AF904557

The Recent and Fossil meet
Kempf Database Ostracoda

CRUSTACEANA MONOGRAPHS constitutes a series of books on carcinology in its widest sense. Contributions are handled by the Series Editor(s) and may be submitted through the office of KONINKLIJKE BRILL Academic Publishers N.V., P.O. Box 9000, NL-2300 PA Leiden, The Netherlands.

Series Editor for the present volume: CHARLES H.J.M. FRANSEN, c/o Naturalis Biodiversity Center, P.O. Box 9517, NL-2300 RA Leiden, The Netherlands; e-mail: charles.fransen@naturalis.nl

Founding Editor: J.C. VON VAUPEL KLEIN, Bilthoven, The Netherlands.

Editorial Committee: N.L. BRUCE, Wellington, New Zealand; Mrs. M. CHARMANTIER-DAURES, Montpellier, France; Mrs. D. DEFAYE, Paris, France; H. DIRCKSEN, Stockholm, Sweden; R.C. GUIAŞU, Toronto, Ontario, Canada; R.G. HARTNOLL, Port Erin, Isle of Man; E. MACPHERSON, Blanes, Spain; P.K.L. NG, Singapore, Rep. of Singapore; H.-K. SCHMINKE, Oldenburg, Germany; F.R. SCHRAM, Langley, WA, U.S.A.; C.D. SCHUBART, Regensburg, Germany; G. VAN DER VELDE, Nijmegen, Netherlands; H.P. WAGNER, Leiden, Netherlands; D.I. WILLIAMSON, Port Erin, Isle of Man.

Published in this series:

CRM 001 - *Stephan G. Bullard*	Larvae of anomuran and brachyuran crabs of North Carolina
CRM 002 - *Spyros Sfenthourakis et al. (eds.)*	The biology of terrestrial isopods, V
CRM 003 - *Tomislav Karanovic*	Subterranean Copepoda from arid Western Australia
CRM 004 - *Katsushi Sakai*	Callianassoidea of the world (Decapoda, Thalassinidea)
CRM 005 - *Kim Larsen*	Deep-sea Tanaidacea from the Gulf of Mexico
CRM 006 - *Katsushi Sakai*	Upogebiidae of the world (Decapoda, Thalassinidea)
CRM 007 - *Ivana Karanovic*	Candoninae (Ostracoda) from the Pilbara region in Western Australia
CRM 008 - *Frank D. Ferrari & Hans-Uwe Dahms*	Post-embryonic development of the Copepoda
CRM 009 - *Tomislav Karanovic*	Marine interstitial Poecilostomatoida and Cyclopoida (Copepoda) of Australia
CRM 010 - *Carrie E. Schweitzer et al.*	Systematic list of fossil decapod crustacean species
CRM 011 - *Peter Castro et al. (eds.)*	Studies on Brachyura: a homage to Danièle Guinot
CRM 012 - *Patricio R. De los Ríos-Escalante*	Crustacean zooplankton communities in Chilean inland waters
CRM 013 - *Katsushi Sakai*	Axioidea of the world and a reconsideration of the Callianassoidea (Decapoda, Thalassinidea, Callianassida)
CRM 014 - *Charles H.J.M. Fransen et al. (eds.)*	Studies on Malacostraca: Lipke Bijdeley Holthuis Memorial Volume
CRM 015 - *Akira Asakura et al. (eds.)*	New Frontiers in Crustacean Biology: Proceedings of the TCS Summer Meeting, Tokyo, 20-24 September 2009
CRM 016 - *Danielle Defaye et al. (eds.)*	Studies on Freshwater Copepoda: a Volume in Honour of Bernard Dussart
CRM 017 - *Hironori Komatsu et al. (eds.)*	Studies on Eumalacostraca: a homage to Masatsune Takeda
CRM 018 - *Masahiro Dojiri & Ju-Shey Ho*	Systematics of the Caligidae, copepods parasitic on fishes
CRM 019 - *Darren C.J. Yeo et al. (eds.)*	Advances in Freshwater Decapod systematics and biology

Editors' addresses:

Finn Viehberg, Institute of Geology and Mineralogy, University of Cologne, Zülpicher Strasse 49A, 50674 Cologne, Germany; e-mail: finn.viehberg@uni-koeln.de

Renate Matzke-Karasz, Palaeontology & Geobiology; GeoBio Center, Ludwig Maximillian University Munich, Richard-Wagner-Strasse 10, 80333 Munich, Germany; e-mail: matzke-karasz@lmu.de

Lisa Park Boush, Center for Integrative Geosciences, University of Connecticut, Storrs, CT 06269, USA; e-mail: Lisa.park_boush@uconn.edu

Alison Smith, Department of Geology, Kent State University, South Lincoln Street, Kent, OH 44242, USA; e-mail: alisonjs@kent.edu

Cover: Scottia pseudobrowniana Pincer

The Recent and Fossil meet Kempf Database Ostracoda Festschrift Eugen Karl Kempf — Proceedings of the 15th International German Ostracodologists' Meeting

By

Finn Viehberg, Renate Matzke-Karasz, Lisa Park Boush and Alison Smith
(Editors)

Crustaceana Monographs, 20

TUTA SUB AEGIDE PALLAS · 1683 ·

BRILL

LEIDEN • BOSTON

This book is printed on acid-free paper.

Library of Congress Control Number: 2014954649

ISBN13: 978 90 04 28759 4
E-ISBN: 978 90 04 28760 0

© 2015 by Koninklijke Brill NV, Leiden, The Netherlands.
Koninklijke Brill NV incorporates the imprints Brill, Brill Nijhoff, Global Oriental and Hotei Publishing.

All rights reserved. No part of this publication may be reproduced, translated, stored in a retrieval system, or transmitted in any form or by any means, electronic, mechanical, photocopying, recording or otherwise, without prior written permission from the publisher.

Authorization to photocopy items for internal or personal use is granted by Koninklijke Brill NV provided that the appropriate fees are paid directly to The Copyright Clearance Center, 222 Rosewood Drive, Suite 910, Danvers, MA 01923, USA.
Fees are subject to change.

PRINTED IN THE NETHERLANDS

CONTENTS

The page numbers in the above Table of Contents refer to the bracketed page numbers in this volume.

[When citing this chapter, refer to Crustaceana 87 (2014) 897–900]

PREFACE

FOSSIL AND RECENT MEET KEMPF DATABASE

BY

FINN A. VIEHBERG[1,5]), RENATE MATZKE-KARASZ[2]), LISA PARK BOUSH[3]) and ALISON J. SMITH[4])

[1]) Institute of Geology and Mineralogy, Universität zu Köln, Zülpicher Strasse 49a, D-50674 Cologne, Germany

[2]) Department of Earth and Environmental Sciences, GeoBio-Center^{LMU}, Ludwig-Maximilians-Universität München, Richard-Wagner-Strasse 10, D-80333 Munich, Germany

[3]) Center for Integrative Geosciences, University of Connecticut, Storrs, CT 06269, U.S.A.

[4]) Department of Geology, Kent State University, Kent, OH 44242, U.S.A.

The majority of the papers in this issue was presented at the 14th International German Ostracodologists' Meeting (IGOM 2012; 11-14 October 2012) with over 60 participants registered from 15 countries (Viehberg & Gromig, 2012). The International German Ostracodologists' Meeting (Deutschsprachiges Ostracodologen Treffen) returned to Cologne where Prof. Dr. Eugen Kempf initiated it in 1988. On the occasion of his 80th birthday the conference participants honoured his major achievements in ostracod research. The motto of the meeting, "Fossil and Recent meet Kempf Database", was linked to his invaluable database, the Kempf Database Ostracoda (KDO). The ostracod taxonomy of fossil and Recent taxa is comprehensible primarily due to his untiring enthusiasm and prescient decision to use computer data processing from the early years. This eventually led to an extensive database for marine ostracods (Kempf, 1986a, 1986b, 1987a, 1987b, 1994, 1995a, 1995b, 1995c, 1996, 2002a, 2004, 2008a, 2008b, 2008c, 2008d) and non-marine ostracods (Kempf 1980a, 1980b, 1980c, 1980d, 1991, 1997a, 1997b, 1997c, 1997d, 2002b, 2006, 2013). Unacceptably, this work is often not cited in international journals, although the majority of ostracodologists consult the database during their research projects.

A lively and comprehensive biography of Eugen Karl Kempf is presented by Matzke-Karasz (2014), which was written partly in close collaboration with Kempf himself. The subsequent papers are case studies that focus on taxonomic groups in various aspects or ecological studies that all benefit more or less from his work.

[5]) Corresponding author; e-mail: finn.viehberg@uni-koeln.de

Similar to the KDO, Mestre et al. (2014) summarised updated information on the neglected ostracod family Entocytheridae and present a relational database with an extensive reference list.

Namiotko et al. (2014) review the genus *Typhlocypris* Vejdovský, 1882 that is under debate to be a senior synonym of the genus *Pseudocandona* Kaufmann, 1900. The authors identify specific characteristics in the valve morphology and separate both genera distinctly using limb traits.

In a small yet significant study Zhai & Zhao (2014) contribute to the knowledge of the geographical distribution of several ostracod species in Palaearctic China. Several species are recorded for the first time in this area.

A new species of the genus *Bathyconchoecia* is sampled in the bathypelagic zone of Nansha Trough (North-West Borneo Trough), southern South China Sea. Yin et al. (2014) describe and illustrate the species with differential diagnosis to *B. angeli* George, 1977.

Mohammed & Keyser (2014) identify Yemen ostracods new to the scientific community in the tidal flats close to Aden City, and thus erect a new genus within the ostracod family Trachyleberididae for these individuals.

Wrozyna et al. (2014) present a detailed quantitative morphological approach of carapace and limbs variations in *Cytheridella ilosvayi* Daday, 1905 to reveal differences in individuals from sexual and asexual reproduction lines.

Peculiar morphological features in *Tanycypris* enables individuals of an amazing escape behaviour that is described and video-recorded by Matzke-Karasz et al. (2014).

An extensive investigation of lacustrine ostracods in Lake Geneva, Switzerland, by Decrouy & Vennemann (2014) presents the results of a monthly sampling campaign at different water depths. The ontogeny and phenology of various species reveal also ecological preferences of various species.

A smaller case study is presented from the town moat of Bremen, Germany by Scharf & Viehberg (2014). Population dynamics are monitored in a highly variable environment of different salinities.

Because ostracods successfully inhabit various environments, sampling methods need to be modified to extract individuals for scientific purpose. Scharf et al. (2014) present a new method to isolate ostracods from stony substrates and eventually to separate living ostracods from the sampled sediment.

The above listed contributions highlight the value of ostracods in interdisciplinary research projects covering an array of different topics.

ACKNOWLEDGEMENTS

We thank the authors of the papers in this special issue for their time and effort they spent in preparing the manuscripts. Additionally, we greatly acknowledge the financial support of the meeting by the Deutsche Forschungsgemeinschaft (DFG) (Vi 532 1/1).

REFERENCES

DECROUY, L. & T. W. VENNEMANN, 2014. Life histories and distribution of ostracods with depth in western Lake Geneva (Petit Lac), Switzerland: a reconnaissance study. Crustaceana, **87**: 1095-1123.

KEMPF, E. K., 1980a. Index and bibliography of nonmarine Ostracoda. Index A. Part 1. Sonderveröffentl. Geolog. Inst. Univ. Köln, **35**: 1-188.

— —, 1980b. Index and bibliography of nonmarine Ostracoda. Index B. Part 2. Sonderveröffentl. Geolog. Inst. Univ. Köln, **36**: 1-180.

— —, 1980c. Index and bibliography of nonmarine Ostracoda. Index C. Part 3. Sonderveröffentl. Geolog. Inst. Univ. Köln, **37**: 1-204.

— —, 1980d. Index and bibliography of nonmarine Ostracoda. Bibliography A. Part 4. Sonderveröffentl. Geolog. Inst. Univ. Köln, **38**: 1-186.

— —, 1986a. Index and bibliography of marine Ostracoda. Index A. Part 1. Sonderveröffentl. Geolog. Inst. Univ. Koln, **50**: 1-762.

— —, 1986b. Index and bibliography of marine Ostracoda. Index B. Part 2. Sonderveröffentl. Geolog. Inst. Univ. Köln, **51**: 1-712.

— —, 1987a. Index and bibliography of marine Ostracoda. Index C. Part 3. Sonderveröffentl. Geolog. Inst. Univ. Köln, **52**: 1-774.

— —, 1987b. Index and bibliography of marine Ostracoda. Bibliography A. Part 4. Sonderveröffentl. Geolog. Inst. Univ. Köln, **53**: 1-454.

— —, 1991. Index and bibliography of nonmarine Ostracoda. Bibliography B. Part 5. Sonderveröffentl. Geolog. Inst. Univ. Köln, **77**: 1-238.

— —, 1994. Index and bibliography of marine Ostracoda. Bibliography B. Part 5. Sonderveröffentl. Geolog. Inst. Univ. Köln, **88**: 1-361.

— —, 1995a. Index and bibliography of marine Ostracoda. Index A Supplement 1. Part 6. Sonderveröffentl. Geolog. Inst. Univ. Köln, **100**: 1-239.

— —, 1995b. Index and bibliography of marine Ostracoda. Index B Supplement 1. Part 7. Sonderveröffentl. Geolog. Inst. Univ. Köln, **101**: 1-203.

— —, 1995c. Index and bibliography of marine Ostracoda. Index C Supplement 1. Part 8. Sonderveröffentl. Geolog. Inst. Univ. Köln, **102**: 1-230.

— —, 1996. Index and bibliography of marine Ostracoda. Bibliography C. Part 9. Sonderveröffentl. Geolog. Inst. Univ. Köln, **103**: 1-198.

— —, 1997a. Index and bibliography of nonmarine Ostracoda. Index A Supplement. Part 6. Sonderveröffentl. Geolog. Inst. Univ. Köln, **109**: 1-142.

— —, 1997b. Index and bibliography of nonmarine Ostracoda. Index B Supplement 1. Part 7. Sonderveröffentl. Geolog. Inst. Univ. Köln, **110**: 1-134.

— —, 1997c. Index and bibliography of nonmarine Ostracoda. Index C Supplement 1. Part 8. Sonderveröffentl. Geolog. Inst. Univ. Köln, **111**: 1-152.

— —, 1997d. Index and bibliography of nonmarine Ostracoda. Bibliography C. Part 9. Sonderveröffentl. Geolog. Inst. Univ. Köln, **112**: 1-144.

— —, 2002a. Index and bibliography of marine Ostracoda. Bibliography D. Part 10, CD-ROM.

— —, 2002b. Index and bibliography of nonmarine Ostracoda. Bibliography D. Part 10, CD-ROM.

— —, 2004. Index and bibliography of marine Ostracoda. Index D. Recent marine Ostracoda of the World. Part 16, CD-ROM.

— —, 2006. Index and bibliography of nonmarine Ostracoda. Index D. Recent nonmarine Ostracoda of the World. Part 16, CD-ROM.

— —, 2008a. Index and bibliography of marine Ostracoda. Index A Supplement 2. Part 11, CD-ROM.

— —, 2008b. Index and bibliography of marine Ostracoda. Index B Supplement 2. Part 12, CD-ROM.

— —, 2008c. Index and bibliography of marine Ostracoda. Index C Supplement 2. Part 13, CD-ROM.

— —, 2008d. Index and bibliography of marine Ostracoda. Bibliography E. Part 14, CD-ROM.

— —, 2013. Index and bibliography of nonmarine Ostracoda. Bibliography F. Part 15, CD-ROM.

MATZKE-KARASZ, R., 2014. Eugen Karl Kempf, the man behind the Kempf Database Ostracoda. Crustaceana, **87**: 901-922.

MATZKE-KARASZ, R., C. NAGLER & S. HOFMANN, 2014. The ostracod springtail — camera recordings of a previously undescribed high-speed escape jump in the genus *Tanycypris* (Cypridoidea, Ostracoda). Crustaceana, **87**: 1072-1094.

MESTRE, A., J. S. MONRÓS & F. MESQUITA-JOANES, 2014. A review of the entocytheridae (Ostracoda) of the world: updated bibliographic and species checklists and global georeferenced database, with insights into host specificity and latitudinal patterns of species richness. Crustaceana, **87**: 923-951.

MOHAMMED, M. & D. KEYSER, 2014. *Adencythere kodalnamerensis*, a new genus and a new species (Ostracoda) from Kod Al-Namer Beach, Aden City, Yemen. Crustaceana, **87**: 1036-1042.

NAMIOTKO, T., D. L. DANIELOPOL, C. MEISCH, M. GROSS & N. MORI, 2014. Redefinition of the genus *Typholocypris* Vejdovský, 1882 (Crustacea, Ostracoda, Candonidae). Crustaceana, **87**: 952-984.

SCHARF, B., A. ADLER & F. A. VIEHBERG, 2014. New methods for collecting Ostracoda (Crustacea) in stony sediments with remarks to the separation of ostracods from the collected sediment. Crustaceana, **87**: 1136-1147.

SCHARF, B. & F. A. VIEHBERG, 2014. Living Ostracoda (Crustacea) from the town moat of Bremen, Germany. Crustaceana, **87**: 1124-1135.

VIEHBERG, F. A. & R. GROMIG, 2012. Abstract Book. 14th International German Ostracodologists' Meeting. Cologne, October 11-14, 2012. Kölner Forum für Geologie und Paläontologie, **21**: 1-67.

WROZYNA, C., W. E. PILLER & M. GROSS, 2014. Morphotypes of *Cytheridella ilosvayi* (Ostracoda, Crustacea) detected by soft and hard part analyses. Crustaceana, **87**: 1043-1071.

YIN, Q., Q. CHEN & K. LI, 2014. *Bathyconchoecia liui* n. sp., a new species of ostracod (Myodocopa, Halocyprididae) from the South China Sea. Crustaceana, **87**: 1027-1035.

ZHAI, D. & W. ZHAO, 2014. On some Recent non-marine ostracods from Northern China, with description of one new species. Crustaceana, **87**: 985-1026.

[When citing this chapter, refer to Crustaceana 87 (2014) 901–922]

This paper is dedicated to Prof. Dr. Eugen Kempf on the occasion of his 80th birthday (16 April 2012)

EUGEN KARL KEMPF, THE MAN BEHIND THE KEMPF DATABASE OSTRACODA

BY

RENATE MATZKE-KARASZ[1])

Ludwig-Maximillians-Universität, Department of Earth and Environmental Sciences, Palaeontology and Geobiology, GeoBio-Center LMU, Richard-Wagner-Strasse 10, D-80333 Munich, Germany

ABSTRACT

ITIS, WoRMS, EOL, iBOL, GBIF, ANTABIF, OBIS. FlyBase, FishBase, FosFARbase, Fauna Europaea. World Modern Foraminifera Database, EDNA Fossil Insect Database, European Diatom Database, The Reptile Database, Mammalian Species of the World. Today's biologists are blessed with databases, and there is no doubt about their relevance for scientific work. (Palaeo-)Biologists put a lot of time and efforts into establishing and maintaining a multitude of databases, not to mention their endeavours to raise funds for this work. Many databases are produced and edited by a multitude of authors and editors, providing a great amount of data being supplied to the project within a short period of time — all too often at the expense of quality. The recent success of databases is intrinsically tied to the increasing access to, and use of, the World Wide Web, and it comes as no surprise that only few such biological databases have been initiated prior to the late 1990s. A rare example of an evidentially long-standing, relational database, authored and managed by an individual expert of a crustacean class, is the Kempf Database Ostracoda. Prof. Eugen Karl Kempf, now celebrating his eightieth anniversary, started data compilation in the 1960s as a card index and innovatively switched to a machine punch card system in the 1970s. Ever since the first publication from his taxonomic-bibliographic database in 1980, ostracod researchers worldwide benefitted from an ineffably high amount of, currently, more than 285 000 datasets of highest integrity. It is time to honour the person behind the Kempf Database Ostracoda.

Key words. — Relational database, ostracodology, taxonomic, bibliographic, University of Cologne

ZUSAMMENFASSUNG

ITIS, WoRMS, EOL, iBOL, GBIF, ANTABIF, OBIS. FlyBase, FishBase, FosFARbase, Fauna Europaea. World Modern Foraminifera Database, EDNA Fossil Insect Database, European Diatom Database, The Reptile Database, Mammalian Species of the World. Heutzutage sind Biologen gesegnet mit Datenbanken und es gibt keinen Zweifel über deren Relevanz für die wissenschaftliche

[1]) e-mail: matzke-karasz@lmu.de

Arbeit. (Paläo-)Biologen stecken viel Zeit und Mühe in die Erstellung und Erhaltung einer Vielfalt von Datenbanken — ganz zu Schweigen von den Bemühungen um deren Finanzierung. Viele Datenbanken werden von einer großen Zahl von Autoren und Editoren erstellt und gepflegt, wodurch schnell eine große Datenmenge generiert wird — leider oft mit dem Nachteil einer geringen Qualität. Der heutige Erfolg von Datenbanken ist direkt mit dem immer einfacher werdenden Zugang und Gebrauch des Internets verbunden, so dass es nicht verwundert, dass nur wenige dieser Datenbanken vor den späten 1990er Jahren entstanden. Eine seltene Ausnahme einer solchen offensichtlich altbewährten, relationalen Datenbank, erstellt und gepflegt von einem Experten einer Crustaceen-Klasse, ist die *Kempf Database Ostracoda*. Prof. Eugen Karl Kempf, heute achtzig Jahre jung, begann seine Datensammlung in den 1960er Jahren als Kartei, um sie bereits in den 1970er Jahren als elektronische Datenbank im Lochkartensystem weiterzuführen. Seit der ersten Publikation aus seiner taxonomisch-bibliographischen Datenbank im Jahr 1980 haben Ostracodologen weltweit in einem unbeschreiblichen Maße von den nun mehr als 285 000 Datensätzen höchster Integrität profitiert. Es ist höchste Zeit, den Menschen hinter der *Kempf Database Ostracoda* zu ehren.

Key words. — Relational database, ostracodology, taxonomic, bibliographic, University of Cologne

INTRODUCTION

Cologne University, Department of Geo- and Environmental Sciences, winter term 1984. A biology student in her third year sits in the lecture hall, listening to her first lecture on micropalaeontology, the subsidiary subject she had picked for her diploma studies.

Today's topic: O-S-T-R-A-C-O-D-A. Crustaceans on a microscopic scale with calcified, bivalved carapace, enclosing the soft body completely. A continuous fossil recorded from at least the Ordovician, including countless biostratigraphic relevant index fossils; more than 80 000 fossil and Recent named species known; representatives in all kinds of water bodies, even in humid terrestrial habitats; indicator organisms for a variety of environmental factors — this sounds like more than enough reasons to have a closer look at these beasts! She starts enjoying her excursion into palaeontology.

The professor, an energetic, wiry, athletic person, moves in front of the audience with an elegant spring in his step. His thoughtful, precise and clear speech grips the audience. At the end of today's lesson, she hears him stating with some indignation that given "this high importance of ostracods in palaeontology, palaeoecology and biostratigraphy, it is incomprehensible that research on Recent ostracods is widely neglected!"

CHILDHOOD

City of Cologne, 16 April 1932. Time for celebration in the house of baker Eugen Kempf senior! Little Eugen Karl Kempf Jr. was born, first son and heir,

Fig. 1. Kempf family in 1944, with Eugen standing between father and younger brother Alfred.

determined to follow his father's steps and to become a baker in the foreseeable future.

It is the time of economic depression, with 12% unemployment in the city of 'Eau de Cologne' and The Cathedral. It is also the last year of popular Mayor Konrad Adenauer (who was to become German Chancellor in 1949), who had to escape from threatened violence by the Nazis in 1933 (Dietmar, 1997). Eugen's already difficult life as a schoolboy (fig. 1) before and during times of the Second World War was impeded by the destruction of the family's house during a bomb attack in 1943. The homeless family found shelter and rest in a small town in Baden-Württemberg, homeland of Eugen Sr., where junior for the first time enjoyed recreation in the country side and started to discover the secrets of nature (fig. 2).

Reunited with Eugen Sr., who had been called to war in 1940, the family returned to the entirely devastated hometown in 1945, and withstood poverty and misery by putting all forces and motivation into rebuilding a destroyed bakery. It was only natural that Eugen Jr. started his apprenticeship as a baker after middle school, thus fulfilling the expectations of the family.

EDUCATION AND STUDIES

Traditionally, German people have high demands on the quality of bread and cake — ask a German abroad about what is missed the most! The baker's

Fig. 2. Rural idyll prior to final combats of WWII: the Kempf family found shelter from massive bombing of Cologne in the town of Meßkirch (Baden-Württemberg). This figure is published in colour in the online edition of this journal, which can be accessed via http://booksandjournals.brillonline.com/content/journals/15685403.

profession is, therefore, especially challenging in Germany, both mentally and physically (the author feels competent on this behalf, also being born into a baker's family). Nevertheless, young journeyman Eugen Karl soon felt intellectually underchallenged. In 1952, he decided to join classes of an evening high school for a time span of three and a half years in order to complete his school career, while at the same time working hard days in the bakehouse. At that school, biology was taught by Dr. M. Heinrich, who had written a Ph.D. thesis on Devonian fossils (Heinrich, 1914). Also, he took part in geological excursions offered by the adult education centre. Here, he first met Ulrich Jux, an early geologist at the University of Cologne, later replacing Martin Schwarzbach as director of the Geological Institute.

In 1956, Eugen Kempf received the university-entrance diploma, his golden ticket to an academic career. He immediately enrolled in the faculty of natural sciences at Cologne University, fulfilling the desire to study geological sciences that had emerged under the influence of his biology teacher and from the excursions guided by Ulrich Jux. However, he was much too reasonable and dutiful to simply discard his professional education; in contrast, he followed it up until he could pass the examination for his master craftsman's certificate in 1957. As a fully approved 'Bäckermeister', he could finally focus on what he had become passionate about: learning more about earth and its history.

The University of Cologne has a history: founded in 1388 by the citizens of Cologne (and not by the clergy), reputedly the fourth-oldest European university, closed down by the French occupying forces in 1798 and re-founded in 1919. In 1938, the geological institute had moved into a remarkable round building originating from the former city fortification. Partly destroyed by bombs during WWII, it was rededicated as an urgently needed hospital for infectious diseases and the institute had to 'hibernate' in a single, small room of the main university building. Only in the 1950s could the geological institute move back and slowly expand. Now headed by Prof. Martin Schwarzbach (1907-2003) the 'father' of palaeoclimatology, the Cologne Geological Institute cooperated with that of Munich University and that of nearby University of Bonn. Cologne students passed 'external' semesters in Munich and/or Bonn, while reconstruction and teaching capacity in Cologne was moved forward. Especially instructive and skilful excursions with Schwarzbach (fig. 3) and his wife — a botanist who took the opportunity to teach the students during long travels in the university bus — shaped young palaeontologist Eugen Karl.

An ostracod species named in dedication to Prof. Schwarzbach in 1967, *Ilyocypris schwarzbachi*, expresses the high esteem Eugen Karl felt towards his teacher and doctoral supervisor. After passing his first exam, it was time for a semester 'abroad' in Bavaria, together with his fellow student Friedrich Strauch (future founder and president of the Alfred-Wegener-Stiftung). Impressive names were listed as readers for geology and palaeontology in the 1958 calendar of Munich's Ludwig-Maximilian-University: Werner Quenstedt (a universal naturalist, grandson of early expert in Jurassic geology F. A. Quenstedt), Richard Dehm (mammalian palaeontology), Herbert Hagn (micropalaeontology), Werner Zeil (alpine geology), Karl Mägdefrau (botany and palaeobotany), Hans-Georg Fischer (petrography, grandson of Sebastian Fischer, author of 31 ostracod species in the 1850s) and Walter Wüst (ornithology).

Fig. 3. Professor Martin Schwarzbach (left) during a students' field-trip to study Palaeozoic deposits of southern Belgium in the range of river Maas in 1957, guided by Professor Marius Lecompte of the Catholic University of Louvain (right).

Strengthened intellectually by courses and excursions in the beautiful alpine landscapes and physically by nourishing Bavarian food, Eugen Karl returned to the Rhineland to study one semester at the University of Bonn. There he learned a lot from readers like Roland Brinkmann (geology), Heinrich K. Erben (palaeontology), Hans-Joachim Schweitzer (palaeobotany), Paul Woldstedt (fundamentals of the Pleistocene) and Alfred Neuhaus (mineralogy).

Thereafter he returned to Cologne University to finalise his studies with a diploma degree in 1963. Soon he became research assistant of Ulrich Jux, in charge of the palaeobotanical collection and later of electron microscopy. He enticed an exceptionally skilled mechanic away from the car building industry, W. Mackowiak, who put enormous efforts in optimising techniques and developing accessories for scanning and, first of all, transmission electron microscopy. Among others, the team successfully aimed at producing distortion-free TEM images, even at low magnifications. Beside these tasks, Eugen took over the responsibility for the institute's library matters, thus becoming familiar with a field that would become crucial for the future database work. In the 1970s, the editorial and financial responsibilities of the institute's budget-free journal, *Sonderveröffentlichungen*, were also put in Eugen's hands. The previously rather unimpressive booklets became provokingly colourful designed on the outside and scrupulously edited on the inside; the latter being a process painfully experienced by many authors of subsequently published doctoral theses.

Scientifically, his palaeontological work on the Holstein interglacial of Tönisberg in the lower Rhine area led to the doctoral degree in natural sciences in 1965. Further work analysed various fossil plant remains and ostracods. With up-to-date electron microscopy available, Eugen Karl targeted innovation: he elaborated TEM methods to investigate fossil sporoderms of megaspores from various periods of the earth column and thus contributed to their use as biostratigraphic indicators. Above all, however, for the first time it could be documented that the fine structure of fossil megaspore walls is often excellently preserved in Cenozoic, Mesozoic, and even in Palaeozoic specimens.

RESEARCH AND ACADEMIC TEACHING

Eugen Karl presented his habilitation monograph to the faculty and received the *venia legendi* in 1971, marking the start of an independent micropalaeontological research group. Scientifically and socially settled within a growing institute (fig. 4), he was appointed adjunct professor (1974) and lifetime professor (1980).

In this period of his career, many research travels were undertaken to various corners of the world, often together with the new head of department, Ulrich Jux. Siberia, Afghanistan and Egypt (fig. 5) were the most distant destinations for their expeditions. Eugen Karl's determination at work, his diplomatic sensitiveness, his patience — if not persistence — proved exceptionally helpful during expeditions. Equipped with a distinct objectivity in all his statements, combined with a friendly and positive humour, Eugen often de-escalated precarious situations that occurred when on geological expedition, especially in Africa and Asia. His typical Rhineish warm-heartedness and philanthropy frequently 'paved the way', where there was not even a dirt track before. Beside a whole lot of fossil material sampled for the institute's collection during these expeditions, several publications emerged from them.

Internationally, Eugen Kempf has always been intensively interconnected with ostracodologists. At the 10th International Symposium on Ostracoda in 1988, he

Fig. 4. Social life at the Cologne Geological Institute. Top, one of the famous Cologne Carnival parties in the 1980s. From left to right: Prof. Dr. H. U. Kasper, W. Mackowiak ("master technician" of electron microscopes), late Dr. R. H. Weiss, Prof. Dr. E. K. Kempf and late D. Rohde, a student working on Cretaceous Ostracoda. Bottom, celebratory colloquium at the occasion of Prof. Dr. U. Jux' 60th birthday in 1989 (with late Mrs. Jux). This figure is published in colour in the online edition of this journal, which can be accessed via http://booksandjournals.brillonline.com/content/journals/15685403.

was elected chair of the International Research Group on Ostracoda (IRGO), the representation of ostracod workers worldwide since 1963. Besides, he initiated regular meetings of German speaking ostracodologists and organised the first such meeting 1988 in Cologne. Ever since, this meeting has taken place almost every year, offering a platform for newcomers to meet with senior researchers in an informal surrounding (fig. 6).

Fig. 5. Ulrich Jux, Egyptian policeman, Jamal and Eugen Kempf (left to right). The scientists are temporarily under arrest due to 'unofficial' entry from the wilderness of the desert. This figure is published in colour in the online edition of this journal, which can be accessed via http://booksandjournals.brillonline.com/content/journals/15685403.

When teaching students, Eugen Karl always avoided wasted words, but convinced with clear and bluntly phrased statements, often spiced with a dry wit. Excursions with this professor were very popular, because they were known to be very instructive and at the same time relaxed. The friendly atmosphere made long excursion days pass quickly (fig. 7) and, whenever possible, they were terminated by a powerful swim in the sea. Several Ph.D. theses were carefully supervised, always demanding a superior level of specimen and data documentation.

PRIVATE LIFE

Although profession and professionalism have always been the guiding stars in Eugen's life, he did not neglect the other, private and amateur part of it. Being a genuinely athletic person, Eugen always engaged in sports. In line with expectations, he did not just try to avoid rusting by exercise, but to obtain perfection and to explore his metes and bounds. Instead of leisure biking, Eugen competed in bicycle races like the traditional race around Cologne. His more artistic side was

Fig. 6. Field trip on the occasion of the first meeting of ostracodologists of German tongue, organised by Eugen Kempf in 1988. W. Mackowiak, Heinz Malz, unidentifiable colleague, Friedrich Plumhoff, Alexander Liebau and Eugen Kempf (left to right). This figure is published in colour in the online edition of this journal, which can be accessed via http://booksandjournals.brillonline.com/content/journals/15685403.

developed by ballroom dancing, which he brought to perfection together with his graceful wife Elisabeth (fig. 8). The couple married in 1961 and two children later completed the family. Dancing remained their passion until today, after succeeding countless tournaments as both, participants and officials.

Kempf's long term mentor, superior, colleague and expedition companion, Prof. Ulrich Jux, together with Kempf's first Ph.D. student, the late Roseline Weiss, authored an honouring consideration of his work at the occasion of his 65th anniversary. It finishes wishing the fresh retiree 'a retirement in health and happy creativeness'. Fifteen years later it is clear that this wish proved to be very effective. The retirement did not prevent Kempf from working tirelessly on the project of his life, the Kempf Database Ostracoda, and on responding to any taxonomic inquiry brought to him privately by e-mail or through the ostracodologists' e-mail discussion list OSTRACON. The sometimes ludicrous quality of taxonomic information on ostracods in internet databases made him start entering (currently some 2000) reliable datasets into the information system Wikispecies, including links to the original literature if available online, e.g., in the Biodiversity Heritage

Fig. 7. Hands-on teaching during a student excursion to the Isle of White in the 1980s. This figure is published in colour in the online edition of this journal, which can be accessed via http://booksandjournals.brillonline.com/content/journals/15685403.

Library. Thus, his family sees him restless in the word's most positive sense; motivation to push further the ostracod mission seems unbreakable.

THE KEMPF DATABASE OSTRACODA

In the last decade, multi-proxy studies in palaeontology have proved to be an efficient tool in the study of past environments (Birks & Birks, 2006). Their increase in numbers nicely indicates the increasing interest in responses of biota to past environmental changes. Nowadays, such multi-proxy studies are usually authored by a group of specialists, all treating their very field of expertise. However, multi-proxy studies were already around much earlier, and in those

Fig. 8. Left: Snap shot of Eugen Kempf passing by on his racing bike during Germany's second-oldest bicycle race 'Rund um Köln' (in 1950). Right: Eugen and his wife Elisabeth in pure elegance on the dance floor (in the 1980s).

days the various proxies were usually investigated by a single author. A perfect example is Kempf's first publication on the Holstein-interglacial of Tönisberg (Lower Rhine), with analyses of the mollusc and ostracod fauna, combined with the study of pollen, fruits, seeds and megaspores (Kempf, 1966).

Becoming acquainted with the taxonomy of such diverse organisms requires an intensive literature study and, thus, a good literature management. Almost unimaginable in times of Endnote, Mendeley and Zotero, hundreds of index cards were written on the widespread specialist publications — unwittingly forming the basis of the future Kempf Database Ostracoda (KDO).

Further micropalaeontological studies directed Eugen Kempf more and more towards ostracods and the growing 'uni-directional' card index became less and less manageable, when, in 1974, the faculty-independent computation centre at the University of Cologne opened up the opportunity of data computation to all university staff. However, the first euphoria quickly calmed down when Eugen Kempf and the computation centre staff tried to work out ways for processing the vast taxonomic and literature data. The staff was unable to cope with expectations. Rules of taxonomy are not straightforwardly translated into computational algorithms.

Months of consideration and trials followed, until the basic structure of the KDO was formed — a structure which has proven functional and successful until today. The key feature of this structure is the manifold access to the core of the database, the literature. Three indexes were created, a species-based index, a genera-based index and a chronological index, through which the nomenclatural history of single species and genera can be traced, finally leading to the citations of the publications in question. The titles of these publications are compiled in a bibliography of taxonomically relevant literature. Additionally, a second bibliographic index lists all publications on ostracods without taxonomically relevant information.

Equipped with this theoretical system, a second attempt to transfer the card index into digital dataset was successful. The heart of the computation centre in those days was the Standard Elektrik Lorenz ER 56 transistor computer (core storage capacity of 4000 words of 7 decimal digits each). A magnetic memory storage allowed storage of 10 000 35-bit-words. Data encoding was done by punch cards — standardized pieces of stiff paper, with 10 lines of digits printed in standardized positions, each line containing 80 digits (fig. 9). Information was represented by holes that were punched on predefined positions, using special punching machines that much resembled early typewriters. Thus, each data set had to be punched manually into a card, before it could be read and further processed by the computers.

Eugen started punching his datasets. Card by card was read into the computer during long nights, when the computation centre was less busy. Information was short-term stored on the core memory, and finally transferred to magnetic tapes. Over the early years, more than 40 000 datasets were treated this way; the collection of punch cards bears witness (fig. 9). In the mid-1970s, a computer terminal was set up in the palaeontological institute, connected to the computation centre via leased line. Here, the magnetic tapes could be controlled and datasets could be sorted.

A drawback in those days, however, was the impossibility to produce a clearly arranged, printable output, a spin-off so to say, of the database. The problem was caused by the line-at-a-time printers neither allowing for different fonts or font sizes, nor for a neat column-structured print image. Thus, the RRZK acquisition of a daisy wheel printer terminal in 1979 changed the situation drastically: for the first time, a setting copy could be prepared, paving the way for the first published parts of the KDO in 1980 on non-marine ostracods (Kempf, 1980a-d). The corresponding set on marine ostracods followed some years later (Kempf, 1986c-d, 1987, 1988). They were incorporated into the special publication series of the Cologne Geological Institute, with gaudy soft covers provided by the then series editor Eugen Kempf (fig. 10).

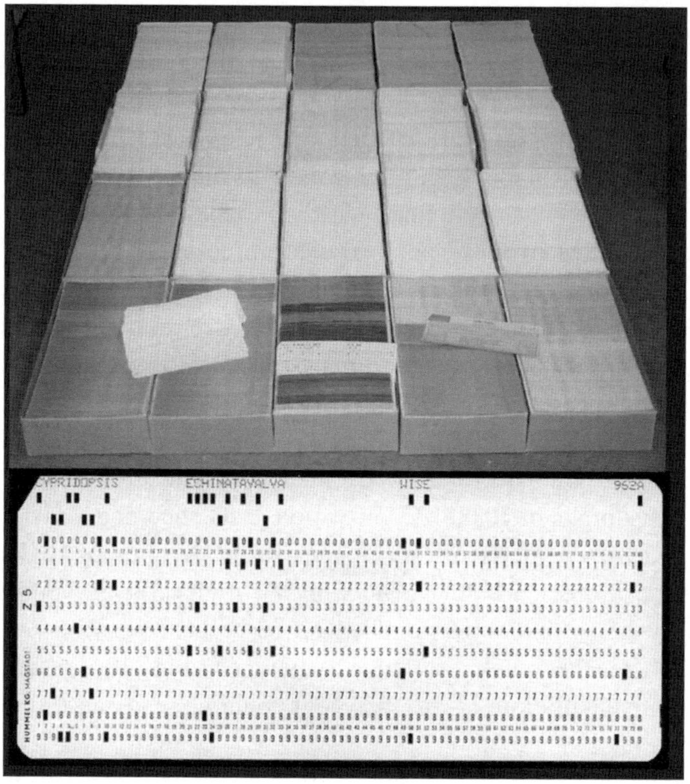

Fig. 9. In the early 1980s, for the first index of marine Ostracoda, Eugen Kempf punched 40 000 datasets into punch cards prior to reading them into the computer of the Cologne University Computation Center RRZK. This figure is published in colour in the online edition of this journal, which can be accessed via http://booksandjournals.brillonline.com/content/journals/15685403.

Two years later, the universe of computation experienced a quantum leap forward: the first portable microcomputer was produced by the Osborne Computer Corporation, the Osborne 1 (fig. 11). Eugen afforded the horrendous sum for hard- and software to become more independent of the university's computation centre. The flexibility of this portable machine (weighing a bit more than 10 kg) was liberation, outweighing the diminutiveness of the screen.

Other computer systems have since been coming and going: IBM, Atari, Apple, Microsoft, Linux, etc, etc. Luckily, right from the early days, the database has been kept independent of operating systems, computer languages and commercial software. Thus, Eugen Kempf continuously fed the database with taxonomical data from literature he gathered over the years, always following the quality precept that no entry will be made if the original publication is not in his hands. To date, more than 20 000 publications on ostracods form the private Kempf library.

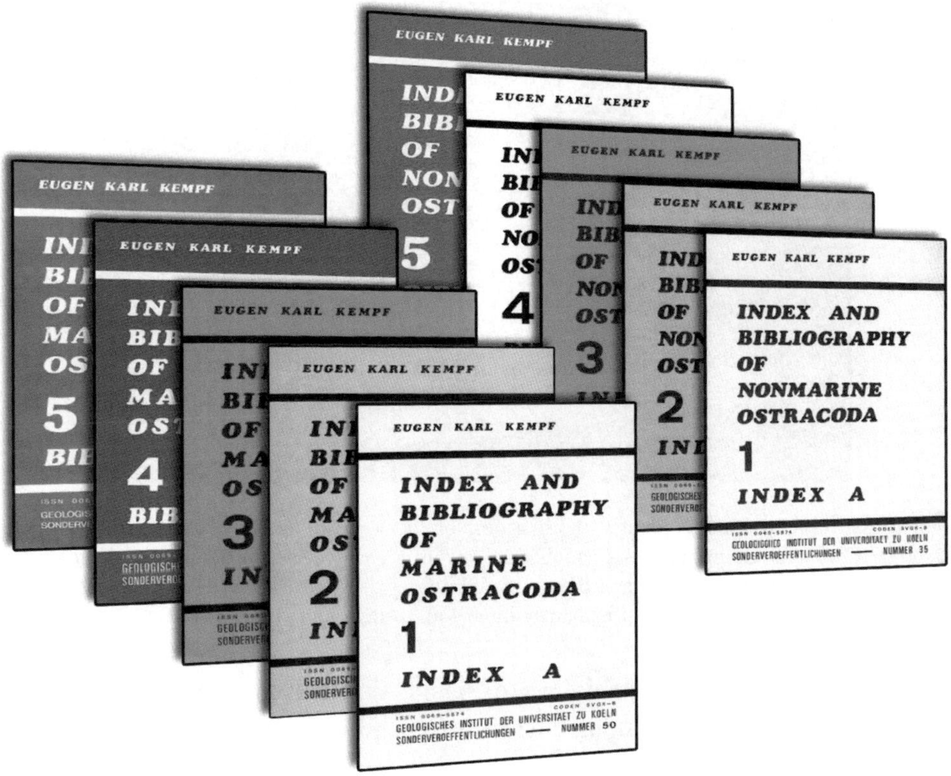

Fig. 10. The first set of indexes and bibliographies of the Kempf Database Ostracoda — marine and non-marine. This figure is published in colour in the online edition of this journal, which can be accessed via http://booksandjournals.brillonline.com/content/journals/15685403.

The most convenient way to acquire literature items was, and still is, receiving them as reprints from authors; however, the vast majority of papers was taken directly from journals and books. These had often to be searched in far-off libraries, ordered for as inter-library loans, or accepted as small microfiche-copies. Titles of Russian publications were transliterated to make them compatible with the database. The Kempf library has meanwhile become the white knight for all ostracodologists who failed in locating a certain rare publication. The last resort is an inquiry to Eugen Kempf, which will always be answered in a friendly, helpful and extensive way. If the desired publication is not present in the Kempf library, it probably does not exist any more — or has never been written!

By following up the history of species descriptions and subsequent taxonomical rearrangements, the KDO is a perfect tool to avoid the publication of homonyms if it is used by authors of new species. If not, such dissonances of nomenclature are noticed at the latest, when entered into the KDO. In these cases, authors are given notice by Eugen Kempf, usually leading to a new, corrective publication.

Fig. 11. Osborne 01 microcomputer, the first portable personal computer, released in 1982 by the Osborne Computer Corporation. Its acquisition was an important step in the history of the KDO. This figure is published in colour in the online edition of this journal, which can be accessed via http://booksandjournals.brillonline.com/content/journals/15685403.

If this did not happen for a couple of years, Eugen Kempf himself published replacement names, thus making the species scientifically available (Kempf, 2010, 2011b, 2012).

In addition to taxonomical data, the KDO also contains biostratigraphical and biogeographical data, albeit not at the same degree of completeness. Input of taxonomical data always had first priority and even for Eugen Kempf the day could not be extended to more than 24 hours. Lack of funding prevented him from hiring assistants.

Beginning in 1980, Kempf made the taxonomical data accessible to users by publishing a set of spin-offs for both marine and non-marine ostracods. Each set consists of three index versions and a compilation of taxonomic literature references. The carefully considered basic structure of the database allowed for producing a chronological index, a generic index and a species-name index, all pointing to the correct literature reference provided separately. A fourth index represents the first phase of spin-offs of the stratigraphical part of the database, this being the Holocene and Living Ostracoda in "Recent Marine Ostracoda of the World" and "Recent Non-Marine Ostracoda of the World". Distinguishing feature of these indexes is the compression of stratigraphic data into a single-line code. This concept allows for manifold sorting processes, which are a prerequisite to service varied queries. On top, a second literature reference compiled publications on ostracods without taxonomical relevance.

These spin-offs and their regular supplements have been published as books and CD-ROMs between 1980 and 2008 (26 volumes). Using the indexes, the nomenclatural history of species can be easily tracked, type taxa can be identified, current species inventories of genera can be authoritatively compiled, authorships of species can be firmly approved, and citations of original descriptions can be verified. If the original literature cannot be acquired, the Kempf library behind the database is the emergency helper. None of the taxonomic internet databases and scientific search engines will ever be able to provide anything similar.

A key feature of the KDO is the foresightful technique of allocating each species a Universal Taxonomic Identification Number (UTIN), using a system that allows for tracking taxonomical changes on genus and species levels. Fifty years after Kempf's decision to code species names numerically for the KDO, modern online taxonomical databases work quite the same in giving species their own primary key, resulting in a number of species key systems identical to the number of databases considered. Why not use the same, universal taxon identification system in times where online databases are increasingly cross-linked and becoming dependent on each other? This and similar issues are of great concern to Eugen Kempf and he is always ready to share his experience and expertise in the field of database development, be it on conferences or online.

Today, the KDO comprises almost 300 000 datasets, all handpicked and controlled with highest scientific expertise. For more than 30 years, researchers have been profiting from these datasets, which is not only reflected in a great number of personal acknowledgment letters sent to Eugen Kempf. The major impact the database has, and will have, is one that cannot be quantified: it is concealed in the severe reduction of taxonomic errors in published papers and in those countless work hours that researchers worldwide could save by using the KDO indexes.

Through all the years, Eugen Kempf kept a vision he had at a time when others could not imagine how a database would be useful. Applications for financial support were as fruitless as they could be in times when terms like 'relational database' raised pure blankness. A lot of private funds were invested to maintain the database and in the end of the 1990s, financial pressures led to the decision to publish further spin-offs on CD-ROMs in private edition. These are eagerly awaited by ostracod workers, be it oceanographers or limnologists, biostratigraphers, palaeoecologists, climate researchers, isotope specialists or reproduction biologists — the KDO serves them all. New trends in taxonomically relevant fields, like 'genetic barcoding' of faunas, will not outdate the KDO. In contrast, even a higher frequency of taxonomically relevant literature is to be expected and reliable references for alignment with older literature will be of steady interest. Thus, the need of a high quality, reliable, platform- and project-independent taxonomic database will even increase in the future.

CONCLUSION

Back to 1984, lecture hall of palaeontology, University of Cologne, micropalaeontology course. The professor states again that given the "high importance of ostracods in palaeontology, palaeoecology and biostratigraphy it is incomprehensible that research on Recent ostracods is widely neglected by biologists!" He is not aware that this time, a young student of biology sitting in the audience will not take it as a disappointed side-note, but as an invitation.

She takes part in Prof. Kempf's micropalaeontology practicals and excursions to become more familiar with the fossil remains of these microcrustaceans. She starts testing scanning electron microscopy techniques on dissected soft bodies of Recent representatives. She trains in techniques for sampling living ostracods during joint field trips with Eugen Kempf, provisioned by the inevitable professorial shortbread and accompanied by shared stories and frequent guffaws.

Thus caringly supervised by palaeontologist Kempf, she takes her first individual scientific steps during graduation and dissertation in the Zoology Department. Owing to her inexperience, their long, intensive meetings frequently turned out to be valuable 'instructive monologues' rather than discussions. Eugen Karl shares with her his scientific knowledge together with his tea, as well as countless difficult-to-obtain offprints, Xerox or microfilm copies of relevant literature that were gathered during decades of database creation. Whenever the development of the Kempf Database Ostracoda takes another serious step, pre-prints in spiral binding are handed over to her (until today) as ultimate update of literature and taxa entered. Her accepted doctoral thesis, like many other theses before, receives his final editing touch before publication in the institutional series.

Almost 30 years later, it may be allowed to declare Eugen Karl's 1984 appeal to do more research on living ostracods successful in the case of this 'young student'. And as plainly any other ostracodologist of the last four decades, she picked the books or the CDs of the Kempf database a thousand times during her work to look for reliable data on species, genera, publication dates, new combinations and their authorships, papers of a certain author, publications in a certain period of time, and so forth. What a gain of time in favour of other tasks to be carried out! What a gain of scientific certitude to know that these data are reliable! Thus, like any other ostracodologist of the last four decades, she knows that her discipline profited massively from the Kempf database and certainly will do so in the future.

At the occasion of his 80th birthday, it is time to honour Prof. Eugen Karl Kempf (fig. 12), the man behind the ultimate ostracod database, the Kempf Database Ostracoda.

Fig. 12. Prof. Dr. Eugen Karl Kempf. This figure is published in colour in the online edition of this journal, which can be accessed via http://booksandjournals.brillonline.com/content/journals/15685403.

ACKNOWLEDGEMENTS

During the phase of writing this biography, discussions with Finn Viehberg helped me a lot. Alison J. Smith polished my English grammar and spelling. Thank you, Alison and Finn!

REFERENCES

BIRKS, H. H. & H. J. B. BIRKS, 2006. Multi-proxy studies in palaeolimnology. Vegetation History and Archaeobotany, **15**: 235-251.

DIETMAR, C., 1997. Chronik Köln: 1-596. (Chronik Verlag im Bertelsmann Lexikon Verlag, Gütersloh/Munich).

HEINRICH, M., 1914. Studien in den Riffkalken des rheinischen oberen Mitteldevons. (Ph.D. Thesis, University Bonn, Bonn).

BIBLIOGRAPHY OF EUGEN KARL KEMPF

BRUNNACKER, K., W. BOENIGK, B. DOLEZALEK, E. K. KEMPF, A. KOCI, H. MENTZEN, M. RAZI RAD & K.-P. WINTER, 1978. Die Mittelterrassen am Niederrhein zwischen Köln und Moenchengladbach. Fortschritte in der Geologie von Rheinland und Westfalen, **28**: 277-324.

JUX, U. & E. K. KEMPF, 1971. Microstructures of the Mesozoic megaspore *Tasmanitriletes* n. g. Grana, **11**(2): 95-100.

— — & — —, 1983. Regional geology of Sistan (South-west Afghanistan). In: M. TOSI (ed.), Prehistoric Sistan, **1**: 5-60. (Istituto Italiano per il Medio ed Estremo Oriente — IsMEO, Roma).

JUX, U., E. K. KEMPF & U. MANZE, 1971. Schichtenfolge der marinen Oberkreide bei Bande Amir (Zentral-Afghanistan). Neues Jahrbuch für Geologie und Paläontologie, Monatshefte, **1971**(12): 712-733.

— —, — — & — —, 1971b. Stauseen durch Travertinabsatz im zentralafghanischen Hochgebirge. Zeitschrift für Geomorphologie, Neue Folge, Supplementband, **12**: 107-137.

KEMPF, E. K., 1966. Das Holstein-Interglazial von Tönisberg im Rahmen des niederrheinischen Pleistozäns. Eiszeitalter und Gegenwart, **17**: 5-60.

— —, 1967a. *Ilyocypris schwarzbachi* n. sp. (Crustacea, Ostracoda) und ein vorläufiges Ostrakoden-Diagramm aus dem pleistozänen Löß von Kärlich (Neuwieder Becken). Sonderveröffentlichungen des Geologischen Instituts der Universität zu Köln, **13** (Schwarzbach-Festschrift): 65-79.

— —, 1967b. Ostrakoden aus dem Holstein-Interglazial von Tönisberg (Niederrheingebiet). Monatsberichte der Deutschen Akademie der Wissenschaften zu Berlin, **9**(2): 119-139.

— —, 1968a. Die Mastixioideen-Flora von Adendorf (Miozän, Rheinland). Zeitschrift der Deutschen Geologischen Gesellschaft, **118**(1966)(1): 182-185.

— —, 1968b. Wasserfarne im rheinischen Braunkohlentertiär und ihr stratigraphischer Wert. Zeitschrift der Deutschen Geologischen Gesellschaft, **118**(1966)(1): 161.

— —, 1968c. Mollusken aus dem Holstein-Interglazial des Niederrheingebietes. Archiv für Molluskenkunde, **98**(1/2): 1-22.

— —, 1968d. Geologische und paläontologische Ergebnisse der Abteufung von Schacht Tönisberg. Der Niederrhein, **35**(1): 12-17.

— —, 1969a. Elektronenmikroskopie der Megasporen von Azolla tegeliensis aus dem Altpleistozän der Niederlande. Palaeontographica, Abteilung B, **128**(3/6): 167-179.

— —, 1969b. Elektronenmikroskopie der Sporodermis von känozoischen Megasporen der Wasserfarn-Gattung *Azolla*. Paläontologische Zeitschrift, **43**(1/2): 95-108.

— —, 1970. Elektronenmikroskopie der Sporodermis von Megasporen der Gattung Selaginella (Pteridophyta). Review of Palaeobotany and Palynology, **10**(2): 99-116.

— —, 1971a. Elektronenmikroskopie der Sporodermis von Mega- und Mikrosporen der Pteridophyten-Gattung *Salvinia* aus dem Tertiär und Quartär Deutschlands. Palaeontographica, Abteilung B, **136**(1/4): 47-70.

— —, 1971b. Ostrakoden-Forschung als Hilfswissenschaft der Ur- und Frühgeschichte. Informationsblätter zu Nachbarwissenschaften der Ur- und Frühgeschichte, **2** (Zoologie 3): 1-5.

— —, 1971c. Electron microscopy of the megaspore *Horstisporites semireticulatus* from Liassic Strata of Germany. Grana, **11**(1): 18-22.

— —, 1971d. Ökologie, Taxonomie und Verbreitung der nichtmarinen Ostrakoden-Gattung Scottia im Quartär von Europa. Eiszeitalter und Gegenwart, **22**: 43-63.

— —, 1972. Electron microscopy of Mesozoic megaspores from Denmark. Grana, **11**(1971)(3): 151-163.

— —, 1973a. Ostrakoden aus dem Jungpleistozän des Damaskus-Beckens. Zeitschrift für Geomorphologie, Neue Folge, **17**(3): 336-347.

— —, 1973b. Geowissenschaftliche Institute der Universität Köln erhalten modernstes Rasterelektronenmikroskop. Mitteilungen der Universität zu Köln, **1973**(11): 4-8.

— —, 1973c. Transmission electron microscopy of fossil spores. Palaeontology, **16**(4): 787-797.

— —, 1974. Niedrige Vergrösserungen — ein Grenzgebiet der Elektronenmikroskopie. ZEISS Informationen, **21**(83): 57-60.
— —, 1975. On *Ilyocypris schwarzbachi* Kempf. Stereo-Atlas of Ostracod Shells, **2**(38): 239-246.
— —, 1976. Low magnifications — a marginal area of electron microscopy. ZEISS Information, **21**(83): 57-60.
— —, 1977. Das mittelpleistozäne Interglazial von Berstadt. Geologische Abhandlungen Hessen, **75**: 68-73.
— —, 1980a. Index and bibliography of nonmarine Ostracoda 4: Bibliography A. Sonderveröffentlichungen des Geologischen Instituts der Universität zu Köln, **38**: 1-186.
— —, 1980b. Index and bibliography of nonmarine Ostracoda 3: Index C. Sonderveröffentlichungen des Geologischen Instituts der Universität zu Köln, **37**: 1-204.
— —, 1980c. Index and bibliography of nonmarine Ostracoda 2: Index B. Sonderveröffentlichungen des Geologischen Instituts der Universität zu Köln, **36**: 1-180.
— —, 1980d. Index and bibliography of nonmarine Ostracoda 1: Index A. Sonderveröffentlichungen des Geologischen Instituts der Universität zu Köln, **35**: 1-188.
— —, 1981. Zur Erforschungsgeschichte der nichtmarinen Ostrakoden. Sonderveröffentlichungen des Geologischen Instituts der Universität zu Köln, **41** (Brunnacker-Festschrift): 341-351.
— —, 1986a. Late Quaternary environmental changes in the eastern Sahara (Libyan Desert) documented by an ostracode diagram. In: Changements globaux en Afrique durant le Quaternaire: Passé — Present — Futur. Travaux et Documents, Institut Français de Recherche Scientifique pour le Développement en Cooperation (ORSTOM), **197**: 235-238.
— —, 1986b. Index and bibliography of marine Ostracoda 2: Index B. Sonderveröffentlichungen des Geologischen Instituts der Universität zu Köln, **51**: 1-712.
— —, 1986c. Index and bibliography of marine Ostracoda 1: Index A. Sonderveröffentlichungen des Geologischen Instituts der Universität zu Köln, **50**: 1-762.
— —, 1987. Index and bibliography of marine Ostracoda 3: Index C. Sonderveröffentlichungen des Geologischen Instituts der Universität zu Köln, **52**: 1-774, Köln.
— —, 1988. Index and bibliography of marine Ostracoda 4: Bibliography A. Sonderveröffentlichungen des Geologischen Instituts der Universität zu Köln, **53**: 1-454.
— —, 1989. Data on geographical and stratigraphical occurrence in publications dealing with Ostracoda. Cypris, International Ostracoda Newsletter, **7**: 118-121.
— —, 1991. Index and bibliography of nonmarine Ostracoda 5: Bibliography B. Sonderveröffentlichungen des Geologischen Instituts der Universität zu Köln, **77**: 1-232.
— — (ed.), 1993a. Festschrift Ulrich Jux. Sonderveröffentlichungen des Geologischen Instituts der Universität zu Köln, **70**: 1-600.
— —, 1993b. Ulrich Jux 60 Jahre: Biographische Skizze mit Bibliographie 1954-1989. Sonderveröffentlichungen des Geologischen Instituts der Universität zu Köln, **70** (Festschrift Ulrich Jux): 11-32.
— —, 1993c. Hydropteriden-Floren als Zeitmarken im Rheinischen Braunkohlen-Tertiär. Sonderveröffentlichungen des Geologischen Instituts der Universität zu Köln, **70** (Festschrift Ulrich Jux): 527-596.
— —, 1994a. Case 2896 — *Scottia* Brady & Norman, 1889 (Crustacea, Ostracoda): proposed designation of *Scottia pseudobrowniana* Kempf, 1971 as the type species. Bulletin of Zoological Nomenclature, **51**(4): 304-305.
— —, 1994b. Index and bibliography of marine Ostracoda 5: Bibliography B. Sonderveröffentlichungen des Geologischen Instituts der Universität zu Köln, **88**: 1-366.
— —, 1995a. Index and bibliography of marine Ostracoda 8: Index C, Supplement 1. Sonderveröffentlichungen des Geologischen Instituts der Universität zu Köln, **102**: 1-230.
— —, 1995b. Index and bibliography of marine Ostracoda 7: Index B, Supplement 1. Sonderveröffentlichungen des Geologischen Instituts der Universität zu Köln, **101**: 1-206.

— —, 1995c. Index and bibliography of marine Ostracoda 6: Index A, Supplement 1. Sonderveröffentlichungen des Geologischen Instituts der Universität zu Köln, **100**: 1-244.
— —, 1996. Index and bibliography of marine Ostracoda 9: Bibliography C. Sonderveröffentlichungen des Geologischen Instituts der Universität zu Köln, **103**: 1-198.
— —, 1997a. Index and bibliography of nonmarine Ostracoda 9: Bibliography C. Sonderveröffentlichungen des Geologischen Instituts der Universität zu Köln, **112**: 1-150.
— —, 1997b. Index and bibliography of nonmarine Ostracoda 8: Index C, Supplement 1. Sonderveröffentlichungen des Geologischen Instituts der Universität zu Köln, **111**: 1-152.
— —, 1997c. Index and bibliography of nonmarine Ostracoda 7: Index B, Supplement 1. Sonderveröffentlichungen des Geologischen Instituts der Universität zu Köln, **110**: 1-134.
— —, 1997d. Index and bibliography of nonmarine Ostracoda 6: Index A, Supplement 1. Sonderveröffentlichungen des Geologischen Instituts der Universität zu Köln, **109**: 1-148.
— —, 2002a. Index and bibliography of marine Ostracoda **10**: Bibliography D. CD-ROM edition only, with some 1500 comprehensive references to non-taxonomic publications, Frechen.
— —, 2002b. Index and bibliography of nonmarine Ostracoda **10**: Bibliography D. CD-ROM edition only, with some 1500 comprehensive references to non-taxonomic publications, Frechen.
— —, 2004. Index and bibliography of marine Ostracoda **16**: Index D, Recent marine Ostracoda of the World. CD-ROM edition only, with over 20 000 data sets, Frechen.
— —, 2006. Index and bibliography of nonmarine Ostracoda **16**: Index D, Recent nonmarine Ostracoda of the World. CD-ROM edition only, with over 27 500 data sets, Frechen.
— —, 2008a. Index and bibliography of marine Ostracoda **14**: Bibliography E. CD-ROM edition only, with some 800 comprehensive references to taxonomic publications, Frechen.
— —, 2008b. Index and bibliography of marine Ostracoda **13**: Index C, Supplement 2. CD-ROM edition only, with some 5200 new taxa in chronological order, Frechen.
— —, 2008c. Index and bibliography of marine Ostracoda **12**: Index B, Supplement 2. CD-ROM edition only, with some 5200 new taxa, species in alphabetical order, Frechen.
— —, 2008d. Index and bibliography of marine Ostracoda **11**: Index A, Supplement 2. CD-ROM edition only, with some 5200 new taxa, genera in alphabetical order, Frechen.
— —, 2009a. Prägende Jugendjahre während des Zweiten Weltkriegs in Messkirch. In: E. BÜCHIN (ed.), Denken — Gedanken — Andenken, Zum 90. Geburtstag von Elsbeth Büchin, 87-89, 107-130, 3 Abb., Messkirch.
— —, 2009b. *Paraparchites eskridgensis*, new name for the ostracode *Paraparchites punctatus* Watabe and Kaesler, 2004. Journal of Paleontology, **83**(3): 501.
— —, 2010. New names for homonym species and genera of Ostracoda (Arthropoda: Crustacea). Munis Entomology & Zoology, **5**(2): 551-558.
— —, 2011a. Nomenclature of Ostracoda, responsibility of authors, and the code of ethics. Joannea Geologie und Paläontologie, **11**: 96-98.
— —, 2011b. *Juxilyocypris* gen. nov. and replacement names for homonym species or genera of Ostracoda (Arthropoda: Crustacea). Munis Entomology & Zoology, **6**(2): 955-969.
— —, 2012. Substitute names for homonym species of Cytherellidae (Ostracoda: Podocopa: Platycopida). Munis Entomology & Zoology, **7**(2): 1260-1267.
KEMPF, E. K. & C. NINK, 1993. *Henryhowella asperrima* (Ostracoda) aus der Typusregion (Miozän: Badenian; Wiener Becken). Sonderveröffentlichungen des Geologischen Instituts der Universität zu Köln, **70** (Festschrift Ulrich Jux): 95-114.
KEMPF, E. K. & B. W. SCHARF, 1981. Lebende und fossile Muschelkrebse (Crustacea: Ostracoda) vom Laacher See. Mitteilungen der Pollichia, **68**(1980): 205-236.
SYLVESTER-BRADLEY, P. C. & E. K. KEMPF, 1973. On *Ilyocypris monstrifica* (Norman). Stereo-Atlas of Ostracod Shells, **1**(28): 149-152.

First received 6 July 2013.
Final version accepted 15 December 2013.

[When citing this chapter, refer to Crustaceana 87 (2014) 923–951]

This paper is dedicated to Prof. Dr. Eugen Kempf on the occasion of his 80th birthday (16 April 2012)

A REVIEW OF THE ENTOCYTHERIDAE (OSTRACODA) OF THE WORLD: UPDATED BIBLIOGRAPHIC AND SPECIES CHECKLISTS AND GLOBAL GEOREFERENCED DATABASE, WITH INSIGHTS INTO HOST SPECIFICITY AND LATITUDINAL PATTERNS OF SPECIES RICHNESS

BY

ALEXANDRE MESTRE[1,2]), JUAN S. MONRÓS[2]) and FRANCESC MESQUITA-JOANES[1,2,3])

[1]) Department of Microbiology and Ecology, University of Valencia, Dr Moliner 50, E-46100 Burjassot, Spain

[2]) "Cavanilles" Institute of Biodiversity and Evolutionary Biology, University of Valencia, Polígono La Coma s/n, E-46980 Paterna, Valencia, Spain

ABSTRACT

The creation of biodiversity datasets freely available for the scientific community is a valuable task to stimulate global research on biodiversity. Among others, the Global Biodiversity Information Facility (GBIF) is a remarkable resource providing free online access to biodiversity data on many diverse taxonomic groups (including Ostracoda) from both terrestrial and marine ecosystems. More specific databases for Ostracoda georeferenced data have been built (e.g., NACODe and OMEGA), some with freely available data. However, the Entocytheridae, a family of ostracods living commensal on other crustaceans, with 220 living species, representing the third non-marine ostracod family in number of species, has been remarkably under-represented in the currently available biodiversity databases. To cover this gap, we present here a free-access world database of Entocytheridae published in GBIF and review the current knowledge of the group by updating a bibliographic and species checklists of the Entocytheridae. We also analyse the host specificity of the group and the latitudinal species richness pattern in North and Central America (including 186 spp.). The current database includes 3509 georeferenced records from 220 species, in contrast to just 44 entocytherid georeferenced records previously published in GBIF. In addition, the updated species list accounts for 43 species and 2 genera that were not included in the previous compendium on Entocytheridae published by Hart & Hart in 1974, so as 40 species not included in the 2013 Checklist provided by the Catalogue of Life. We show that the specialisation in one unique host species is not the rule in Entocytheridae, and evidence an unusual latitudinal pattern of species richness in North and Central American entocytherids, most probably related to the biogeographic history of their hosts.

[3]) Corresponding author; e-mail: mezquita@uv.es

RESUMEN

La creación de bases de datos de biodiversidad libres para la comunidad científica es una tarea valiosa para estimular la investigación en biodiversidad. Entre otros repositorios, la Infraestructura Mundial de Información en Biodiversidad (GBIF) es un destacado recurso libre y gratuito que provee acceso a datos de biodiversidad de grupos taxonómicos muy diversos (incluyendo Ostracoda) de ecosistemas terrestres y marinos. Anteriormente ya se habían construido bases de datos específicas para datos georreferenciados de Ostracoda (p.e. NACODe y OMEGA), algunas de acceso libre. Sin embargo, los Entocytheridae, una familia de ostrácodos comensales de otros crustáceos que, con 220 especies vivas, representan la tercera familia de ostrácodos no marinos en número de especies, han estado notablemente subrepresentados en la bases de datos disponibles. Con el fin de cubrir esta carencia, presentamos una base de datos mundial de la familia Entocytheridae de acceso libre publicada en GBIF, y revisamos el conocimiento actual del grupo mediante la actualización de las listas de las especies conocidas de Entocytheridae y de su bibliografía. También analizamos la especificidad de hospedadory el patrón latitudinal de riqueza de especies en América del Norte y Central (incluyendo 186 spp.). La base de datos actual contiene 3509 registros de 220 especies, que contrasta con los 44 únicos registros georreferenciados previamente publicados en GBIF. Además, la lista actualizada contiene 43 especies y 2 géneros que no aparecían en la última revisión sobre Entocytheridae publicada por Hart & Hart en 1974, así como 40 especies no incluidas en la lista de 2013 del Catálogo de la Vida. Por otro lado, mostramos que la especialización en una única especie de hospedador no parece ser la regla en los Entocytheridae, y descubrimos un patrón de riqueza de especies inusual en los entocitéridos de América del Norte y Central, probablemente relacionado con la historia biogeográfica de sus hospedadores.

INTRODUCTION

The integration of biodiversity information datasets covering all taxonomic groups to create a global biodiversity database freely available to the scientific community is a valuable task to stimulate global research on biodiversity. To be able to attain such an aim, the existence of a consistent, reliable and updated list of the known biological species is a previous and essential step. With this aim, the Integrated Taxonomy Information System (ITIS) partnership (http://itis.gov) and the Species 2000 federation (http://sp2000.org/) planned the Catalogue of Life (http://catalogueoflife.org/), in order to build a list of all known species of organisms on Earth from multiple peer-reviewed taxonomic databases, periodically providing updated species checklist editions, such as the Annual Checklist (e.g., Roskov et al., 2013). Wikispecies (http://en.wikipedia.org/wiki/Wikispecies) is another project with similar aims to provide an open access and dynamic catalogue of all species directed at scientists, with high potential to store useful information related to each species. Another important step is the creation of standard formats to store biodiversity information, in order to facilitate its transference and sharing. Along this line, the Biodiversity Information Standards (TDWG, Berendsohn et al., 2011) created the Darwin Core (Darwin Core Task Group & TDWG, 2009), a body of standards to store information about taxa, their occurrence in nature as documented by observations, specimens, and samples, and related information.

Taxonomically supported by the Catalogue of Life and based on Darwin Core standard formats, the Global Biodiversity Information Facility (GBIF, http://data.gbif.org) is an international organisation to provide open free access to biodiversity data via internet through a global network of countries and organizations. The Ocean Biogeographic Information System (OBIS, http://iobis.org), World Register of Marine Species (WoRMS, Appeltans et al., 2012) and the Freshwater Animal Diversity Assessment (FADA, Balian et al., 2008) are similar projects involving marine and freshwater species databases.

More specific databases restricted to a given taxonomic group are more or less available to scientists. There are various examples of this type of database for Ostracoda. Indeed, the interest of ostracodologists in the construction of Ostracoda databases is remarkable. One of the first works on this area is the Kempf Database on Ostracoda (KDO, Kempf, 1980a-d, 1991, 1997a-d, 2002, 2006), a bibliographic compendium for the ostracodologists. In addition to this, a series of ostracod database projects rose with a biogeographic approach: the North American Combined Ostracode Database (NACODe, Curry & Smith, 2012), the Nonmarine Ostracod Distribution in Europe (NODE, Horne et al., 1998), the Ostracod Metadatabase of Environmental and Geographical Attributes (OMEGA, Horne et al., 2011) and the World Ostracoda Database (Brandão et al., 2013). The NACODe is a fusion of two previous databases, the Delorme Ostracode Database, that contains data from Canadian freshwater bodies, accessible via the Canadian Museum, and the North American Nonmarine Ostracode Database (NANODe), built from North American freshwater collections and with free internet access; NODE is a database restricted to freshwater European data mainly from the literature, but not freely accessible; OMEGA, also privately hosted, is a more recent database that includes the above-mentioned two databases with additional South African and Japanese records, and was designed to calibrate non-marine ostracods for palaeoclimate and palaeoenvironmental applications; finally, the World Ostracoda Database incorporates at present a free-access world database of marine ostracods integrated into WoRMS.

The entocytherids are a peculiar family of ostracods because of the obligate epibiont (symbiont) mode of life of all its component species. Their hosts are crustaceans belonging to Decapoda, Amphipoda and Isopoda. Commensalism is the most accepted hypothesis about the relationships between the entocytherids and their hosts (Hart & Hart, 1974). The ostracod family Entocytheridae, represented by 178 species, 31 genera and 5 subfamilies according to Hart & Hart (1974), the last review of this family hitherto, is the third non-marine Ostracoda family (out of 13 living families) with the highest number of species after the Cyprididae and

Candonidae (Martens et al., 2008). However, despite we can find a high number of publications dealing with the presence of entocytherids (e.g., Hobbs & Walton, 1975, 1976; Hobbs & Peters, 1977; Andolshek & Hobbs, 1986), they are not included in major ostracod biogeographic databases, with the exception of NODE. However, this resource is restricted to European records, while the two main entocytherid subfamilies (Entocytherinae and Notocytherinae), representing almost 95% of the entocytherid species, are native to North and Central America (Entocytherinae) and Australasia (Notocytherinae) (Hart & Hart, 1974). Moreover, the GBIF online resource only included 44 entocytherid occurrences hitherto, mainly corresponding to sampling localities of specimens preserved in Museum collections (e.g., National Museum of Natural History, Australian Museum). Another example of the reduced attention paid to the entocytherids by most ostracodologists are recent reviews of the world non-marine ostracod fauna that only consider free-living ostracods (Martens et al., 2008; Martens & Savatenalinton, 2011), disregarding this commensal group, or describing it in a very succinct way (e.g., Karanovic, 2012).

Host specificity is an important parameter to understand the symbiont-host relationships, and it is closely related to the evolution of their association. Generally, parasites tend to be strongly host-specific (Poulin, 2007). In contrast, following the theory about the forces involved in host specificity proposed by Combes (1991), we expect lower host specificity in commensals than in parasites because the "compatibility filter" that the biont has to overcome to colonize a host should be less restrictive for commensals than parasites. Nevertheless, compared with studies carried out on parasites, the commensal-host relationship has been under-treated. Entocytheridae is a group that has been found in association with a wide variety of hosts composed by crustacean species from three different orders: Decapoda, Amphipoda and Isopoda. Due to the wide range of host species in the group and their commensal relationships with their host, the entocytherids are an interesting group to analyse issues on host specificity in symbiotic interactions.

Georeferenced databases with large-scale geographic ranges are essential in testing one of the most recognised patterns in biogeography: the latitudinal gradient of species richness, which decrease from the tropics toward the poles (Hillebrand, 2004). This is a very general pattern, but there are some exceptions (e.g., Rivadeneira et al., 2011). Non-free-living species represent special systems that may be strongly cued to particular factors such as their host specificity, dispersal abilities and coevolution (Poulin, 2007). In this sense, the entocytherids could be useful for checking this type of patterns in non-free-living species.

The main aim of this work was to build a free-access biogeographic database of the family Entocytheridae with all the published world occurrences of the group hitherto. We also aimed at compiling updated bibliographic and species checklists of this ostracod family, not updated since the review of Hart & Hart (1974). Additionally, we analyse the host specificity of the group and the latitudinal species richness pattern of the North and Central American species.

MATERIAL AND METHODS

In order to build our biogeographic database, we previously carried out an intensive bibliographic search to obtain all the publications containing in the text body (excluding the references section) a direct mention to the family Entocytheridae or any member of the group. Afterwards, we checked these publications by looking for all the entocytherid occurrences. Then, we created an entocytherid database using the Database Management System PostgreSQL v. 9.1 (The PostgreSQL Global Development Group, 2012) where we transferred the occurrence data obtained from the bibliographic search. The database's variables followed the standard Darwin Core. When finished, the "Entocytheridae (Ostracoda) World Database" (http://www.gbif.es/gbif/ic_colecciones.php?ID_Coleccion=10114) was converted to a Darwin Core archive by using the Integrated Publishing Toolkit (IPT) v. 2.0.5 (http://www.gbif.es:8080/ipt/) in GBIF, where a copy of the free-access database (Mestre & Mesquita-Joanes, 2013) has been hosted.

To analyse the host specificity of Entocytheridae species, we used the information obtained from the literature regarding to the host species cited in association with the entocytherids in the localities where they were found together. We only considered those citations where the host species was identified. Therefore, the records without identified hosts were removed, leaving a total of 203 entocytherid species for analysis. With this information, we did a histogram for the distribution of entocytherid species according to the number of associated host species. To analyse the latitudinal changes in species richness for the North and Central American entocytherids we represented the entocytherid species richness per 5° latitudinal interval on the American continent between 15-55° of latitude. The species richness of each latitudinal interval was estimated using the Margalef's diversity index ($D_{MG} = (S - 1)/\ln(N)$, where S is the number of species and N the number of occurrences; Magurran, 2004) in order to minimize the effects of sampling bias between latitude intervals. All the analyses were done using PostgreSQL v. 9.1 and R-cran v. 3.0.1 (R Core Team, 2013) through the RPostgreSQL package v. 0.3 (Conway et al., 2012) connecting both software programmes.

RESULTS

As a result of our bibliographic search, we found a total of 215 publications containing direct citations to entocytherids (Appendix A). The temporal variation

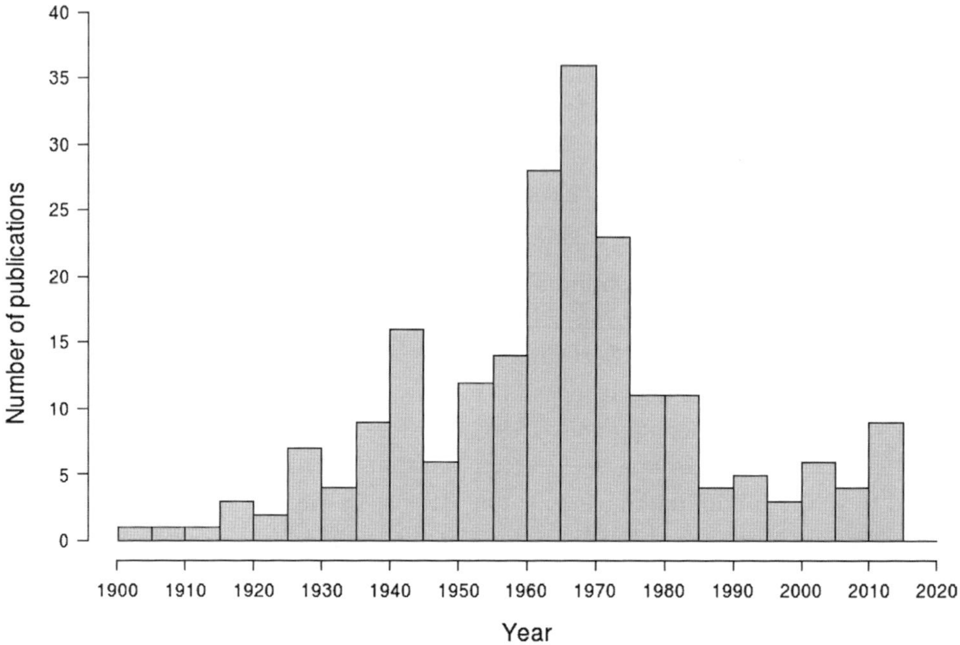

Fig. 1. Temporal evolution of the number of publications on entocytherids. Publications that have a direct mention of the family Entocytheridae or any component of the group in their text body (excluding the references section) are included (Appendix A), summing up to a total of 215 publications.

of the number of publications on entocytherids (fig. 1) shows that the main effort was made during the 1960s and early 1970s with a maximum of 35 publications in 5 years (this is, 7 publications per year on average). During the late 1970s and in the 1980s there was a reduction in the number of publications until it reached a mean number below one publication per year until present, with a slight increasing trend during the last few years.

The present database contains entocytherid records for 220 species, 35 genera and 5 subfamilies (included in the species checklist in Appendix B) with a total of 3509 occurrences. The main body of the data, with 3429 occurrences, comes from North and Central America, dominated by members of the subfamily Entocytherinae (with 183 species), together with some records of three Sphaeromicolinae species of the genus *Hobbsiella*. Oceania, represented by the subfamily Notocytherinae, contains 37 records. The European entocytherid records, with also 37 occurrences, correspond to the subfamilies Sphaeromicolinae and Hartiellinae, with the exception of some exotic American entocytherids (Entocytherinae), mainly located in the Iberian Peninsula. Finally, there are isolated records of Entocytherinae in Hawaii (1 occurrence) and Japan (2 occurrences), and Microsyssitrinae in southern India (2 occurrences) and South Africa (1 occurrence) (fig. 2).

Fig. 2. World distribution of the ostracod family Entocytheridae based on the free-access Entocytheridae (Ostracoda) World Database published in GBIF (Mestre & Mesquita-Joanes, 2013) and presented in this study, composed by 3509 occurrences and 220 species.

Regarding the analysis of host specificity, we identified 248 host species in association with the entocytherids (234 Decapoda, 13 Isopoda and 1 Amphipoda). The species of the main host group, Decapoda, were composed by 217 Cambaridae, 12 Parastacidae, 4 Astacidae and 1 Pseudothelphusidae. The distribution of the entocytherid species according to the number of hosts (fig. 3) shows a truncated pattern over lower numbers of host species, with 104 species (approx. 50% of the 203 species analysed) having less than three hosts. However, only 73 entocytherid species had only one associated host species. Moreover, 32 entocytherids were cited in association with more than 20 host species, with some cases exceeding 40 host species (i.e., 47 for *Ankylocythere sinuosa*, 48 for *A. ancyla*, 50 for *Uncinocythere simondsi* and 64 for *Entocythere elliptica*).

The data selected for the analysis of latitudinal variation in species richness of North and Central American entocytherids included 186 entocytherid species, recorded in a total of 2410 American locations between 15 and 55° of latitude. Our results show a non-linear latitudinal pattern of entocytherid species richness characterised by a maximum peak at intermediate latitudes (around 30-40°) with a strong decrease towards higher and lower latitudes (fig. 4).

DISCUSSION

Research on the Entocytheridae, with a pronounced peak in productivity around the 1960s and the first half of the 1970s, that culminated in the review by Hart & Hart (1974), suffered a negative trend in number of publications since the

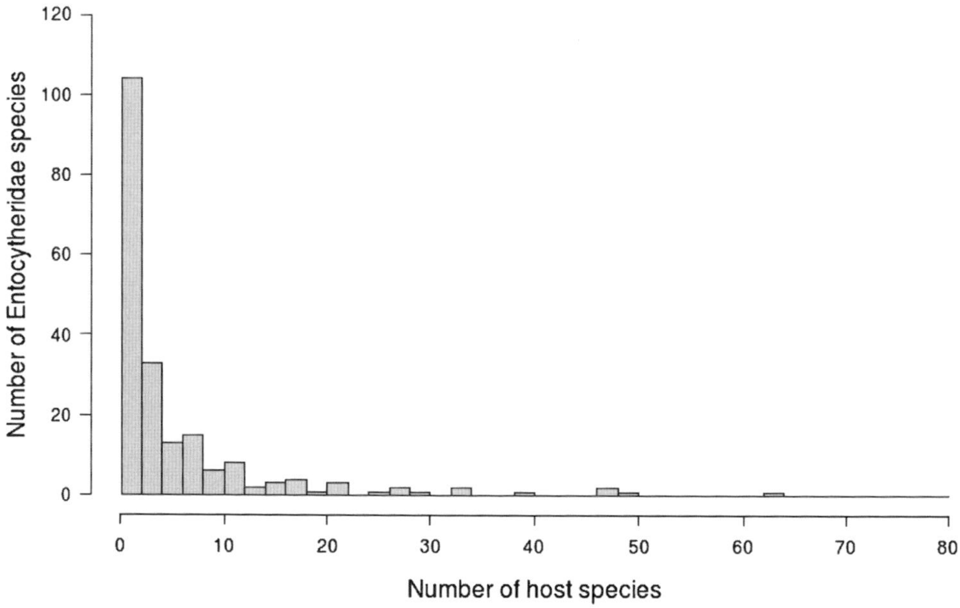

Fig. 3. Frequency distribution of Entocytheridae species in relation to the number of associated host species. The data were obtained from the literature about the group (Annex I), by only considering the citations of identified host species found in association with entocytherids, including 203 entocytherid species and 248 host species.

late 1970s to date, probably due to the decrease of scientific activity of some authors that have considerably contributed to the study of entocytherids (e.g., H. H. Hobbs, Jr. and C. W. Hart, Jr.). The slight increase of publications in recent years are mainly related to the discovering of well-established exotic entocytherid populations in Europe and Japan, associated with the invasive American crayfishes *Procambarus clarkii* (Girard, 1852) and *Pacifastacus leniusculus* (Dana, 1852) (Smith & Kamiya, 2001; Grabow & Martens, 2009; Grabow et al., 2009; Mestre et al., 2011, 2012, 2013; Aguilar-Alberola et al., 2012; Castillo-Escrivà et al., 2012, 2013). In this sense, the Entocytheridae could represent a new model of exotic ostracod dispersal, more related to the invasive success of their hosts and in which the entocytherid-host relationships have an important role, unlike other ostracod dispersal processes such as the passive dispersal by human activities (McKenzie & Moroni, 1986; Escrivà et al., 2012) or bird dispersal (Figuerola & Green, 2002; Frisch et al., 2007), where other different types of factors are involved, such as geographic parthenogenesis and desiccation-resistant eggs (McKenzie & Moroni, 1986).

The Entocytheridae (Ostracoda) World Database (Mestre & Mesquita-Joanes, 2013) increases the number of georeferenced entocytherid sites in GBIF with 3509 new records, compared to the previous number of just 44 records. The database

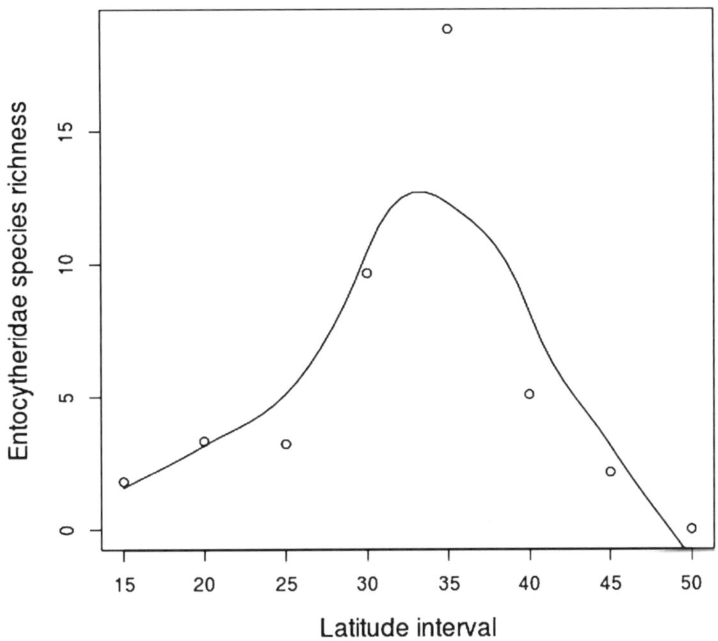

Fig. 4. Species richness of North and Central American Entocytheridae (186 spp.) per 5° latitudinal interval on the American region between 15-55° of latitude. The species richness was estimated using the Margalef's diversity index ($D_{MG} = S - 1/\ln(N)$ where S is the number of species and N the number of occurrences). The fitted curve corresponds to a LOESS model of the Gaussian family with a smoothing parameter $\alpha = 2/3$.

and species checklist also integrate 43 species and 2 genera that were not included in the last review of the family published by Hart & Hart (1974), because they were described later. The two new genera are *Psittocythere*, erected to name a new species discovered by Hobbs & Walton (1975), and *Hobbsiella*, a name proposed by Danielopol & Hart (1985) to group three American species belonging to Sphaeromicolinae: *Hobbsiella cirolanae* (Rioja, 1951), *Hobbsiella coahuiltecae* (Hobbs & Hobbs, 1973) and *Hobbsiella moria* (Hart, 1978). Moreover, the updated species checklist (Appendix B) contains 40 species not present in the 2013 Annual Checklist of the Catalogue of Life (Roskov et al., 2013): *Ankylocythere barbouri*, *Ankylocythere spargosis*, *Dactylocythere scotos*, *Entocythere prisma*, *Plectocythere odelli* and all the species belonging to Notocytherinae (except *Laccocythere aotearoa*), Sphaeromicolinae (except *Hobbsiella moria*), Hartiellinae and Microsyssitrinae. Besides the new taxa, we incorporated some other posterior taxonomic changes proposed for the group. These are the cases of the synonymy of *Ankylocythere sinuosa* and *Ankylocythere tiphophila* (Crawford, 1959) in a unique species, *Ankylocythere sinuosa*, as established by Peters & Pugh (1999), and the substitution of the generic name *Aphelocythere* by *Waltoncythere*, as proposed by Hobbs & Peters (1978).

Our results confirm that low host specificity is well established in the family Entocytheridae (approx. 64% of the 203 species analysed have been found associated with more than one host species), as suggested by Hart & Hart (1974). The lower restrictions in host compatibility related to the host defences compared with parasites might facilitate this establishment. Nevertheless, probably other factors involved on the evolution of the group's host specificity, such as interspecific competition, promoted the specialisation of some entocytherid species in a unique host, as evidenced by the 73 entocytherid species with only one associate host. In this sense, there are some evidences of spatial segregation between entocytherid species within the host to avoid the interspecific competition and allowing them to share the same host (Hobbs et al., 1967).

The peculiar latitudinal pattern in species richness followed by the American entocytherids, with a peak around 30-40°N may be related to the similar observed pattern in the American crayfishes, their main host group, by France (1992). This pattern could be explained by the radiation in diversity of the American crayfishes in the Southeast of North America that occurred from the last Cretaceous and through the Cenozoic (Hobbs, 1988), favoured by the great diversity of semi-aquatic habitats in the region. The decrease of species richness at higher latitudes is probably due to the effect of the recent glaciation events that preclude a long history of colonisation and evolution. The absence of crayfishes and entocytherids in lower latitudes is thought to be a result of competitive exclusion by the freshwater crabs inhabiting these regions (Scholtz, 2002).

In this survey on the ostracod family Entocytheridae, we present a contribution to the GBIF project with a free access world Entocytheridae database (Mestre & Mesquita-Joanes, 2013) with 3509 records from 220 different species. We also contribute with both updated reference and species checklists of Entocytheridae, with significant changes from the last review of the family by Hart & Hart (1974). Moreover, we show that the high specialization in one unique host species is not the rule in Entocytheridae, but there is a high variability in the number of host species found in association with an entocytherid species, with extreme cases of some species with more than 40 observed associated host species. Finally, we evidence an unusual latitudinal pattern of species richness in the North and Central American entocytherids, probably related to their link with the biogeographic history of their hosts, through the frame of coevolution. We expect that the increasing availability of ostracod databases such as the one presented here will allow further studies on crustacean biogeography and a better assessment of world biodiversity patterns.

ACKNOWLEDGEMENTS

This work would not have been made possible without the previous efforts of many researchers, cited as authors of the publications included in Appendix B, who sampled crayfishes, isopods or amphipods, or checked crayfish collections, from a great number of localities around the world, looking for entocytherids, and published their results allowing the availability of data for this work. Our work was partially funded by the Spanish Ministry of Science and Innovation project ECOINVADER (CGL2008-01296/BOS) and the University of Valencia ("V-Segles" predoctoral grant to A.M.).

REFERENCES

AGUILAR-ALBEROLA, J. A., F. MESQUITA-JOANES, S. LÓPEZ, A. MESTRE, J. C. CASANOVA, J. RUEDA & A. RIBAS, 2012. An invaded invader: infection of the red swamp crayfish *Procambarus clarkii* (Girard 1852) by entocytherid ostracods in the Iberian Peninsula. Hydrobiologia, **688**: 63-73.

ANDOLSHEK, M. & H. H. HOBBS JR., 1986. The entocytherid ostracod fauna of Southeastern Georgia. Smithsonian Contributions to Zoology, **424**: 1-43.

APPELTANS, W., P. BOUCHET, G. A. BOXSHALL, C. DE BROYER, N. J. DE VOOGD, D. P. GORDON, B. W. HOEKSEMA, T. HORTON, M. KENNEDY, J. MEES, G. C. B. POORE, G. READ, S. STÖHR, T. C. WALTER & M. J. COSTELLO (eds.), 2012. World Register of Marine Species, available online at http://www.marinespecies.org (accessed 12 September 2013).

BALIAN, E. V., H. SEGERS, C. LÉVVÊQUE & K. MARTENS, 2008. An introduction to the Freshwater Animal Diversity Assessment (FADA) project. Hydrobiologia, **595**: 3-8.

BERENDSOHN, W., A. GÜNTSCH, N. HOFFMANN, A. KOHLBECKER, K. LUTHER & A. MÜLLER, 2011. Biodiversity information platforms: from standards to interoperability. ZooKeys, **150**: 71-87.

BRANDÃO, S. N., M. V. ANGEL & I. KARANOVIC, 2013. World Ostracoda Database, available online at http://www.marinespecies.org/ostracoda (accessed 12 September 2013).

CASTILLO-ESCRIVÀ, A., A. MESTRE, J. S. MONRÓS & F. MESQUITA-JOANES, 2012. Population dynamics of *Ankylocythere sinuosa* (Rioja, 1942) (Ostracoda, Entocytheridae), epibiont on the red swamp crayfish (*Procambarus clarkii*) in Pego-Oliva Wetland (Eastern Spain). In: F. A. VIEHBERG & R. GROMIG (eds.), Abstracts, 14th International German Ostracodologists' Meeting, October 11-14, 2012, Cologne. Kölner Forum für Geologie und Paläontologie, **21**: 8.

— —, — —, — — & — —, 2013. Population dynamics of an epibiont Ostracoda on the invasive red swamp crayfish *Procambarus clarkii* in a western Mediterranean wetland. Hydrobiologia, **714**: 217-228.

COMBES, C., 1991. Evolution of parasite life cicles. In: C. A. TOFT, A. AESCHLIMANN & L. BOLIS (eds.), Parasite-host associations: coexistence or conflict?: 62-82. (Oxford University Press, Oxford).

CONWAY, J., D. EDDELBUETTEL, T. NISHIYAMA, S. K. PRAYAGA & N. TIFFIN, 2012. R interface to the PostgreSQL database system: 1-36. (PostgreSQL Global Development Group and The Regents of the University of California).

CURRY, B. B. & A. J. SMITH, 2012. NACODe (North American Combined Ostracode Database): introduction and application. Geological Society of America Abstracts with Programs, **44**(7): 39.

DANIELOPOL, D. L. & C. W. HART, JR., 1985. Notes of the center of origin and antiquity of the Sphaeromicolinae, with description of *Hobbsiella*, new genus (Ostracoda, Entocytheridae). Stygologia, **1**: 54-70.

DARWIN CORE TASK GROUP & BIODIVERSITY INFORMATION STANDARDS — TDWG, 2009. Darwin Core, available online at http://rs.tdwg.org/dwc (accessed 12 September 2013).

ESCRIVÀ, A., R. SMITH, J. A. AGUILAR-ALBEROLA, T. KAMIYA, I. KARANOVIC, J. RUEDA, E. I. SCHORNIKOV & F. MESQUITA-JOANES, 2012. Global distribution of *Fabaeformiscandona subacuta*: an exotic invasive Ostracoda on the Iberian Peninsula? Journal of Crustacean Biology, **32**(6): 949-961.

FIGUEROLA, J. & A. J. GREEN, 2002. Dispersal of aquatic organisms by waterbirds: a review of past research and priorities for future studies. Freshwater Biology, **47**: 483-494.

FRANCE, R., 1992. The North American latitudinal gradient in species richness and geographical range of freshwater crayfish and amphipods. The American Naturalist, **139**(2): 342-354.

FRISCH, D., A. J. GREEN & J. FIGUEROLA, 2007. High dispersal capacity of a broad spectrum of aquatic invertebrates via waterbirds. Aquatic Sciences, **69**: 568-574.

GRABOW, K. & A. MARTENS, 2009. Neozoische Aufsitzer auf in Deutschland eingeschleppten Signalkrebsen. Internationales Flußkrebsforum 07.-10.Mai 2009 Gersfeld/Rhön, Deutschland: 38-42.

GRABOW, K., G. SCHOOLMANN & A. MARTENS, 2009. Epizoen auf in Deutschland eingeschleppten Flusskrebsen. Deutsche Gesellschaft für Limnologie (DGL) Erweiterte Zusammenfassungen der Jahrestagung 2008 (Konstanz), Hardegsen 2009: 386-389.

HART, D. G. & C. W. HART, JR., 1974. The ostracod family Entocytheridae: 1-239. (Fulton Press, Lancaster, PA).

HILLEBRAND, H., 2004. On the generality of the latitudinal diversity gradient. The American Naturalist, **163**: 192-211.

HOBBS JR., H. H., 1988. Crayfish distribution, adaptative radiation and evolution. In: D. M. HOLDICH & R. S. LOWERY (eds.), Freshwater crayfish: biology, management and exploitation: 52-82. (Croom Helm., London).

HOBBS JR., H. H., P. C. HOLT & M. WALTON, 1967. The crayfishes and their epizootic ostracod and branchiobdellid associates of the Mountain Lake, Virginia, region. Proceedings of the United States Natural Museum, **123**(3602): 1-84.

HOBBS JR., H. H. & D. J. PETERS, 1977. The entocytherid ostracods of North Carolina. Smithsonian Contributions to Zoology, **247**: 1-73.

— — & — —, 1978. A substitute name for the homonym *Aphelocythere* Hobbs and Peters (Ostracoda Entocytheridae). Proceedings of the Biological Society of Washington, **91**(4): 1037.

HOBBS JR., H. H. & M. WALTON, 1975. New entocytherid ostracods from Tennessee with a key to the species of the genus *Ascetocythere*. Proceedings of the Biological Society of Washington, **88**(2): 5-20.

— — & — —, 1976. New entocytherid ostracods from Kentucky and Tennessee. Proceedings of the Biological Society of Washington, **89**(33): 393-404.

HORNE, D. J., A. BALTANÁS & G. PARIS, 1998. Geographical distribution of reproductive modes in living non-marine ostracods. In: K. MARTENS (ed.), Sex and parthenogenesis: 77-99. (Backhuys, Leiden).

HORNE, D. J., B. B. CURRY, L. D. DELORME, K. MARTENS, A. J. SMITH & R. J. SMITH, 2011. OMEGA: the ostracod metadatabase of environmental and geographical attributes. Joannea-Geologie und Paläontologie, **11**: 80-84.

KARANOVIC, I., 2012. Recent freshwater ostracods of the world. Crustacea, Ostracoda, Podocopida: 1-608. (Springer, New York, NY).

KEMPF, E. K., 1980a. Index and bibliography of nonmarine Ostracoda. Bibliography A. Part 4. Sonderveröffentlichungen des Geologischen Institut der Universität zu Köln, **38**: 1-186.

— —, 1980b. Index and bibliography of nonmarine Ostracoda. Index A. Part 1. Sonderveröffentlichungen des Geologischen Institut der Universität zu Köln, **35**: 1-188.

— —, 1980c. Index and bibliography of nonmarine Ostracoda. Index B. Part 2. Sonderveröffentlichungen des Geologischen Institut der Universität zu Köln, **36**: 1-180.

— —, 1980d. Index and bibliography of nonmarine Ostracoda. Index C. Part 3. Sonderveröffentlichungen des Geologischen Institut der Universität zu Köln, **37**: 1-204.

— —, 1991. Index and bibliography of nonmarine Ostracoda. Bibliography B. Part 5. Sonderveröffentlichungen des Geologischen Institut der Universität zu Köln, **77**: 1-238.

— —, 1997a. Index and bibliography of nonmarine Ostracoda. Bibliography C. Part 9. Sonderveröffentlichungen des Geologischen Instituts der Universität zu Köln, **112**: 1-144.

— —, 1997b. Index and bibliography of nonmarine Ostracoda. Index A Supplement. Part 6. Sonderveröffentlichungen des Geologischen Institut der Universität zu Köln, **109**: 1-142.

— —, 1997c. Index and bibliography of nonmarine Ostracoda. Index B Supplement 1. Part 7. Sonderveröffentlichungen des Geologischen Institut der Universität zu Köln, **110**: 1-134.

— —, 1997d. Index and bibliography of nonmarine Ostracoda. Index C Supplement 1. Part 8. Sonderveröffentlichungen des Geologischen Institut der Universität zu Köln, **111**: 1-152.

— —, 2002. Index and bibliography of nonmarine Ostracoda. Part 10, Frechen.

— —, 2006. Index and bibliography of nonmarine Ostracoda. Index D, Recent nonmarine Ostracoda of the World. Part 16, Frechen.

MAGURRAN, A. E., 2004. Measuring biological diversity: 1-256. (Blackwell Publishing, Oxford).

MARTENS, K. & S. SAVATENALINTON, 2011. A subjective checklist of the Recent, free-living, non-marine Ostracoda (Crustacea). Zootaxa, **2855**: 1-79.

MARTENS, K., I. SCHÖN, C. MEISCH & D. J. HORNE, 2008. Global diversity of ostracods (Ostracoda, Crustacea) in freshwater. Hydrobiologia, **595**: 185-193.

MCKENZIE, K. G. & A. MORONI, 1986. Man as an agent of crustacean passive dispersal via useful plants — exemplified by Ostracoda *Ospiti esteri* of the Italian ricefields ecosystem — and implications arising therefrom. Journal of Crustacean Biology, **6**(2): 181-198.

MESTRE, A. & F. MESQUITA-JOANES, 2013. Entocytheridae (Ostracoda) World Database, available online at http://www.gbif.es/ic_colecciones.php?ID_Coleccion=10114 (accessed 16 September 2013).

MESTRE, A., J. S. MONRÓS & F. MESQUITA-JOANES, 2011. Comparison of two chemicals for removing an entocytherid (Ostracoda: Crustacea) species from its host crayfish (Cambaridae: Crustacea). International Review of Hydrobiology, **96**(4): 347-355.

— —, — — & — —, 2012. Building a database of the entocytheridae ostracods of the world. In: F. A. VIEHBERG & R. GROMIG (eds.), Abstracts, 14th International German Ostracodologists' Meeting, October 11-14, 2012, Cologne. Kölner Forum für Geologie und Paläontologie, **21**: 45-46.

— —, — — & — —, 2013. Do Rapoport effects apply to epibiont aquatic invertebrates? The case of ostracod Entocytheridae commensal on freshwater crustaceans. Frontiers of biogeography, **4**(Suppl. 1): 114-115.

PETERS, D. J. & J. E. PUGH, 1999. On the entocytherid ostracods of the Brazos River basin and adjacent coastal region of Texas. Proceedings of the Biological Society of Washington, **112**: 338-351.

POULIN, R., 2007. Evolutionary ecology of parasites: 1-332. (Princeton University Press, Princeton, NJ).

R CORE TEAM, 2013. R: a language and environment for statistical computing: 1-409. (R Foundation for Statistical Computing, Vienna).

RIVADENEIRA, M. M., M. THIEL, E. R. GONZÁLEZ & P. A. HAYE, 2011. An inverse latitudinal gradient of diversity of peracarid crustaceans along the Pacific Coast of South America: out of the deep south. Global Ecology and Biogeography, **20**: 437-448.

ROSKOV, Y., T. KUNZE, L. PAGLINAWAN, T. ORRELL, D. NICOLSON, A. CULHAM, N. BAILLY, P. KIRK, T. BOURGOIN, G. BAILLARGEON, F. HERNANDEZ & A. DE WEVER (eds.), 2013. Species 2000 & ITIS Catalogue of Life, 11th March 2013. Digital resource at www.catalogueoflife.org/col/. (Species 2000, Reading).

SCHOLTZ, G., 2002. Phylogeny and evolution. In: D. M. HOLDICH (ed.), Biology of freshwater crayfish: 30-52. (Blackwell Science, London).

SMITH, R. J. & T. KAMIYA, 2001. The first record of an entocytherid ostracod (Crustacea: Cytheroidea) from Japan. Benthos Research, **56**: 57-61.

THE POSTGRESQL GLOBAL DEVELOPMENT GROUP, 2012. PostgreSQL 9.2.1 Documentation: 1-2773. (The PostgreSQL Global Development Group).

APPENDIX A

List of publications with a direct mention of the family Entocytheridae or any member of the group in the main text (excluding the references section).

AGUILAR-ALBEROLA, J. A., F. MESQUITA-JOANES, S. LÓPEZ, A. MESTRE, J. C. CASANOVA, J. RUEDA & A. RIBAS, 2012. An invaded invader: infection of the red swamp crayfish *Procambarus clarkii* (Girard 1852) by entocytherid ostracods in the Iberian Peninsula. Hydrobiologia, **688**: 63-73.

ALLEN, S. A., 1933. Parasites and commensals of North Carolina crayfishes. Journal of Elisha Mitchell Scientific Society, **49**: 119-121.

ANDOLSHEK, M. & H. H. HOBBS, JR., 1986. The entocytherid ostracod fauna of Southeastern Georgia. Smithsonian Contributions to Zoology, **424**: 1-43.

BAKER, J. H., 1969. On the relationship of *Ankylocythere sinuosa* (Rioja 1942) (Ostracoda, Entocytheridae) to the crayfish *Procambarus simulans simulans* (Faxon 1884). Transactions of the American Microscopical Society, **88**(2): 293-294.

BARR JR., T. C., 1967. Ecological studies in the Mammoth Cave System of Kentucky. I: the biota. International Journal of Speleology, **3**(1-2): 147-204.

BEDINGER, M. S. & H. H. HOBBS, JR., 1965. Observations on a new troglobitic crayfish (with notes on the distribution of troglobitic crayfishes in the Ozark region). Bulletin of the National Speleological Society, **27**(3): 93-96.

BOWMAN, T. E., 1988. *Nitokra sphaeromatra*, a new harpacticoid copepod crustacean associated with the wood-boring isopod, *Sphaeroma peruvianum* in Costa Rica. Proceedings of Biological Society of Washington, **101**(1): 171-175.

BRONSTEIN, Z. S., 1947. Rakoobraznye 2, Ostracoda presnykh vod. Trudi Zoologicheskogo Instuta Akademii Nauk SSSR, **3**: 1-339.

CASTILLO-ESCRIVÀ, A., A. MESTRE, J. S. MONRÓS & F. MESQUITA-JOANES, 2012. Population dynamics of *Ankylocythere sinuosa* (Rioja, 1942) (Ostracoda, Entocytheridae), epibiont on the red swamp crayfish (*Procambarus clarkii*) in Pego-Oliva Wetland (Eastern Spain). In: F. A. VIEHBERG & R. GROMIG (eds.), Abstracts, 14th International German Ostracodologists' Meeting, October 11-14, 2012, Cologne. Kölner Forum für Geologie und Paläontologie, **21**: 8.

— —, — —, — — & — —, 2013. Population dynamics of an epibiont Ostracoda on the invasive red swamp crayfish *Procambarus clarkii* in a western Mediterranean wetland. Hydrobiologia, **714**: 217-228.

CHAPPUIS, P. A., 1927. Die Tierwelt der Unterirdischen Gewässer. Die Binnengewässer, Stuttgart, **3**: 1-175.
CRAWFORD, E. A., JR., 1959. Five new ostracods of the genus *Entocythere* (Ostracoda, Cytheridae) from South Carolina. University of South Carolina Publications, Biology, (3) **2**(4): 149-189.
— —, 1960. The genus *Entocythere*, an ostracod epizooic on crayfishes (Abstract). The Association of Southeastern Biologists Bulletin, **7**(2): 26.
— —, 1961. Three new species of the genus *Entocythere* (Ostracoda, Cytheridae) from North and South Carolina. American Midland Naturalist, **65**(1): 236-245.
— —, 1965. Three new species of epizoic ostracods (Ostracoda, Entocytheridae) from North and South Carolina. American Midland Naturalist, **74**(1): 148-154.
DANIELOPOL, D. L., 1970. Une nouvelle espèce de genre *Darwinula* des eaux souterraines de Roumania et quelques remarques sur la morphologie des Darwinulidae (Ostracoda-Podocopida). Travaux de l'Institut de Spéologie "Émile Racovitza", **9**: 135-149.
— —, 1971a. Définition de la tribu Hartiellini nov. et remarques sur le poecilogynie de *Hartiella dudichi* (Ostracoda, Entocytheridae, Sphaeromicolinae). Travaux de l'Institut de l'Spéologie "Emile Racovitza", **10**: 189-207.
— —, 1971b. Quelques remarques sur le peuplement ostracodologique des eaux douces souterraines d'Europe. In: H. J. OERTLI (ed.), Colloque sur la Paléoécologie des Ostracodes. Bulletin du Centre de Recherche S.N.P.A., supplement 5: 179-190. (Société nationale des pétroles d'Aquitaine, Pau).
— —, 1977a. Recherches sur les Ostracodes Entocytheridae. Donnees sur *Sphaeromicola cebennica juberthiei* nov. ssp. et *Sphaeromicola cirolanae* Rioja. International Journal of Speleology, **9**: 21-41.
— —, 1977b. On the origin and diversity of European freshwater interstitial ostracods. In: H. LÖFFLER & D. DANIELOPOL (eds.), Aspects of ecology and zoogeography of recent and fossil ostracoda: 295-305. (Dr. W. Junk Publishers, The Hague).
— —, 1982. The preparation of ostracods limbs for optical microscopy (dissection, staining and mounting procedures). In: R. F. MADDOCKS (ed.), Applications of Ostracoda to economic and scientific problems (program and abstracts): 43-44. (Department of Geosciences, University of Houston, Houston, TX).
DANIELOPOL, D. L. & C. W. HART, JR., 1985. Notes of the center of origin and antiquity of the Sphaeromicolinae, with description of *Hobbsiella*, new genus (Ostracoda, Entocytheridae). Stygologia, **1**: 54-70.
DANIELOPOL, D. L., P. MARMONIER, A. J. BOULTON & G. BONADUCE, 1994. World subterranean ostracod biogeography, dispersal or vicariance. Hydrobiologia, **287**(1): 119-129.
DEBOUTTEVILLE, C. D., 1960. Biologie des eaux souterraines littorales et continentales: 1-740. (Hermann, Paris).
DELORME, L. D., 1970. Freshwater ostracods of Canada. Part IV. Families Ilyocypridae, Notodromadidae, Darwinulidae, Cytherideidae, and Entocytheridae. Canadian Journal of Zoology, **48**(6): 1251-1259.
DOBBIN, C. N., 1941. Fresh-water ostracoda from Washington and other western localities. University of Washington Publications, Biology, **4**: 174-246.
EDGERTON, B. F., L. H. EVANS, F. J. STEPHENS & R. M. OVERSTREET, 2002. Synopsis of freshwater crayfish diseases and commensal organisms. Aquaculture, **206**: 57-135.
EVENSON, C. D., 1969. Designation of lectotypes of fresh-water species described by Dobbin, 1941 (Ostracoda, Crustacea). In: J. W. NEALE (ed.), The taxonomy, morphology and ecology of recent Ostracoda: 491-494. (Oliver & Boyd, Edinburgh).
FERGUSON JR., E., 1944. Studies of seasonal life history of three species of freshwater Ostracoda. American Midland Naturalist, **32**(3): 713-727.
— —, 1958. A supplementary list of species and records of distribution for North American Ostracoda. Proceedings of the Biological Society of Washington, **71**: 197-202.

— —, 1965. Morphology and taxonomy of freshwater Ostracoda. Symposium on Crustacea: 12 (Marine Biological Association of India, Mandapam Camp, South India).

— —, 1968. Recently described species and distributional records for North American freshwater Ostracoda. American Midland Naturalist, **79**(2): 499-506.

FRANTZ, T. C. & A. J. CORDONE, 1966. A preliminary checklist of invertebrates collected from Lake Tahoe, 1961-1964. Occasional Papers of the Biological Society of Nevada, **8**: 1-12.

FRANZ, R., J. BAUER & T. MORRIS, 1994. Review of biollogically significant caves and their faunas in Florida and South Georgia. Brimleyana, **20**: 1-109.

GALL, W. K. & R. R. JEZERINAC, 1998. Commensal ostracod (Ostracoda: Entocytheridae) provides evidence for the postglacial dispersal of the burrowing crayfish, *Cambarus diogenes* (Decapoda: Cambaridae), into Western New York. Bulletin of the Buffalo Society of National Sciences, **36**: 203-213.

GRABOW, K. & A. MARTENS, 2009. Neozoische Aufsitzer auf in Deutschland eingeschleppten Signalkrebsen. Internationales Flußkrebsforum 07.-10.Mai 2009 Gersfeld/Rhön, Deutschland: 38-42.

GRABOW, K., G. SCHOOLMANN & A. MARTENS, 2009. Epizoen auf in Deutschland eingeschleppten Flusskrebsen. Deutsche Gesellschaft für Limnologie (DGL) Erweiterte Zusammenfassungen der Jahrestagung 2008 (Konstanz), Hardegsen 2009: 386-389.

HART JR., C. W., 1959. The ostracods of the genus *Entocythere* from the lower Chattahoochee-Flint basin: with a review of the occurrence of the genus in Florida, and descriptions of two new species. Bulletin of the Florida State Museum, Biological Sciences, **4**(6): 193-204.

— —, 1960. A new ostracod (Crustacea, Ostracoda) of the genus *Entocythere* from Pennsylvania. Notulae Naturae of the Academy of Natural Sciences of Philadelphia, **335**: 1-4.

— —, 1962. A revision of the ostracods of the family Entocytheridae. Proceedings of the Academy of Natural Sciences of Philadelphia, **114**(3): 121-147.

— —, 1963. The status of entocytherid ostracod research in the United States and Mexico. Proceedings of the XVI International Congress of Zoology, Washington, DC, **1**: 51.

— —, 1964. Two new entocytherid ostracods from the vicinity of Washington, D.C. Proceedings of the Biological Society of Washington, **77**(28): 243-246.

— —, 1965a. New entocytherid ostracods and distribution records for five midwestern states. Transactions of the American Microscopical Society, **84**(2): 255-259.

— —, 1965b. Three new entocytherid ostracods from the western United States with new locality data for two previously described western entocytherids. Crustaceana, **8**(2): 190-196.

— —, 1971. A new species of parasitic ostracod of the genus *Acetabulastoma* (Paradoxostomatidae, Paradoxostominae) with a discussion of the copulatory appendage homologies. Notulae Naturae, Academy of Natural Sciences of Philadelphia, **442**: 1-11.

— —, 1978. A new species of the genus *Sphaeromicola* (Ostracoda: Entocytheridae: Sphaeromicolinae) from Texas, with notes on relationships between European and North American species. Proceedings of the Biological Society of Washington, **91**: 724-730.

HART JR., C. W. & J. CLARK, 1984. A new commensal ostracod of the genus *Microssyssitria* from South Africa (Ostracoda: Entocytheridae: Microssyssitriinae). Proceedings of the Biological Society of Washington, **97**(1): 217-220.

HART JR., C. W. & D. G. HART, 1966a. Four new entocytherid ostracods from Kentucky, with notes on the troglobitic *Sagittocythere barry*. Notulae Naturae of the Academy of Natural Sciences of Philadelphia, **388**: 1-10.

— — & — —, 1966b. New Australian entocytherid ostracods and their relation to North American species. American Zoologist, **6**(4): 564.

— — & — —, 1967. The entocytherid ostracods of Australia. Proceedings of the Natural Sciences of Philadelphia, **119**(1): 1-51.

— — & — —, 1968. The sorting, storage and preparation of microscopic crustaceans, with special reference to entocytherid ostracods. Crustaceana, **14**(3): 315-318.

— — & — —, 1969a. Evolutionary trends in the ostracod family Entocytheridae, with notes on the distributional patterns in the Southern Appalachians. In: P. C. HOLT, R. L. HOFFMAN & C. W. HART, JR. (eds.), The distributional history of the biota of the Southern Appalachians. Part 1: Invertebrates: 179-190. (Virginia Polytechnic Institute, Blacksburg, VA).

— — & — —, 1969b. The functional morphology of entocytherid ostracod copulatory appendages, with a discussion of possible homologues in other ostracods. In: J. W. NEALE (ed.), The taxonomy, morphology and ecology of recent ostracods: 154-167. (Oliver and Boyd, Edinburgh).

— — & — —, 1971. A new ostracod (Entocytheridae, Notocytherinae) commensal on New Zealand crayfishes. Proceedings of the Biological Society of Washington, **83**(50): 579-584.

— — & — —, 1975. Ostracoda, Podocopa: Family Entocytheridae: 1-64. (Crustaceorum Catalogus, Amsterdam).

HART JR., C. W., L. HAYEK, J. CLARK & W. H. CLARK, 1985. The life history and ecology of the entocytherid ostracod *Uncinocythere occidentalis* (Kozloff and Whitman) in Idaho. Smithsonian Contributions to Zoology, **419**: 1-22.

HART JR., C. W. & H. H. HOBBS, JR., 1961a. Eight new troglobitic ostracods of the genus *Entocythere* (Crustacea, Ostracoda) from the Eastern United States. Proceedings of the Academy of Natural Sciences of Philadelphia, **113**(8): 173-185.

— — & — —, 1961b. Observations on the distribution and relationships of the troglobitic ostracods of North America (Abstract). The Association of Southeastern Biologists Bulletin, **8**(2): 25.

HART JR., C. W., N. B. NAIR & D. G. HART, 1967. A new ostracod (Ostracoda: Entocytheridae) commensal on a wood-boring marine isopod from India. Notulae Naturae of the Academy of Natural Sciences of Philadelphia, **409**: 1-11.

HART, D. G. & C. W. HART, JR., 1970. A new ostracod (Entocytheridae, Notocytheridae) commensal on New Guinea crayfishes. Zoologische Mededelingen, **44**(19): 279-283.

— — & — —, 1971. New entocytherid ostracods of the genera *Ankylocythere*, *Dactylocythere*, *Entocythere*, *Geocythere*, and *Uncinocythere* — with a new diagnosis of the genus *Entocythere*. Proceedings of the Academy of Natural Sciences of Philadelphia, **123**(5): 105-125.

— — & — —, 1974. The ostracod family Entocytheridae: 1-239. (Fulton Press, Lancaster, PA).

HARTMANN, G., 1963. Zur phylogenie und systematik der ostracoden. Zeitschrift für Zoologische Systematik und Evolutionsforschung, **1**(1-2): 1-204.

— —, 1964. Neontological and paleontological classification of Ostracoda. Publicazioni della Stazione Zoologica di Napoli, **33**(Suppl.): 550-587.

HARTMANN, G. & H. S. PURI, 1974. Summary of neontological and paleontological classification of Ostracoda. Mitteilungen aus dem Hamburgischen Zoologischen Museum und Institut, **82**: 263-267.

HIRUTA, S. & R. J. SMITH, 2001. Living freshwater and shallow marine Ostracoda in Eastern Hokkaido, Japan (Field excursion E). In: N. IKEYA (ed.), Guidebook of excursions of the 14th International Symposium on Ostracoda: 107-125. (Shizuoka, Japan).

HOBBS JR., H. H., 1941. A new crayfish from San Luis Potosi, Mexico (Decapoda, Astacidae). Zoologica, Scientific Contribution of the New York Zoological Society, **26**(1): 1-4.

— —, 1953. The epizootic associates of the crayfishes of the New River system with particular reference to the ostracods. Journal of the Tennessee Academy of Sciences, **28**(3): 180-181.

— —, 1955. Ostracods of the genus *Entocythere* from the New River system in North Carolina, Virginia, and West Virginia. Transactions of the Microscopical Society, **74**(4): 325-333.

— —, 1957. Observaciones acerca de las especies del género *Entocythere* (Crustáceos, ostrácodos) de Cuba. Anales del Instituto de Biología, Mexico, **27**(2): 431-436.

— —, 1966. An illustrated key to the species of the genus *Ankylocythere* with a description of a new species from Louisiana (Ostracoda, Entocytheridae). Proceedings of the Louisiana Academy of Sciences, **29**: 67-75.

— —, 1967. A new genus and three new species of ostracods with a key to the genus *Dactylocythere* (Ostracoda: Entocytheridae). Proceedings of the United States National Museum, **122**(3587): 1-10.

— —, 1971. The entocytherid ostracods of Mexico and Cuba. Smithsonian Contributions to Zoology, **81**: 1-55.

— —, 1982. A new crayfish (Decapoda: Cambaridae) from the State of Puebla, Mexico, with new locality records for *Procambarus* (*Villalobosus*) *xochitlanae* and entocytherid ostracod symbionts. Association for Mexican Cave Studies Bulletin, **28**: 39-44.

— —, 1983. *Distocambarus* (*Fitzcambarus*) *carlsoni*, and new subgenus and species of crayfish (Decapoda: Cambaridae) from South Carolina. Proceedings of the Biological Society of Washington, **96**(3): 429-439.

HOBBS JR., H. H. & T. C. BARR, JR., 1972. Origins and affinities of the troglobitic crayfishes of North America (Decapoda: Astacidae) II. Genus *Orconectes*. Smithsonian Contributions to Zoology, **105**: 1-84.

HOBBS JR., H. H. & M. S. BEDINGER, 1964. A new troglobitic crayfish of the genus *Cambarus* (Decapoda, Astacidae) from Arkansas with a note on the range of *Cambarus cryptoides* Hobbs. Proceedings of the Biological Society of Washington, **77**: 9-16.

HOBBS JR., H. H. & R. FRANZ, 1991. A new troglobitic crayfish, *Procambarus* (*Lonnbergius*) *morrisi*, (Decapoda: Cambaridae) from Florida. Proceedings of the Biological Society of Washington, **104**: 55-63.

HOBBS JR., H. H. & C. W. HART, JR., 1966. On the entocytherid ostracod genera *Ascetocythere*, *Plectocythere*, *Phymocythere* (gen. nov.) and *Cymocythere* with descriptions of new species. Proceedings of the Academy of Natural Sciences of Philadelphia, **118**(2): 35-61.

HOBBS JR., H. H. & H. H. HOBBS, III, 1970. New entocytherid ostracods with a key to the genera of the subfamily Entocytherinae. Smithsonian Contributions to Zoology, **47**: 1-19.

— — & — —, 1973. The genus *Sphaeromicola* (Ostracoda, Entocytheridae) in Mexico. Studies on the cavernicole fauna of Mexico and adjacent regions. Association for Mexican Cave Studies, **5**: 39-42.

HOBBS JR., H. H., H. H. HOBBS, III & M. A. DANIEL, 1977. A review of the troglobitic decapod crustaceans of the Americas. Smithsonian Contributions to Zoology, **244**: 1-183.

HOBBS JR., H. H., P. C. HOLT & M. WALTON, 1967. The crayfishes and their epizootic ostracod and branchiobdellid associates of the Mountain Lake, Virginia, region. Proceedings of the United States Natural Museum, **123**(3602): 1-84.

HOBBS JR., H. H. & A. C. MCCLURE, 1983. On a small collection of entocytherid ostracods with the descriptions of three new species. Proceedings of the Biological Society of Washington, **96**(4): 770-779.

HOBBS JR., H. H. & F. O. PERKINS, 1967. A new burrowing crayfish from North Carolina (Decapoda, Astacidae). Proceedings of the Biological Society of Washington, **80**(21): 141-146.

HOBBS JR., H. H. & D. J. PETERS, 1977. The entocytherid ostracods of North Carolina. Smithsonian Contributions to Zoology, **247**: 1-73.

— — & — —, 1978. A substitute name for the homonym *Aphelocythere* Hobbs and Peters (Ostracoda Entocytheridae). Proceedings of the Biological Society of Washington, **91**(4): 1037.

— — & — —, 1982. The entocytherid ostracod fauna of Northern Georgia. Proceedings of the Biological Society of Washington, **95**(2): 297-318.

— — & — —, 1989. New records of the entocytherid ostracods infesting burrowing crayfishes, with the description of a new species, *Ascetocythere stockeri*. Proceedings of the Biological Society of Washington, **102**(2): 324-330.

— — & — —, 1991. Additional records of entocytherid ostracods infesting burrowing crayfishes, with descriptions of 5 new species. Proceedings of the Biological Society of Washington, **104**(1): 64-75.

— — & — —, 1993. New records of entocytherid ostracods infesting burrowing and cave-dwelling crayfishes, with descriptions of 2 new species. Proceedings of the Biological Society of Washington, **106**(3): 455-466.

HOBBS JR., H. H. & A. VILLALOBOS, 1958. The exoesqueleton of a freshwater crab as a microhabitat for several invertebrates. Virginia Journal of Science, **9**(4): 395-396.

HOBBS JR., H. H. & M. WALTON, 1960. Three new ostracods of the genus *Entocythere* from the Hiwassee drainage system in Georgia and Tennessee. Journal of the Tennessee Academy of Sciences, **35**(1): 17-23.

— — & — —, 1961. Additional new ostracods from the Hiwasse drainage system in Georgia, North Carolina, and Tennessee. Transactions of the American Microscopical Society, **80**(4): 379-384.

— — & — —, 1962. New ostracods of the genus *Entocythere* from the Mountain Lake region, Virginia (Ostracoda, Entocytheridae). Virginia Journal of Science, **13**(2): 42-48.

— — & — —, 1963a. Four new species of the genus *Donnaldsoncythere* (Ostracoda, Entocytheridae) from Virginia with a key to the species of the genus. Transactions of the American Microscopical Society, **82**(4): 363-370.

— — & — —, 1963b. Three new ostracods (Ostracoda, Entocytheridae) from the Duck River drainage in Tennessee. American Midland Naturalist, **69**(2): 456-461.

— — & — —, 1966. A new genus and six new species of entocytherid ostracods (Ostracoda, Entocytheridae). Proceedings of the United States National Museum, **119**(3542): 1-12.

— — & — —, 1968. New entocytherid ostracods from Southern United States. Proceedings of the Academy of Natural Sciences of Philadelphia, **120**(6): 237-252.

— — & — —, 1970. New entocyterid ostracods from Tennessee and Virginia. Proceedings of the Biological Society of Washington, **82**(68): 851-854.

— — & — —, 1975. New entocytherid ostracods from Tennessee with a key to the species of the genus *Ascetocythere*. Proceedings of the Biological Society of Washington, **88**(2): 5-20.

— — & — —, 1976. New entocytherid ostracods from Kentucky and Tennessee. Proceedings of the Biological Society of Washington, **89**(33): 393-404.

— — & — —, 1977. New entocytherid ostracods of the genus *Dactylocythere*. Proceedings of the Biological Society of Washington, **90**(3): 600-614.

HOBBS III, H. H., 1965. Two new genera and species of the ostracod family Entocytheridae, with a key to the genera. Proceedings of the Biological Society of Washington, **78**(19): 159-164.

— —, 1968. The possibility of host specificity in the genus *Ornithocythere* (Ostracoda, Entocytheridae). Association of Southeastern Biologists Bulletin, **15**(2): 40-41.

— —, 1969a. A new genus and two new species of entocytherid ostracods from Alabama and Mississippi. Proceedings of the Biological Society of Washington, **82**(10): 167-170.

— —, 1969b. Studies in ecological and host specificity in entocytherid ostracods (Ostracoda, Entocytheridae): 1-93. (Unpublished Master's Thesis, Mississippi State University, Jackson, MS).

— —, 1970. New entocytherid ostracods of the genus *Ornithocythere* and the description of a new genus. Proceedings of the Biological Society of Washington, **83**(15): 171-182.

— —, 1971. New entocytherid ostracods of the genera *Ankylocythere* and *Dactylocythere*. Proceedings of the Biological Society of Washington, **84**(17): 137-146.

— —, 1973. The population dynamics of cave crayfishes and their commensal ostracods from southern Indiana: 1-247. (Unpublished Ph.D. Thesis, Indiana University, Bloomington, IN).

— —, 1975. Distribution of Indiana cavernicolous crayfishes and their ecto-commensal ostracods. International Journal of Speleology, **7**: 273-302.

— —, 1978. New species of ostracods from Gulf coastal plain (Ostracoda, Entocytheridae). Transactions of the American Microscopical Society, **97**(4): 502-511.

— —, 1982. Host-commensal relationships of crayfishes and ostracods of the gulf coast plain, U.S.A. Ohio Journal of Science, **82**: 2.

HOFF, C., 1942a. The ostracods of Illinois: their biology and taxonomy. Illinois Biological Monographs, **19**(1-2): 1-196.

— —, 1942b. The subfamily Entocytherinae, a new subfamily of freshwater cytherid ostracods, with descriptions of two new species of the genus *Entocythere*. The American Midland Naturalist, **27**: 63-73.

— —, 1943. Two new ostracods of the genus *Entocythere* and records of the previously described species. Journal of the Washington Academy of Sciences, **33**(9): 275-286.

— —, 1944a. New American species of the ostracod genus *Entocythere*. American Midland Naturalist, **32**(2): 327-357.

— —, 1944b. The origin of neartic fresh-water ostracods. Ecology, **25**(3): 369-372.

HORNE, D. J., A. COHEN & K. MARTENS, 2002. Taxonomy, morphology and biology of Quaternary and living ostracoda. In: J. A. HOLMES & A. R. CHIVAS (eds.), The Ostracoda. Applications in Quaternary research. Geophysical Monograph, **131**: 5-36. (American Geophisical Union, Washington, DC).

HOWE, H. V., 1955. Handbook of ostracod taxonomy: 1-386. (Louisiana State University Studies, Physical Sciences, Baton Rouge, LA).

— —, 1962. Ostracod taxonomy: 1-346. (Louisiana State University Press, Baton Rouge, LA).

HUBAULT, E., 1937. *Monolistra hercegoviniensis* Absolon; spheromien cavernicole d'Herzegovine et *Sphaeromicola stammeri* Klie, son commensal. Bioespeologica LXVI: Archives de Zoologie Experimentale et Generale, **78**(9): 357-373.

— —, 1938. *Sphaeromicola sphaeromidicola* nov. sp., commensal de *Sphaeromides virei* Valle, en Istrie et considerations sur l'origine de diverses especes cavernicoles perimediterraneennes. Archives de Zoologie Experimentale et Generale, **80**: 11-24.

JEANNEL, R., 1926. Faune cavernicole de la France. Encyclopedie Entomologique, Paris, **7**: 1-334.

KAESTNER, A., 1970. Invertebrate zoology, **III**: 1-523. (Wiley-Interscience, New York, NY).

KARANOVIC, I., 2012. Recent freshwater ostracods of the world. Crustacea, Ostracoda, Podocopida: 1-608. (Springer, New York, NY).

KESLING, R. V., 1951. The morphology of ostracod molt stages. Illinois Biological Monographs, **21**(1-3): 1-324.

KLIE, W., 1926. Ostracoda. Biologie der Tiere Deutschlands by Paul Schultz, Berlin, **22**(16): 1-56.

— —, 1930. Uber eine neue Art der Ostracodengattung *Sphaeromicola*. Zoologischer Anzeiger, **88**(9-10): 272-276.

— —, 1931. Campagne spéologique de C. Bolivar et R. Jeannel dans l'Amérique du Nord (1928). 3. Crustacés Ostracodes. Biospeleologica: Archives Zoologie Expérimentale et Générale, **71**(3): 333-344.

— —, 1938a. Ostracoda. (Dahl. Tierwelt Deutschlands, T.34, Crustacea, III, Jena).

— —, 1938b. *Sphaeromicola dudichi* n. sp. (Ostr.) ein kommensale des Bohramphipoden *Chelura terebrans*. Zoologischer Anzeiger, **121**(11-12): 317-322.

KOZLOFF, E. N., 1955. Two new species of *Entocythere* (Ostracoda, Cytheridae) commensal on *Pacifastacus gambelii* (Girard). American Midland Naturalist, **53**(1): 156-161.

KOZLOFF, E. N. & D. C. WHITMAN, 1954. *Entocythere occidentalis* sp. nov., a cytherid ostracod commensal on western species of *Pacifastacus*. American Midland Naturalist, **52**(1): 159-163.

LALANA, R. & M. ORTIZ, 2001. Lista actualizada de especies y bibliografía de los ostrácodos (Crustacea, Ostracoda) de aguas cubanas. Revista de Investigaciones Marinas, **22**(2): 103-106.

LEVINSON, S., 1959. Bibliography and index to new genera and species of ostracods for 1957. Micropaleontology, **5**(2): 241-260.

— —, 1962. Bibliography and index to new genera and species of ostracods for 1958-1959. Micropaleontology, **8**(1): 77-105.

MARSHALL, W. S., 1903. *Entocythere cambaria* n. g. n. sp., a parasitic ostracod. Transactions of the Wisconsin Academy of Sciences, Arts, and Letters, **14**(1): 117-144.

MARTENS, K. & S. SAVATENALINTON, 2011. A subjective checklist of the Recent, free-living, non-marine Ostracoda (Crustacea). Zootaxa, **2855**: 1-79.

MARTENS, K., I. SCHÖN, C. MEISCH & D. J. HORNE, 2008. Global diversity of ostracods (Ostracoda, Crustacea) in freshwater. Hydrobiologia, **595**: 185-193.

MCGREGOR, D. L. & R. V. KESLING, 1969. Copulatory adaptations in ostracods, Part II. Adaptations in living ostracods. Contributions from the Museum of Paleontology, The University of Michigan, **22**(17): 221-239.

MCKENZIE, K. G., 1967. The distribution of Caenozoic marine Ostracoda from the Gulf of Mexico to Australia. In: C. G. ADAMS & D. V. AGER (eds.), Aspects of Tethyan biogeography: 219-238. (Systematics Association Publication No. 7, London).

— —, 1972. Check list of Ostracoda recorded from the Indian subcontinent and Ceylon (1840-1971): 1-59. (Privately printed).

MCKENZIE, K. G., K. J. MÜLLER & M. N. GRAMM, 1983. Phylogeny of ostracoda. In: F. R. SCHRAM (ed.), Crustacean issues I. Crustacean phylogeny: 29-46. (A. A. Balkema, Rotterdam).

MEISCH, C., 2000. Freshwater ostracoda of Western and Central Europe: 1-522. (Spektrum Akademischer Verlag, Heidelberg, Berlin).

MESTRE, A., J. S. MONRÓS & F. MESQUITA-JOANES, 2011. Comparison of two chemicals for removing an entocytherid (Ostracoda: Crustacea) species from its host crayfish (Cambaridae: Crustacea). International Review of Hydrobiology, **96**(4): 347-355.

— —, — — & — —, 2012. Building a database of the entocytheridae ostracods of the world. In: F. A. VIEHBERG & R. GROMIG (eds.), Abstracts, 14th International German Ostracodologists' Meeting, October 11-14, 2012, Cologne. Kölner Forum für Geologie und Paläontologie, **21**: 45-46.

— —, — — & — —, 2013. Do Rapoport effects apply to epibiont aquatic invertebrates? The case of ostracod Entocytheridae commensal on freshwater crustaceans. Frontiers of Biogeography, **4**(Suppl. 1): 114-115.

MOORE, R. C. & C. W. PITRAT, 1961. Treatise on invertebrate paleontology. Part Q, Arthropoda 3: Crustacea, Ostracoda: 1-442. (Geological Society of America and the University of Kansas Press, Lawrence, KS).

MULLER, G. W., 1912. Ostracoda. Das Tierreich, **31**: 1-434.

NEALE, J. W., 1964. Some factors influencing the distribution of recent British Ostracoda. Pubblicazioni della Stazione Zoologica di Napoli, **33**(Suppl.): 247-307.

NICHOLAS, B. G., 1962. Checklist of troglobitic organisms of middle America. American Midland Naturalist, **68**(1): 165-188.

NORDEN, A. W., 1977. A new entocytherid ostracod of the genus *Plectocythere*. Proceedings of the Biological Society of Washington, **90**: 491-494.

— —, 2008. Biological diversity of Plummers Island, Maryland: the crayfishes and their entocytherid ostracod Associates. Bulletin of the Biological Society of Washington, **15**(1): 49-51.

NORDEN, A. W. & B. B. NORDEN, 1985. A new entocytherid ostracod of the genus *Dactylocythere*. Proceedings of the Biological Society of Washington, **98**(3): 627-629.

OKUBO, I., 1988. On the recent *Aurila* species from Japan. In: T. HANAI, N. IKEYA & K. ISHIZAKI (eds.), Evolutionary biology of Ostracoda, its fundamentals and applications: 135-144. (Elsevier, New York, NY).

PARIS, P., 1916a. Note sur quelques Entomostracés récoltés dans le Département de la Côte-d'Or. Bulletin de la Société Zoologique de France, **41**: 112-116.

— —, 1916b. *Sphaeromicola topsenti* n. g. n. sp., Ostracode commensal d'Isopodes troglobies du genre *Caecosphaeroma*. Comptes Rendus Hebdomadaires des Séances de l'Académie des Sciences, Paris, **163**: 307-309.

— —, 1920. Ostracodes (Première Série). Biospeologica: Archives de Zoologie Expérimentale et Générale, **58**: 475-487.

— —, 1925a. La faune cavernicole de la Côte-d'Or. Bulletin de l'Association Française pour l'Avancement des Sciences, **49**: 449-450.

— —, 1925b. Sur la bionomie de quelques Crustacés troglobies de la Côte-d'Or. Bulletin de l'Association Française pour l'Avancement des Sciences, **49**: 451-454.

PENNAK, T. W., 1953. Fresh-water invertebrates of the United States: 1-769. (Ronald Press, New York, NY).

PETERS, D. J., 1971. Entocytherid ostracods of the Lower James-York peninsula, Virginia. Virginia Journal of Science, **22**(3): 100.

— —, 1974. The ecological and geographical distribution of some Virginia ostracods. Association of Southeastern Biologists Bulletin, **21**(2): 74.

— —, 1975. The entocytherid Ostracoda fauna of the James and York River Basins with a description of a new member of the genus *Entocythere*. Virginia Polytechnic Institute and University, Research Division Bulletin, **93**: 1-50.

PETERS, D. J. & J. E. PUGH, 1999. On the entocytherid ostracods of the Brazos River basin and adjacent coastal region of Texas. Proceedings of the Biological Society of Washington, **112**: 338-351.

PRINS, R., 1964. *Attheyella carolinensis* Chappuis (Copepoda, Harpacticoida) on freshwater crayfishes from Kentucky. Transactions of the American Microscopical Society, **83**(3): 370-371.

REDDELL, J. R., 1965. A checklist of the cave fauna of Texas. I. The invertebrates (exclusive of insecta). Texas Journal of Science, **17**(2): 143-187.

REDDELL, J. R. & R. W. MITCHELL, 1971. Studies on the cavernicole fauna of Mexico. Association for Mexican Cave Studies, **4**: 1-239.

REMY, P. A., 1943. Nouvelle station du sphaeromien troglobie *Caecosphaeroma* (*V.*) *burgundum* Dollfus var. *rupis fucaldi* Hubault et de l'ostracode commensal *Sphaeromicola topsenti* Paris. Bulletin de la Société Zoologique de France, **68**: 35-38.

— —, 1946. *Sphaeromicola stammeri* Klie var. *hamiguer* n. var. ostracode commensal du Sphaeromide obscuriole *Caecosphaeroma* (*C.*) *virei* Dollfus. Collections du Musée de Zoologie, Nancy, **1**: 7-9.

— —, 1948a. Sur la distribution géographique des Sphaeromidae, Monolistrini, et des *Sphaeromicola*, leurs ostracodes commensaux. Bulletin du Muséum d'Histoire Naturelle de Marseille, **8**(1): 1-4.

— —, 1948b. Description de *Sphaeromicola cebennica*, n. sp., ostracode cythéride commensal de l'isopode cirolanide cavernicole *Sphaeromides raymondi* Dollfus. Bulletin Mensuel de la Société Linnéenne de Lyon, **17**(7): 129-132.

— —, 1951. Stations de crustacés obscurioles. Biospeologica: Archives de Zoologie Expérimentale et Générale, **88**: 217-230.

RIOJA, E., 1940a. Estudios carcinológicos. V. Morfología de un ostrácodo epizoario observado sobre *Cambarus* (*Cambarellus*) *montezumae* Sauss. de México, *Entocythere heterodonta* n. sp. y descripción de algunos de sus estados larvarios. Anales del Instituto de Biología, Mexico, **11**(2): 593-609.

— —, 1940b. Notas acerca de crustáceos del Lago Patzcuaro. Anales del Instituto de Biología, Mexico, **11**(2): 469-475.

— —, 1940c. Un caso de biocenosis observado sobre *Cambarus* (*Cambarellus*) *montezumae* Saussure, de México. Revista Chilena de Historia Natural, **44**: 57-59.

— —, 1941a. Estudios carcinológicos. VI. Estudio morfológico del esqueleto interno de apodemas quitinoso de *Entocythere heterodonta* Rioja (Crust. Ostrácodos). Anales del Instituto de Biología, Mexico, **12**(1): 177-191.

— —, 1941b. Estudios carcinológicos. VIII. Las especies del género *Entocythere* (Crust. Ostrácodos). Anales del Instituto de Biología, Mexico, **12**(1): 193-198.

— —, 1942a. Descripción de una especie y una subespecie nuevas del género *Entocythere* Marshall, procedentes de la Cueva Chica (San Luis Potosí, México). Ciencia (Mexico), **3**(7): 201-204.

— —, 1942b. Estudios carcinológicos. XIII. Consideraciones y datos acerca del género *Entocythere* (Crust. Ostracodos) y algunas de sus especies, con descripción de una nueva. Anales del Instituto de Biología, Mexico, **13**(2): 685-697.

— —, 1943a. Estudios Carcinológicos. XIV. Nuevos datos acerca de los *Entocythere* (Crust. Ostracoda) de México. Anales del Instituto de Biología, Mexico, **14**(2): 553-566.

— —, 1943b. Estudios carcinológicos. XV. Polimorfismo femenino en los ostrácodos del género *Entocythere*. Anales del Instituto de Biología, Mexico, **14**(2): 567-585.

— —, 1945. Estudios carcinológicos. XVIII. Observaciones acerca de las variaciones de la pieza copuladora en *Entocythere dobbini*, Rioja (Crust. Ostrácodos). Anales del Instituto de Biología, Mexico, **16**(2): 419-423.

— —, 1949. Estudios Carcinológicos. XXI. Contribución al conocimiento de las especies del género *Entocythere* de México. Anales del Instituto de Biología, Mexico, **20**(1 2): 315-329.

— —, 1951. Estudios carcinológicos. XXV. El hallazgo del género *Sphaeromicola* en América (Ostrácodos, Citéridos) y descripción de una nueva especie. Anales del Instituto de Biología, Mexico, **22**(1): 169-179.

— —, 1953. Los crustáceos cavernícolas de México. IV. Centenario de la Universidad de México (1151-1951). Memoria del Congreso Científico Mexicano. Ciencias Biológicas, **7**: 285-298.

— —, 1955. Estudios Carcinológicos. XXXII. Primeros datos acerca de las especies del género *Entocythere* (Crustaceos, Ostracodos) de la Isla de Cuba. Anales del Instituto de Biología, Mexico, **26**(1): 193-197.

ROELOFS, H. M. A., 1966. Verslag over het onderzoek naar biologie en voorkomen van met de mariene, houtborende amphipode *Chelura terebrans* Philippi geassocieerde ostracoden. Verslagen van de gewone vergadering der Afdeeling natuurkunde Koninklijke Nederlandse Akademie van Wetenschappen, Amsterdam, **75**(9): 147-152.

— —, 1968. Etude du developpement de l'ostracode marin *Sphaeromicola dudichi* Klie, 1938. Bulletin Zoologisch Museum Universiteit van Amsterdam, **1**(5): 39-51.

ROELOFS, J., 1966. Redescription de l'ostracode marin *Sphaeromicola dudichi* Klie, 1938, et sa presence dans l'Atlantique. Beaufortia, **13**(161): 207-212.

RUDJAKOV, I. A., 1963. On the fauna of cave ostracods in western Transcaucasia [Russian with English summary]. Zoologicheskii Zhurnal, **42**(1): 32-40.

SARS, G. O., 1926. Fresh-water Ostracoda from Canada and Alaska. Report, Canadian Arctic Expedition, 1913-1918, **7**(1): 1-32.

SMITH, R. J. & T. KAMIYA, 2001. The first record of an entocytherid ostracod (Crustacea: Cytheroidea) from Japan. Benthos Research, **56**: 57-61.

— — & — —, 2005. The ontogeny of the entocytherid ostracod *Uncinocythere occidentalis* (Kozloff and Whitman, 1954) Hart, 1962 (Crustacea). Hydrobiologia, **538**: 217-229.

SOHN, I. G., 1968. Premature citations and zoological nomenclature. Science, **159**(3813): 441-442.

SPANDI, H., 1926. Die Terwelt der Unterirdischen Gewässer. Speläologische Monographien, **11**: 1-235.

STAMMER, H. J., 1930. Eine neue Höhlensphäromide aus dem Karst, *Monolistra* (*Typhlosphaeroma*) *schottlaenderi*, und die Verbreitung des Genus *Monolistra*. Zoologischer Anzeiger, **88**(910): 291-304.

— —, 1932. Die Fauna des Timavo. Zoologische Jahrbücher, (Systematik), **63**(5-6): 521-656.

— —, 1935. Untersuchungen über die Tierwelt der Karsthöhlengewasser. Verhandlungen der Internationalen Vereinigung für Theoretische und Angewandte Limnologie, **7**(1): 92-99.

STAMPER, M. N., 1957. Incubation and hatching of the egg of a commensal ostracod, *Entocythere illinoisensis* Hoff, 1942. Journal of the Cololorado-Wyoming Academy of Science, **4**(9): 50.

STRASKRABA, M., 1969. Lista de los crustáceos dulceacuícolas de Cuba y sus relaciones zoogeográficas. Academia de Ciencias de Cuba, Instituto de Biología, Serie Biológica, **8**: 1-37.

TRESSLER, W. L., 1947. A check list of the known species of North American freshwater ostracods. American Midland Naturalist, **38**: 698-707.

— —, 1954. Fresh-water Ostracoda from Texas and Mexico. Journal of the Washington Academy of Sciences, **44**(5): 138-149.

— —, 1959. Ostracoda. In: H. B. WARD, G. C. WHIPPLE & W. T. EDMONSON (eds.), Fresh-water biology (2nd ed.): 657-734. (Wiley, New York, NY).

VANDEL, A., 1964. Biospéologie, la biologie des animaux cavernicoles: 1-619. (Gauthier-Villars, Paris).

VILLALOBOS, A. & H. H. HOBBS, JR., 1974. Three new crustaceans from La Media Luna, San Luís Potosí, Mexico. Smithsonian Contributions to Zoology, **174**: 1-18.

WALTON, M. & H. H. HOBBS, JR., 1959. Two new eyeless ostracods of the genus *Entocythere* from Florida. Quarterly Journal of the Florida Academy of Sciences, **22**(2): 114-120.

— — & — —, 1971. Distribution of certain entocytherid ostracods on their crayfish hosts. Proceedings of the Academy of Natural Sciences of Philadelphia, **123**(4): 87-103.

WESENBERG-LUND, C., 1939. Biologie der Süsswassertiere; Wirbellose Tiere: 1-817. (Julius Springer, Vienna).

WESTERVELT JR., C. A. & E. N. KOZLOFF, 1959. *Entocythere neglecta* sp. nov., a cytherid ostracod commensal on *Pacifastacus nigrescens* (Stimpson). American Midland Naturalist, **6**(1): 239-244.

WHATLEY, R. C. & J. M. STEPHENS, 1977. Precocious sexual dimorphism in fossil and recent Ostracoda. In: H. LÖFFLER & D. DANIELOPOL (eds.), Aspects of ecology and zoogeography of Recent and fossil Ostracoda: 69-91. (Dr. W. Junk B. V., Publishers, The Hague).

WILLIAMS, B. W., K. L. WILLIAMS, S. R. GELDER & H. C. PROCTOR, 2011. Distribution of Entocytheridae (Crustacea: Ostracoda) in the Northern prairies of North America and reports of opportunistic clitellate annelids on crayfish hosts. Western North American Naturalist, **71**(2): 276-282.

WOLF, B., 1934-1938. Animalium cavernarum catalogus. Volumes 1-3: 1-918. (Dr. W. Junk B. V., Publishers, The Hague).

YOUNG, W., 1971. Ecological studies of Entocytheridae (Ostracoda). American Midland Naturalist, **85**(2): 399-409.

APPENDIX B

Taxonomic list of the 220 extant species of the family Entocytheridae.

Class OSTRACODA Latreille, 1806
Subclass PODOCOPA G. W. Müller, 1894
Order PODOCOPIDA G. O. Sars, 1866
Suborder CYTHEROCOPINA Gründel, 1967
Superfamily CYTHEROIDEA Baird, 1850
Family Entocytheridae Hoff, 1942
Subfamily Entocytherinae Hoff, 1942
Ankylocythere ancyla Crawford, 1965
Ankylocythere barbouri Villalobos & Hobbs, 1974

Ankylocythere bidentata (Rioja, 1949)
Ankylocythere burkeorum Hobbs III, 1971
Ankylocythere carpenteri Hobbs & McClure, 1983
Ankylocythere chipola Hobbs III, 1978
Ankylocythere copiosa (Hoff, 1942)
Ankylocythere cubensis (Rioja, 1955)
Ankylocythere ephydra Hart & Hart, 1971
Ankylocythere freyi Hobbs III, 1978
Ankylocythere hamata (Hobbs, 1957)
Ankylocythere harmani Hobbs, 1966
Ankylocythere heterodonta (Rioja, 1940)
Ankylocythere hobbsi (Hoff, 1944)
Ankylocythere hyba Hobbs & Walton, 1963
Ankylocythere krantzi Hobbs III, 1978
Ankylocythere maya Hobbs, 1971
Ankylocythere prolata Hobbs & Peters, 1991
Ankylocythere sinuosa (Rioja, 1942)
Ankylocythere spargosis Andolshek & Hobbs, 1986
Ankylocythere tallapoosa Hart & Hart, 1971
Ankylocythere talulus (Hoff, 1944)
Ankylocythere telmoecea (Crawford, 1959)
Ankylocythere toltecae Hobbs, 1971
Ankylocythere tridentata Hart, 1964
Ankylocythere villalobosi Hobbs, 1971
Ascetocythere asceta (Hobbs & Walton, 1962)
Ascetocythere batchi Hobbs & Walton, 1968
Ascetocythere bouchardi Hobbs & Walton, 1975
Ascetocythere coryphodes Hobbs & Hart, 1966
Ascetocythere cosmeta Hobbs & Hart, 1966
Ascetocythere didactylata Hobbs & Hart, 1966
Ascetocythere hoffmani Hobbs & Hart, 1966
Ascetocythere holti Hobbs & Walton, 1970
Ascetocythere hyperoche Hobbs & Hart, 1966
Ascetocythere jezerinaci Hobbs & McClure, 1983
Ascetocythere lita Hobbs & Hobbs, 1970
Ascetocythere myxoides Hobbs & Hart, 1966
Ascetocythere ozalea Hobbs & Hart, 1966
Ascetocythere pseudolita Hobbs & Walton, 1975
Ascetocythere riopeli Hobbs & Walton, 1976
Ascetocythere sclera Hobbs & Hart, 1966
Ascetocythere stockeri Hobbs & Peters, 1989
Ascetocythere triangulata Hobbs & Walton, 1975
Ascetocythere veruta Hobbs & Walton, 1975
Cymocythere clavata Crawford, 1965
Cymocythere cyma (Hobbs & Walton, 1960)
Cymocythere gonia Hobbs & Hart, 1966
Dactylocythere amicula Hart & Hart, 1966
Dactylocythere amphiakis Hart & Hart, 1966
Dactylocythere apheles Hobbs & Walton, 1976
Dactylocythere arcuata (Hart & Hobbs, 1961)

Dactylocythere astraphes Hobbs & Walton, 1977
Dactylocythere banana Hart & Hart, 1971
Dactylocythere brachydactylus Hobbs & Walton, 1976
Dactylocythere brachystrix Hobbs & Walton, 1966
Dactylocythere chalaza (Hobbs & Walton, 1962)
Dactylocythere charadra Hobbs III, 1971
Dactylocythere chelomata (Crawford, 1961)
Dactylocythere coloholca Hobbs & Hobbs, 1970
Dactylocythere cooperorum Hobbs & Walton, 1968
Dactylocythere corvus Hobbs & Walton, 1977
Dactylocythere crawfordi Hart, 1965
Dactylocythere crena Hobbs & Walton, 1975
Dactylocythere cryptoteresis Hobbs & Peters, 1993
Dactylocythere daphnioides (Hobbs, 1955)
Dactylocythere demissa Hobbs & Walton, 1976
Dactylocythere enoploholca Hobbs & Walton, 1970
Dactylocythere exoura Hart & Hart, 1966
Dactylocythere falcata (Hobbs & Walton, 1961)
Dactylocythere guyandottae Hobbs & Peters, 1991
Dactylocythere isabelae Hobbs & Peters, 1977
Dactylocythere jeanae Hobbs, 1967
Dactylocythere koloura Hart & Hart, 1971
Dactylocythere lepta Hobbs & Peters, 1991
Dactylocythere leptophylax (Crawford, 1961)
Dactylocythere macroholca Hobbs & Hobbs, 1970
Dactylocythere mecoscapha (Hobbs & Walton, 1960)
Dactylocythere megadactylus Hart & Hart, 1971
Dactylocythere myura Hobbs & Walton, 1970
Dactylocythere pachysphyrata Hobbs & Walton, 1966
Dactylocythere peedeensis Hobbs & Peters, 1977
Dactylocythere phoxa Hobbs, 1967
Dactylocythere prinsi Hobbs & Walton, 1968
Dactylocythere prionata (Hart & Hobbs, 1961)
Dactylocythere prominula Hobbs & Walton, 1977
Dactylocythere pughae (Hobbs & Hobbs, 1970)
Dactylocythere pygidion Hobbs & Peters, 1991
Dactylocythere runki (Hobbs, 1955)
Dactylocythere sandbergi (Hart & Hart, 1971)
Dactylocythere scissura Hobbs & Walton, 1975
Dactylocythere scotos Norden & Norden, 1985
Dactylocythere speira (Hart & Hart, 1971)
Dactylocythere spinata (Hobbs & Walton, 1970)
Dactylocythere spinescens Hobbs & Walton, 1977
Dactylocythere steevesi (Hart & Hobbs, 1961)
Dactylocythere striophylax (Crawford, 1959)
Dactylocythere susanae (Hobbs III, 1971)
Dactylocythere suteri (Crawford, 1959)
Dactylocythere ungulata (Hart & Hobbs, 1961)
Dactylocythere xystroides (Hobbs & Walton, 1963)
Donnaldsoncythere ardis (Hobbs & Walton, 1963)

Donnaldsoncythere cayugaensis (Hobbs & Walton, 1966)
Donnaldsoncythere donnaldsonensis (Klie, 1931)
Donnaldsoncythere hiwasseensis (Hobbs & Walton, 1971)
Donnaldsoncythere humesi (Hoff, 1943)
Donnaldsoncythere ileata Hobbs & Walton, 1963
Donnaldsoncythere leptodrylus Hobbs & Peters, 1977
Donnaldsoncythere pennsylvanica (Hart, 1960)
Donnaldsoncythere scalis Hobbs & Walton, 1963
Donnaldsoncythere truncata Hobbs & Walton, 1963
Donnaldsoncythere tuberosa (Hart & Hobbs, 1961)
Entocythere cambaria Marshall, 1903
Entocythere claytonhoffi Rioja, 1942
Entocythere costata Hobbs & Peters, 1977
Entocythere dentata Crawford, 1965
Entocythere dorsorotunda Hoff, 1944
Entocythere elliptica Hoff, 1944
Entocythere harrisi Peters, 1975
Entocythere illinoisensis Hoff, 1942
Entocythere internotalus Crawford, 1959
Entocythere kanawhaensis Hobbs & Walton, 1966
Entocythere lepta Hart & Hart, 1971
Entocythere mexicana Rioja, 1943
Entocythere prisma Andolshek & Hobbs, 1986
Entocythere reddelli Hobbs & Walton, 1968
Entocythere ruibali Rioja, 1955
Entocythere tyttha Hobbs & Hobbs, 1970
Geocythere acuta Hart & Hart, 1971
Geocythere geophila (Hart, 1959)
Geocythere gyralea Hart, 1965
Geocythere nessoides Hobbs & Hobbs, 1970
Harpagocythere baileyi Hobbs & Peters, 1977
Harpagocythere georgiae Hobbs III, 1965
Harpagocythere tertius Hobbs & Walton, 1968
Hartocythere torreya (Hart, 1959)
Litocythere lucileae Hobbs & Walton, 1968
Lordocythere petersi Hobbs & Hobbs, 1970
Okriocythere cheia Hart, 1964
Ornithocythere aetodes Hobbs III, 1970
Ornithocythere gypodes Hobbs III, 1969
Ornithocythere popi Hobbs III, 1970
Ornithocythere rhea Hobbs III, 1970
Ornithocythere waltonae Hobbs, 1967
Ornitocythere thomai Hobbs & McClure, 1983
Phymocythere lophota Hobbs & Peters, 1993
Phymocythere phyma (Hobbs & Walton, 1962)
Plectocythere crotaphis Hobbs III, 1965
Plectocythere johnsonae Hobbs & Hart, 1966
Plectocythere kentuckiensis Hobbs & Peters, 1991
Plectocythere odelli Norden, 1977
Psittocythere psitta Hobbs & Walton, 1975

Rhadinocythere serrata (Hoff, 1944)
Sagittocythere barri (Hart & Hobbs, 1961)
Sagittocythere stygia Hart & Hart, 1966
Saurocythere rhipis Hobbs III, 1969
Thermastrocythere riojai (Hoff, 1943)
Uncinocythere allenae (Hart & Hart, 1971)
Uncinocythere ambophora (Walton & Hobbs, 1959)
Uncinocythere bicuspide (Rioja, 1943)
Uncinocythere cassiensis Hart, 1965
Uncinocythere caudata (Kozloff, 1955)
Uncinocythere clemsonella (Crawford, 1961)
Uncinocythere columbia (Dobbin, 1941)
Uncinocythere cuadricuspide (Rioja, 1945)
Uncinocythere dobbinae (Rioja, 1943)
Uncinocythere equicurva (Hoff, 1944)
Uncinocythere ericksoni (Kozloff, 1955)
Uncinocythere holti Hart, 1965
Uncinocythere lucifuga (Walton & Hobbs, 1959)
Uncinocythere neglecta (Westervelt & Kozloff, 1959)
Uncinocythere occidentalis (Kozloff & Whitman, 1954)
Uncinocythere pholetera (Hart & Hobbs, 1961)
Uncinocythere simondsi (Hobbs & Walton, 1960)
Uncinocythere spathe Hart & Hart, 1971
Uncinocythere stubbsi Hobbs & Walton, 1966
Uncinocythere thektura Hart, 1965
Uncinocythere warreni Hobbs & Walton, 1968
Uncinocythere xania Hart & Hobbs, 1961
Uncinocythere xena Hart & Hart, 1971
Uncinocythere zancla Hobbs & Walton, 1963
Uncinocythere zaruri Hobbs, 1971
Waltoncythere acuta (Hobbs & Peters, 1977)

Subfamily Hartiellinae Danielopol, 1971

Hartiella dudichi (Klie, 1938)

Subfamily Microsyssitrinae Hart, Nair & Hart, 1967

Microsyssitria indica Hart, Nair & Hart, 1967
Microsyssitria nhlabane Hart & Clark, 1984

Subfamily Notocytherinae Hart & Hart, 1967

Chelocythere kalganensis Hart & Hart, 1967
Elachistocythere merista Hart & Hart, 1970
Herpetocythere acanthoides Hart & Hart, 1967
Herpetocythere australensis Hart & Hart, 1967
Herpetocythere bendora Hart & Hart, 1967
Herpetocythere gnoma Hart & Hart, 1967
Herpetocythere labidioides Hart & Hart, 1967
Herpetocythere mackenziei Hart & Hart, 1967
Hesperocythere klasteroides Hart & Hart, 1967
Hesperocythere tallanalla Hart & Hart, 1967
Hesperocythere xiphoides Hart & Hart, 1967
Laccocythere aotearoa Hart & Hart, 1971
Lichnocythere synethes Hart & Hart, 1967

Lichnocythere tubrabucca Hart & Hart, 1967
Lichnocythere victoria Hart & Hart, 1967
Notocythere antichthon Hart & Hart, 1967
Notocythere blundelli Hart & Hart, 1967
Notocythere erica Hart & Hart, 1967
Notocythere grampians Hart & Hart, 1967
Notocythere mirranatwa Hart & Hart, 1967
Notocythere rieki Hart & Hart, 1967
Notocythere synomodites Hart & Hart, 1967
Notocythere syssitos Hart & Hart, 1967
Notocythere tasmanica Hart & Hart, 1967
Riekocythere cherax Hart & Hart, 1967
Riekocythere xenika Hart & Hart, 1967

Subfamily Sphaeromicolinae Hart, 1962

Hobbsiella cirolanae (Rioja, 1951)
Hobbsiella coahuiltecae (Hobbs & Hobbs, 1973)
Hobbsiella moria (Hart, 1978)
Sphaeromicola cebennica Remy, 1948
Sphaeromicola hamigera Remy, 1946
Sphaeromicola sphaeromidicola Hubault, 1938
Sphaeromicola stammeri Klie, 1930
Sphaeromicola topsenti Paris, 1916

First received 8 January 2013.
Final version accepted 18 November 2013.

[When citing this chapter, refer to Crustaceana 87 (2014) 952–984]

This paper is dedicated to Prof. Dr. Eugen Kempf on the occasion of his 80th birthday (16 April 2012)

REDEFINITION OF THE GENUS *TYPHLOCYPRIS* VEJDOVSKÝ, 1882 (OSTRACODA, CANDONIDAE)

BY

TADEUSZ NAMIOTKO[1,6]), DAN L. DANIELOPOL[2]), CLAUDE MEISCH[3]), MARTIN GROSS[4]) and NATAŠA MORI[5])

[1]) University of Gdańsk, Faculty of Biology, Department of Genetics, Laboratory of Limnozoology, Wita Stwosza 59, PL-80-308 Gdańsk, Poland

[2]) University of Graz, Institute for Earth Sciences (Geology & Palaeontology), Heinrichstrasse 26, A-8010 Graz, Austria

[3]) National Natural History Museum of Luxembourg, 25 rue Munster, L-2160 Luxembourg, Luxembourg

[4]) Universalmuseum Joanneum, Department for Geology & Palaeontology, Weinzöttlstrasse 16, A-8045 Graz, Austria

[5]) National Institute of Biology, Večna pot 111, SI-1000 Ljubljana, Slovenia

ABSTRACT

The taxonomy of the genus *Typhlocypris* Vejdovský, 1882 is reviewed. New morphological information on *Typhlocypris eremita* (Vejdovský, 1882), the type species of the genus, is provided, and a new reference material is presented. The generic diagnosis is emended with details derived from the developmental trajectory of the valves, from the juvenile stage A-3 to the adult. Those criteria clearly differentiate *Typhlocypris* from the related genus *Pseudocandona* Kaufmann, 1900. As here redefined, *Typhlocypris* is a phylogentic lineage of the subfamily Candoninae containing extant species presently living in aquatic subterranean habitats and fossil species recovered from non-marine Late Palaeogene to Neogene and Quaternary deposits in Europe and western Asia. The type species of *Typhlocypris* is considered a metaspecies, taxonomically treated as *T. eremita* (*sensu lato*), which includes populations resembling the newly designated reference material. The homeomorphic triangular valve shape of the Candoninae is discussed. Careful examination of the valve morphology of *Typhlocypris* combined with the analysis of limb traits helps to distinguish representatives of this genus from unrelated phylogenetic groups presenting similar triangularly shaped valves. It is emphasised that for a useful description of *Typhlocypris* taxa both transmitted light and scanning electron microscopy are necessary.

RÉSUMÉ

La taxonomie du genre *Typhlocypris* Vejdovský, 1882 est révisée. Des données nouvelles sur la morphologie de *Typhlocypris eremita* (Vejdovský, 1882), l'espèce type du genre, sont présentées

[6]) Corresponding author; e-mail: tadeusz.namiotko@biol.ug.edu.pl

et du nouveau matériel de référence est désigné. La diagnose du genre est complétée à l'aide de données provenant du développement ontogénétique des valves, du stade A-3 au stade adulte. Ces caractères nouveaux différencient clairement *Typhlocypris* du genre *Pseudocandona* Kaufmann, 1900, qui lui est apparenté. *Typhlocypris*, comme redéfini ici, est une lignée phylogénétique de la sous-famille des Candoninae contenant des espèces actuelles vivant dans les milieux aquatiques souterrains, ainsi que des espèces fossiles trouvées en Europe et en Asie de l'Ouest dans des couches sédimentaires d'origine non-marine s'étendant du Paléogène supérieur au Néogène et au Quaternaire. *Typhlocypris eremita*, l'espèce type du genre, est considérée comme étant une « méta-espèce », traitée taxonomiquement sous la forme de *T. eremita* (*sensu lato*), incluant des populations morphologiquement proches du nouveau matériel de référence désigné dans le présent travail. Le caractère homéomorphique des carapaces de forme triangulaire parmi les Candoninae est discuté. L'examen détaillé de la morphologie des valves de *Typhlocypris*, en combinaison avec l'étude de la morphologie des appendices permet de distinguer *Typhlocypris* des lignées phylogénétiques non apparentées possédant elles-aussi des carapaces de forme triangulaire. Il est montré que la description appropriée des taxons de ce genre doit se baser conjointement sur la microscopie optique par transparence et électronique à balayage.

INTRODUCTION

Candoninae Kaufmann, 1900 is a very species-rich group, widely distributed in non-marine habitats, especially in the Palaearctic Region. In this zoogeographic domain it is documented by a high number of both fossil and extant taxa (Kempf, 1980a, b, c, d, 1997a, b, c, d; Martens et al., 2008; Martens & Savatenalinton, 2011). Unfortunately, a number of groups of the subfamily, like the genera *Candona* Baird, 1845 and *Pseudocandona* Kaufmann, 1900, are poorly defined and need to be reviewed (Danielopol et al., 2011). Some taxa, like *Typhlocypris* Vejdovský, 1882, were used by ostracodologists in such different ways that presently their semantic content has become totally confused (compare *inter alia* the systematics of Candoninae proposed by Krstić & Guan, 2000 with that of Karanovic, 2012). Even worse, it has become practically impossible to use *Pseudocandona* and *Typhlocypris* when dealing with topics of wider interest, e.g., for biogeographic studies at regional or global scales. Therefore ostracodologists (*inter alia* Prof. K. Martens, message on 9 Jan. 2009 to D.L.D. and C.M.) requested taxonomic clarification of such groups.

Recently, for a review of the taxonomy of Candoninae it has been proposed (Danielopol et al., 2011) to re-examine the original diagnoses of critical taxa, the genus *Typhlocypris* included, to look for new criteria in order to better define both the genus- and species-group taxa and to try to harmonise the systematics used by palaeontologists with those of neontologists. The latter ostracodologists are in a better situation to improve critical taxonomies because they could use morphological traits of both the carapace and limbs. In the present study we first present a review of the different ways along which zoologists and palaeontologists integrated the taxonomic position of *Typhlocypris* within the general systematics of Candoninae. We then present new data on the morphology of *T. eremita*

(Vejdovský, 1882), the type species of the genus. We point out the inaccuracy of the original description of the type species and therefore great difficulty in using that description as a stable reference for further comparative taxonomic studies. To improve that situation we present information on a new reference material for the type species *T. eremita*. Once solved the problem of the fixation of an objective reference system we will propose new criteria to differentiate the closely related genera *Typhlocypris* and *Pseudocandona*. We continue with the fossil record of *Typhlocypris* taxa and discuss the homeomorphism of the triangular carapace shape within the Candoninae. All this information is then used to present a redefinition of the genus *Typhlocypris*. The present contribution is intended to provide ostracodologists, both neontologists and palaeontologists, a new and sound basis for future research. One should note that this contribution is an expanded version of our communication to the 14th International German Ostracodologists' Meeting in Cologne (11-14 Oct. 2012), which honoured Professor Eugen Kempf (Namiotko et al., 2012).

Taxonomic problems encountered by zoologists

The taxon *Typhlocypris* was introduced by Vejdovský (1882) as a subgenus of the genus *Cypris* O. F. Müller, 1776 for the unique species *Cypris eremita* Vejdovský, 1880. Šoštarić (1888) raised the subgenus to the generic rank. Neither Vejdovský (1882) nor Šoštarić (1888) presented a diagnosis for the then monotypic (sub)genus *Typhlocypris*. Following the "International Code of Zoological Nomenclature" (ICZN, 1999), in cases of the genus-group names published before 1931 without an accompanying description or a definition, the first species name (in this case we have a monotypic genus) mentioned in combination with the new generic name is considered the type species of the genus and the descriptive traits of the species act as diagnostic features for the genus. Šoštarić (1888), who proposed the generic status for *Typhlocypris*, uncritically translated into Croatian selected parts of the German text of Vejdovský's (1882) publication, including also some of the erroneous traits existing in Vejdovský's text. Šoštarić (1888) presented the description of *T. eremita* from Prague as relying on the material he had sampled in wells at Zagreb (for details see Appendix A). Vávra in his monograph (1891a) offered for the first time a short diagnosis of the genus *Typhlocypris* and showed that *T. eremita* is a representative of the Candoninae closely related to the species of the genus *Candona*. In another publication Vávra (1891b) noted that the only major character which validates the generic rank of *Typhlocypris* is the lack of eyes. It is now known that all candonines lack any black eye pigment (Meisch, 2000). Müller (1912) considered this trait an adaptive convergency occurring in many subterranean organisms and therefore regarded *Typhlocypris* as a junior synonym of the genus *Candona*. Ostracodologists in the first half of the 20th century followed Müller's (1912) opinion. More recently, *T. eremita* and the related species were included in the genus *Pseudocandona* (cf. Danielopol, 1978; Meisch, 2000; Namiotko & Danielopol, 2004; Martens & Savatenalinton, 2011; but see Karanovic, 2005, 2012). *Pseudocandona* comprises some 84 extant and 28 fossil species (Kempf, 1980a, b, c, d, 1997a, b, c, d; Martens & Savatenalinton, 2011). Although the genus has an almost worldwide distribution, the vast majority of species occurs in the Palaearctic Region (Martens & Savatenalinton, 2011). The genera *Typhlocypris*, *Pseudocandona* and *Candona* have a rather tortuous and interwoven taxonomic history, of which only the most significant steps are reported here.

Typhlocypris versus *Pseudocandona*

The genus *Pseudocandona* was established by Kaufmann (1900) for a single species *Candona pubescens* (Koch, 1837), a species in which the males were known to possess sexually non-dimorphic antennae (A2), i.e., (a) with 5 (versus 6) segments (penultimate segment undivided), and (b) lacking the so-called male bristles (those bristles are now known to be transformed t-setae; Danielopol, 1978; Meisch, 2000). Initially, *Pseudocandona pubescens* (Koch, 1837) — now considered an unused older synonym of *Pseudocandona insculpta* (G. W. Müller, 1900) (see Hartwig, 1901 and Meisch, 2000 for synonymy) — was the unique species assigned to *Pseudocandona*.

Subsequently, both *Typhlocypris* and *Pseudocandona* were put into synonymy with *Candona* s. l. by the most influential authors of the first half of the 20th century (e.g., Müller, 1912; Alm, 1915; Klie, 1938). Bronstein (1947) maintained the genus *Pseudocandona* and, relying on Kaufmann's (1900) diagnostic character (i.e., the non-sexually-dimorphic male A2), included eight species collected from Lake Baikal in the previously monospecific genus. The same author (Bronstein, 1947) also put the genus *Metacandona* Bronstein, 1930 in synonymy with *Pseudocandona*.

In a milestone publication, Triebel (1963), a palaeontologist, treated the systematics of the Candoninae within a bright framework including both palaeontological and zoological information. His treatment was largely followed by a series of zoologists (see below). Triebel (1963) brought the taxonomic knowledge of the lineages *Typhlocypris* and *Pseudocandona* a big step forward. First, he correctly pointed out that the male A2-feature is not a useful taxonomic character at the generic and subgeneric levels, this because it leads to the grouping of otherwise very diverse and therefore not closely related species (for instance, *P. insculpta* and Bronstein's Baikal species). Triebel (1963: 165) therefore redefined *Pseudocandona*, the 'reticulate' surface of the juvenile valves becoming the main differentiating character of the taxon. He consequently transferred the species of the *rostrata*- and *compressa*-groups of the genus *Candona* into *Pseudocandona*, which he classified as a subgenus of *Candona*. Triebel (1963: 162) similarly revived *Typhlocypris*, putting forward the triangular shape of the carapace as the main differentiating character, the larger valve (in this case the left one) being approximately triangular and the smaller one approximately trapezoidal in shape. Triebel (1963) assigned *Typhlocypris* the rank of a subgenus of *Candona*. Subsequently, Petkovski (1969a, b) reintroduced *Pseudocandona* as a separate genus to include the *rostrata*- and *compressa*-groups of species. *Typhlocypris eremita*, the type species of *Typhlocypris*, was considered to belong to the *rostrata*-group. *Typhlocypris* was therefore sunk into synonymy with *Pseudocandona*, becoming an unused older synonym of *Pseudocandona*. This view was adopted by the majority of zoologists up to the present time (e.g., Danielopol, 1978; Martens, 1984; Marmonier, 1988; Meisch, 2000; Namiotko & Danielopol, 2004; Martens & Savatenalinton, 2011).

Danielopol (1978: 27) provided an additional diagnostic character of *Pseudocandona*, i.e., the shape of the M-process of the hemipenis, a process which usually appears flat and proximally only weakly sclerotised in the males of the genus. Meisch (1996) introduced two important 'new' differentiating characters which each on its own permits easy distinction of the extant species of *Pseudocandona* from those of the related genera *Candona* s. str., *Fabaeformiscandona* Krstić, 1972 and *Schellencandona* Meisch, 1996: (a) the presence of dense, long, stiff and perpendicularly attached setae on the valves, and (b) the existence of three setae (d1, d2 and dp) on the protopodite of the cleaning leg (L7) (there are only two such setae, d1 and dp in three other genera mentioned above).

Sywula (e.g., 1973, 1974) was the first to draw the attention to the fact that the generic name *Typhlocypris*, as revised by Vávra (1891a), has priority over the name *Pseudocandona*. Therefore Sywula in his monograph (Sywula, 1974) used *Typhlocypris* instead of *Pseudocandona* assigning it the rank of a subgenus within the genus *Candona*. This scheme, though in a slightly different variant, was also used more recently by Karanovic (2005) (see the following subsection).

Meisch (2000: 152) pointed out that *Typhlocypris* is an unused older synonym of *Pseudocandona* and that, if the species of the *eremita*-group of *Pseudocandona* would later be considered to belong to a separate genus, then *Typhlocypris* would be the valid name of that genus. This statement foreshadowed the reassessment and redefinition of the genus *Typhlocypris* as proposed in the present contribution.

The *eremita*-group of *Pseudocandona*

Danielopol (1971: 182) introduced the *eremita*-group of species within *Pseudocandona* s. l. for the species around *Pseudocandona eremita*. More recently Namiotko & Danielopol (2004) and Namiotko et al. (2004) presented a detailed review of the species of the *eremita*-group of *Pseudocandona*, assigning 11 extant species to that group. Iepure et al. (2007) discussed the evolutionary morphology of the species of the *eremita*-group and described two new species. The present work uses many results of the latter three studies. It should be noticed that Karanovic (2005, 2012) proposed an alternative taxonomic scheme: upon recognising *Typhlocypris* as an older, albeit currently unused synonym of *Pseudocandona*, she decided to apply the nomenclatural priority rule. Consequently, Karanovic (2005) put both names into synonymy, with priority given to *Typhlocypris* and subdivided the genus *Typhlocypris* into two subgenera: (1) *Typhlocypris* (*Pseudocandona*) with the species of the *compressa*-group, and (2) the nominotypical subgenus *Typhlocypris* (*Typhlocypris*) for the remaining species groups formerly assigned to *Pseudocandona*. For both taxonomic and phylogenetic reasons presented in the present paper, this view is not adopted here.

Here we recognise the species around *P. eremita* to form a separate evolutionary lineage and we therefore propose to extract the *eremita*-group of species from the genus *Pseudocandona* Kaufmann to elevate it to the genus level under the name *Typhlocypris* Vejdovský. A redefinition of *Typhlocypris* is given and the evolutionary trajectory of the lineage is discussed.

Taxonomic schemes used by palaeontologists

Héjjas (1894) described a new genus and species, *Kochia trigonella*, from Miocene deposits of (palaeo)Lake Pannon, a site which today is situated in Transylvania, Romania. This ostracod has asymmetric valves, the left one being triangular and displaying a strong umbo. With these traits it resembles the species of the genus *Typhlocypris*. Schneider (in Mandelstam & Schneider, 1956) described from Miocene deposits in the Caucasus another candonine genus with triangular left valves — *Advenocypris* Schneider, 1956. Because of the triangular shape of the carapace, Schneider (in Lubimova et al., 1960) created for this taxon the subfamily Advenocyprinae, which still nowadays is accepted by some specialists (cf. Vekua, 1975). Krstić & Guan (2000) and Krstić (2006) lowered Schneider's (1960) subfamily Advenocyprinae to the status of a tribe (Advenocyprini). Mandelstam & Schneider (1963) described a series of new species from the Miocene and Pliocene, which they assigned to *Advenocypris*. Krstić (2000), however, considered those triangular candonines to belong to a new tribe (Typhlocyprini), for which she did not offer any diagnosis.

Triebel (1963) imposed the idea that both *Kochia* and *Advenocypris* are junior synonyms of *Candona* (*Typhlocypris*). This taxonomic scheme was widely followed by palaeontologists from Central and Western Europe and Turkey (Krstić, 1972, 1993; Freels, 1980; Pokorný, 1988; Gross, 2004). *Kochia trigonella* Héjjas,

1894 was also mentioned as belonging to *Typhlocypris* by various specialists, *inter alia* Freels (1980) and Stoica (in ter Borgh et al., 2013). It should be noted that *Kochia* was already a preoccupied name used by botanists for the Chenopodiaceae genus *Kochia* Roth (cf. Chu & Sanderson, 2008). More importantly, the living triangular candonid *Candona trigonella* Klie, 1931 (a potential species for the genus *Typhlocypris*, cf. Namiotko & Danielopol, 2004 and additional information in the section "Notes on living and fossil species assigned to the genus *Typhlocypris* Vejdovský" of this contribution) uses a preoccupied name (see above) and as such is a junior homonym. Moreover, Sokač (1978) assigned a fossil Quaternary ostracod resembling the extant *eremita* species to the genus *Typhlocypris*. Other palaeontologists assigned their fossil triangular candonids either to the genus *Candona* (e.g., Absolon, 1973, 1978) or to *Pseudocandona* (e.g., Pipík & Bodergat, 2008), without the recognition of a subgenus *Typhlocypris*.

Finally, we have to mention the taxonomic scheme proposed by Krstić (2006) for the triangular candonines. This author uses taxonomic names that zoologists pointed out to be synonyms of other taxa. This is the case of the monotypic genus *Cavernocandona* Hartmann, 1964 (type species *C. dispar* Hartmann, 1964 considered to be a representative of the *eremita*-species group of *Pseudocandona* by Namiotko & Danielopol, 2004), which Krstić (2006) assigned to the tribe Advenocyprini. On the other hand, the living species *Candona pannonicola* Löffler, 1960, also considered to be closely related with the *eremita*-species group (Meisch, 2000; Namiotko & Danielopol, 2004), was assigned to the new genus *Pannonocandona* Krstić, 2006 by Krstić (2006), for which the Pliocene species *P. farkasi* Krstić was designated as type species. Unlike *Cavernocandona*, *Pannonocandona* was assigned to the tribe Pseudocandonini described by Krstić & Guan (2000).

MATERIAL AND METHODS

The material used for the redescription of *Typhlocypris eremita* (Vejdovský, 1882) comes from Čečelice, a locality situated about 30 km north of Prague, Czech Republic. It consists of fossil valves collected by Dr. A. Absolon (Prague) and given as a present to D.L.D. in 1971. The sampling site is an outcrop of 'Sumpfkreide' (lake marl deposited in swamps) dated as 'Altholozän' (Early Holocene) in Absolon (1973: 50 and 1978: 40). Our sample contains five valves registered under the four figure photo-catalogue numbers of the Ostracoda collection of D.L.D. housed until January 2013 in the Department of Geology and Palaeontology, Institute of Earth Sciences, Karl-Franzens University Graz, and then transferred and deposited in the Department of Invertebrate Zoology, Natural History Museum, Vienna and registered under five figure NHMW crustacean collection numbers: (1) adult (Ad) left valve (LV) specimen (Spec) 1, here designated as a new valve reference material for *T. eremita* (photo-catalogue no 8354, NHMW-25481); (2) Ad LV Spec 2 (8355, NHMW-25482); (3) Ad right valve (RV) (8356); (4) A-2 juvenile (juv) RV Spec 1 (8357) both kept in one micropalaeontological cell (as NHMW-25483); and (5) A-2 juv RV Spec 2 (8358, NHMW-25484).

We compared our adult valves with the illustrated material of Absolon (1978, fig. 25). Additional valves came from a well at Hotarele, a village located about 30 km south of Bucharest in the Judeţ Ilfov, Romania. The material originates from the sample of living individuals and empty valves used for the study of the developmental trajectory and life cycle of *T. eremita* (*sensu lato*) published by Danielopol (1980). We used the following valves listed with their photo-catalogue numbers: Ad LV (8299), Ad RV (8300), A-1 juv LV (8301), A-1 juv RV (8302), A-2 juv LV (8303), A-2 juv RV (8304), A-3 juv LV Spec 1 (8305), A-3 juv LV Spec 2 (8306), A-3 juv RV Spec 1 (8307) and A-3 juv RV Spec 2 (8308). This material was also deposited at the Natural History Museum of Vienna (Austria), Department of Invertebrate Zoology under numbers NHMW 25485-25489.

Additionally, we used for comparison fossil specimens of Late Miocene age from the Lake Pannon collected at Mataschen, Styrian Basin (regional stage Pannonian, "Zone B") and Hennersdorf, Vienna Basin (regional stage Pannonian, "Zone E"), which two of us (D.L.D. and M.G.) intensively sampled for ostracods during a number of years. For the first site one should consult the monograph of Gross (2004), whereas for the second one the information can be found in Danielopol et al. (2011).

Valves were photographed by transmitted light microscopy (TLM) with a Nikon DS-5 M digital camera fitted to a Nikon E-200 microscope as well as with a JEOL 6610 scanning electron microscope (SEM). In this latter case we examined the external side of the same valve which was previously photographed in TLM. Microphotographs were stored as tiff and jpg files and used for the digitised outline analysis using the software tps-dig, version 1.37 (Rohlf, 2003). In this way we realized a much more detailed analysis of valves as compared with classic descriptions where generally either TLM or SEM is used. In our case we examined for the same valve its shape, the inner and outer surface and structural elements of the calcified layer, the inner lamellae as well as the marginal pore canals.

For comparative analysis of the valve shapes we used the computer software Morphomatica (Linhart et al., 2007) with protocols for geometric morphometrics superposition using two modes, the standardised and non-standardised for surface. The former mode was used to avoid differences due to the valve size. For the mathematical reconstruction of outlines (the approximation of the B-spline curve) we used 24 (12 × 2) control points. This algorithm slightly differs from that used by Iepure et al. (2007) where 16 (8 × 2) control points defined the outline shape. However, in our case an excellent approximation of the mean and maximum error of the fitted mathematical curve to the digitised outline was obtained with 24 control points. For comparative purposes, especially for the geometric morphometrics analysis, we used a number of illustrations taken from the existing literature. The precise origin of the data is given in the captions to the corresponding figures.

The chaetotaxic scheme used here follows Broodbakker & Danielopol (1982) as revised for the second antenna by Martens (1987) and extended to the walking and cleaning legs by Meisch (1996). For a review see Meisch (2000) and for the nomenclature of valves and limbs we followed Horne et al. (2002) as well.

Here we introduce a rectification to the nomenclature of the endopodial segments of the 7th limb (the cleaning leg L7) for the Candoninae. One of us (D.L.D.) noticed that the endopodial segment 2 (E II) is not separated by an intersegmental septum from the next endopodial segment (E III); therefore, one has to consider them as one compound segment, named here E II + III. By combining the observation in TLM with that obtained with SEM it is possible to deduce that the so-called septum between the endopodial segments E II and E III is just an external annulation which varies in its expression depending on the thickness of the outer cuticular layer of the endopodite. Arguments for the present assertion are offered in the Appendix A which is added to this publication.

The following abbreviations are used: A2, second antenna; Ad, adult; juv, juvenile; L5, fifth limb (= first thoracopod transformed into a maxilliped); L7, seventh limb (= third thoracopod forming a cleaning leg); LV, left valve; RV, right valve. For endopodial segments of L7 we use the codes E I, E II + III and E IV following a scheme of Broodbakker & Danielopol (1982) and as mentioned above.

Hierarchical taxonomic position of the genus *Typhlocypris* follows Martens & Savatenalinton (2011).

RESULTS AND DISCUSSION

Morphology of *Typhlocypris eremita* (Vejdovský, 1882), new aspects

On the accuracy of the description of *Typhlocypris eremita* in Vejdovský's (1882) monograph

We mentioned above that Vejdovský (1882) proposed *Typhlocypris* as a subgenus of the at that time well-known but vaguely defined genus *Cypris*. Originally, the subgenus *Typhlocypris* included only one species (*Cypris eremita*) which name was first introduced in Vejdovský's earlier paper (Vejdovský, 1880) without any description or illustration, and as such was a nomen nudum. The same name (*Cypris eremita*) was made available later for the same concept when Vejdovský (1882)

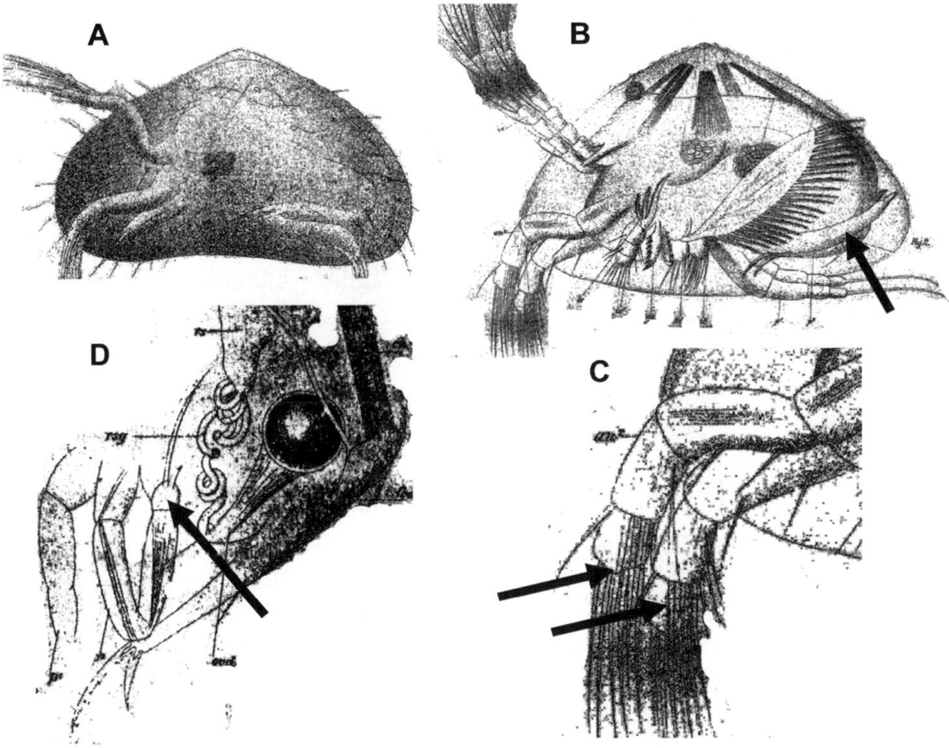

Fig. 1. Reproduction of morphological details of *Typhlocypris eremita* (Vejdovský, 1882) from Vejdovský (1882: Plate 7, figs. 1, 2 and 6). A, Individual with carapace and limbs; B, A-3 juv stage with "Anlage" of L7 (arrow); C, A2 of juv figured in B, with presumed swimming setae (arrows); D, Posterior part of Ad, with details on the distal setae on L7 (arrow). Not to scale.

provided a description, thus the name takes the authorship and date of that act of establishment, not from the earlier publication.

In his description and figures, Vejdovský (1882) starts with the description of the carapace in lateral view (his fig. 1 of Plate VII is reproduced here in fig. 1A) and says that the carapace in lateral view is triangular in shape. However, Vejdovský's (1882) description of limbs is inadequate. As we do not have either the original reference material (no name-bearing type specimen is believed to be extant) or a good diagnosis of the species, ostracodologists encounter serious problems with the accurate identification of *T. eremita*-like individuals coming from populations located outside Prague. This conundrum will be solved here for the first time and inaccuracies of Vejdovský's (1882) description are here clarified.

Vejdovský (1882) figured a juvenile of *T. eremita* with surprising traits (fig. 1 here). First, it is easy to recognise that the juvenile in question is the A-3 stage, as indicated by the "Anlage" of the L7 limb (arrowed in fig. 1B here). Vejdovský's juvenile has anterodorsally a roundly shaped structure, which he suggested be a rudimentary eye (fig. 1B), although he considered the eye to be absent from adult specimens. As pointed out by Namiotko et al. (2012) a similar structure was not seen in controlled experiments with *T. eremita* (*sensu lato*) from the groundwater at Hotarele, southern Romania (details in Danielopol, 1980). Additionally, the juvenile figured by Vejdovský (op. cit.) displays long endopodial setae on A2 (arrow in fig. 1C here), which are equivalent to the swimming setae seen in the genus *Cypris* and many species of the family Cyprididae Baird, 1845 in general. However, as representatives of the subfamily Candoninae are all devoid of swimming setae (Meisch, 2000), in our opinion the supposed swimming setae of *T. eremita* figured by Vejdovský (1882) are most probably not shown inserted in the correct place on the correct podomere, and thus it can be argued that they are in fact not swimming setae and the drawing is mistaken. Finally, in fig. 1D (arrow) we point out the distal setae of L7 illustrated by Vejdovský (1882). In contrast to the Cyprididae, both the adults and A-1 juveniles of the Candoninae usually have three well developed setae there (Kovalenko, 1983). The L7 in Vejdovský's (1882) figure (here reproduced in fig. 1D) belongs to an adult female with a well developed *receptaculum seminis* and an ovocyte inside, but only two distal setae are figured. We consider that Vejdovský did not see the 3rd long seta, typical of the Candoninae.

Typhlocypris eremita from Čečelice and designation of a new reference material

Vejdovský (1882) sampled *T. eremita* in a house-well used for drinking water in Prague at the address 183 Křemencova street (Praha 1, Nové Město, 50°4′43.489″N 14°25′1.582″E). The site is located about 230 m away from the bank of the Vltava River and the well most probably was dug in alluvial sandy-gravel sediments. As this house-well does not exist anymore, on 15 September 1969 one of us (D.L.D.) tried to collect *T. eremita* in alluvial sediments of the so-called hyporheal of the bank of the Vltava River in the southern part of Prague, at Radotin (i.e., at about 11 km from the type location mentioned by Vejdovský, 1882), but only *Fabaeformiscandona wegelini* (Petkovski, 1962) was found. Additionally, D.L.D. addressed an inquiry to the National (Natural History) Museum in Prague, Department of Invertebrate Zoology for a possible location of the holotype and/or other type material left by Vejdovský. The following message was received on 21 February 2013 from Dr. Petr Dolejš: "I checked databases and inspected the collections as well and found no type material of *Cypris* (*Typhlocypris*) *eremita*". Also the well from where Vávra (1891a) got specimens used for his redescription of *T. eremita*, namely Brenntegasse 22, nowadays Spálená 22 (Praha 1, Nové Město), disappeared as we were informed by P. Dolejš, who wrote on 21 February 2013 to D.L.D.: "I am afraid that number 22 does not exist any more because there is a metro-station (currently in reconstruction so I cannot check the situation). The last building has number 18 and the following 30 (numbers 20-22-24-26-28 were probably demolished)". It seems that also a house-well at the address Spálená 64 does not exist anymore, from where *T. eremita* was re-sampled after Vejdovský (1882) by Řeháčková (1956) and illustrated by Jančařík (1952).

Considering the above presented inadequacy of the original description coupled with the lack of the type material, we decided to designate a selected valve of a series of valves from Čečelice (described in detail below) as a new reference material which for the moment is proposed as a standard for *T. eremita*. Although the new reference material meets qualifying conditions for designation of a neotype, in the absence of the soft parts we were disinclined to formally designate the Holocene valve from Čečelice as the name-bearing type of *T. eremita*. The Čečelice valve is the best characterised material we have nowadays which came as nearly as practicable from the original type locality. We decided to adopt this pragmatic solution because it is absolutely necessary to have a fixed reference material for future studies dealing with the taxonomic status of this species. The designed here new reference valve material should be abandoned once new material of living *T. eremita* is collected in Prague, and when valves are correctly described.

Absolon (1978) provided excellent figures of the valves of *Typhlocypris eremita* sampled at Čečelice, ca. 30 km north of the type locality in Prague. His description is useful in order to complement the scarce and sometimes incorrect information existing on this species. This is done here with the help of the material of A. Absolon offered to one of us (D.L.D.) and listed in the Material and Methods section. By the superposition of the valve outlines we compared the material from Čečelice and the illustrations of valves given by Vejdovský (1882), Vávra (1891a), Jančařík (1952) and Absolon (1978) using geometric morphometric techniques in standardised and non-standardised for surface modes.

Vávra's (1891a) and Absolon's (1978) descriptions clearly deal with adult females. When we superposed on these valves the outlines of the valves figured by Vejdovský (1882) and Jančařík (1952) it can be shown that the latter two valves belong to juvenile stages (consult fig. 2A-D). In fig. 3A-B the superposition of the outlines of adult and juvenile valves illustrated by Absolon (1978) allowed us to identify the developmental stage of the valves illustrated by Vejdovský (1882). We also noted (fig. 2A, C) that the adult left valve figured by Vávra (1891a) is similar to that of Absolon (1978). Our left valves No. 8354 and No. 8355 from Čečelice closely fit Absolon's figures (fig. 4). When the outline given by Vejdovský (1882) in his fig. 1 of Plate 7 is superposed on the outline of the A-2 juvenile figured in Absolon (1978, fig. 25, juv.) we notice a close resemblance (fig. 3C). Therefore we can now say with confidence that Vejdovský (1882) and Jančařík

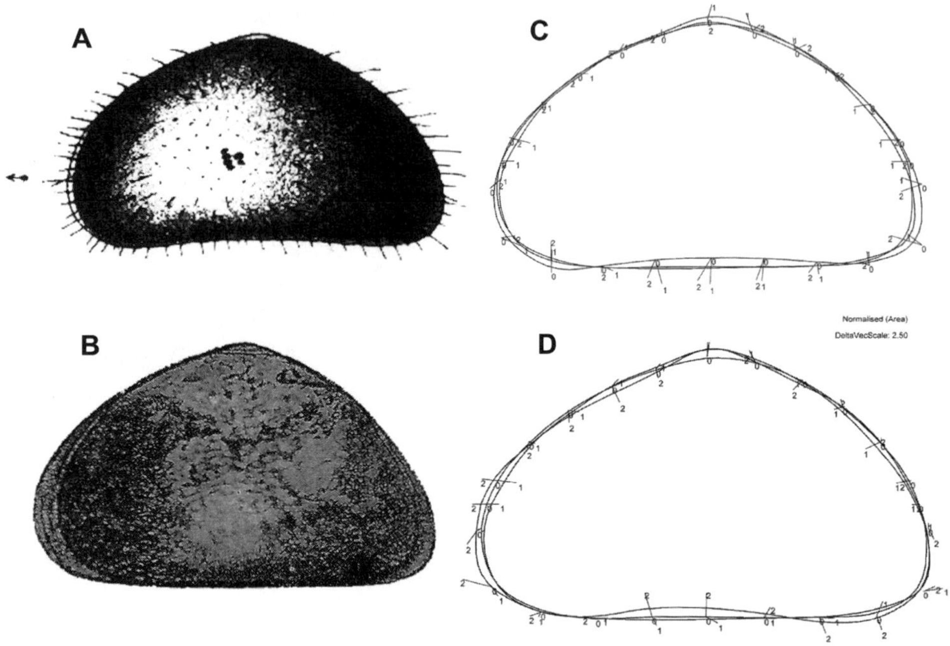

Fig. 2. *Typhlocypris eremita* (Vejdovský, 1882), viewed from the left side. A, Carapace reproduced from Vávra (1891a: fig. 15/1); B, LV in outer lateral view from Jančařík (1952: fig. 2); C, Superposition in standardised for equal surface mode of Ad valve-shapes in lateral views from illustrations of Vávra (1891a: fig. 15/1) — indicated as 0, Jančařík (1952: fig. 2) = 1 and Absolon (1978: fig. 25) = 2; D, Superposition of valve-shapes, standardised for equal surface, from illustrations of Absolon (1978: fig. 25) = 0, Jančařík (1952: fig. 2) = 1 and Vejdovský (1882: pl. 7, fig. 1) = 2. Delta vectors in C and D are 2.5× enlarged.

(1952) documented *T. eremita* only with juvenile valves, what was indeed affirmed in the latter paper (Jančařík, 1952).

From the Čečelice sample (cf. above, Material and Methods) we chose as the new reference material a left adult valve No. 8354, here illustrated in figs. 4A-B and 5A-B, catalogued at the Natural History Museum in Vienna under the number NHMW-25481. The remaining valves of our material from Čečelice offer additional information on *T. eremita*, especially on the diagnostic morphological traits existing on the adult right valve and on the juvenile valves. Hence, we illustrate here the specimens of this series with both TLM and SEM pictures. Figs. 4A and 5A depict the general aspect of the new reference material. They show the triangular shape of the left valve, as well as the reduced extension of the dorsal umbo and of the inner lamella. The anterior marginal pore canals of this valve are short (fig. 4C); the central surface of the valve is covered with foveolae well visible in TLM (fig. 4B) and better documented with SEM pictures

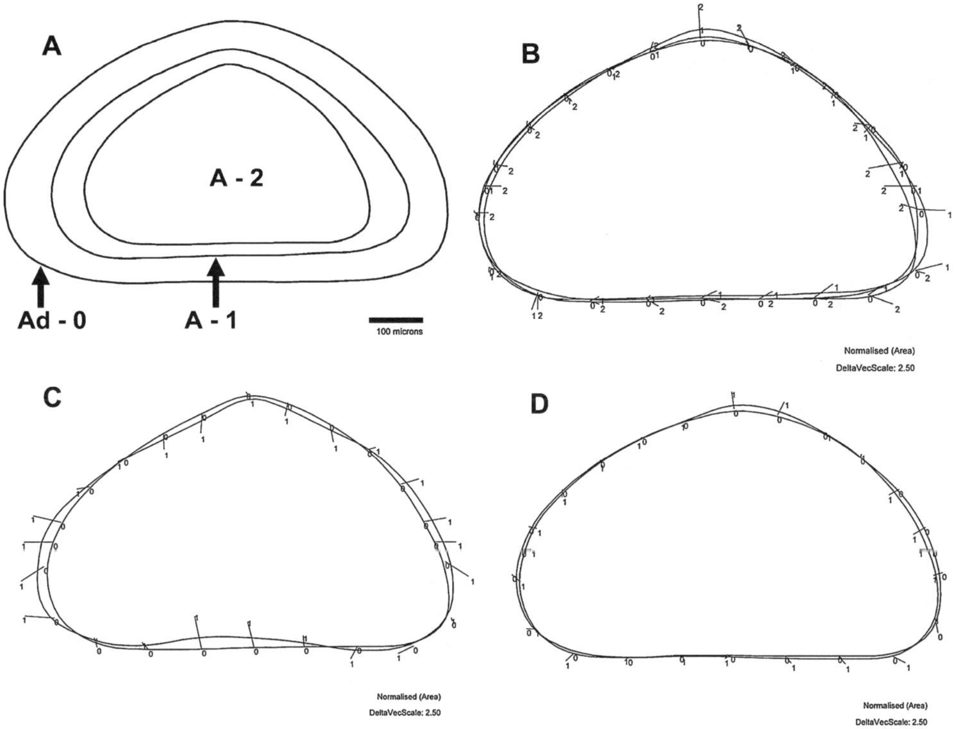

Fig. 3. *Typhlocypris eremita* (Vejdovský, 1882). A-B, Superposition of LV-shapes in lateral views of Ad and juv stages A-1 and A-2. A, non-standardised for equal surface and B, standardised for equal surface outlines from Absolon (1978: fig. 25), Ad indicated as 0, juv A-1 = 1 and A-2 = 2; C, Superposition in the standardised for equal surface mode of A-2 LV from Absolon (1978: fig. 25) = 0 and the outline from Vejdovský (1882: Plate 7, fig. 1) = 1; D, Superposition in the standardised for equal surface mode of Ad LV No. 8354 from Čečelice = 0 and the outline from Farkas (1958: fig. 51/A) = 1. Scale bar in A, 0.1 mm. Delta vectors in B-D are 2.5× enlarged.

(fig. 5A-B). The adult right valve (fig. 4E) is only slightly triangular and the intercardinal dorsal margin has an oblique position. As in the left adult valve, the inner lamella is reduced. The juvenile A-2 right valves (figs. 4F-G and 5C) have a triangular shape with a rounded dorsal margin and a stronger expression of the central ornamentation.

Finally, we have to mention that the modern systematics of *Typhlocypris*, as here discussed, uses mainly details of the valve morphology. Therefore, the valve material from a site near Prague (as near as possible to the original type locality) as the reference for *T. eremita* is the alternative solution we offer. Indeed, groundwater ostracods are hard to sample in the Prague area, this mainly because Vejdovský's and Vávra's wells have disappeared (cf. also Appendix A) and also because no other sites in Prague and surroundings have yielded material of *T. eremita* so far.

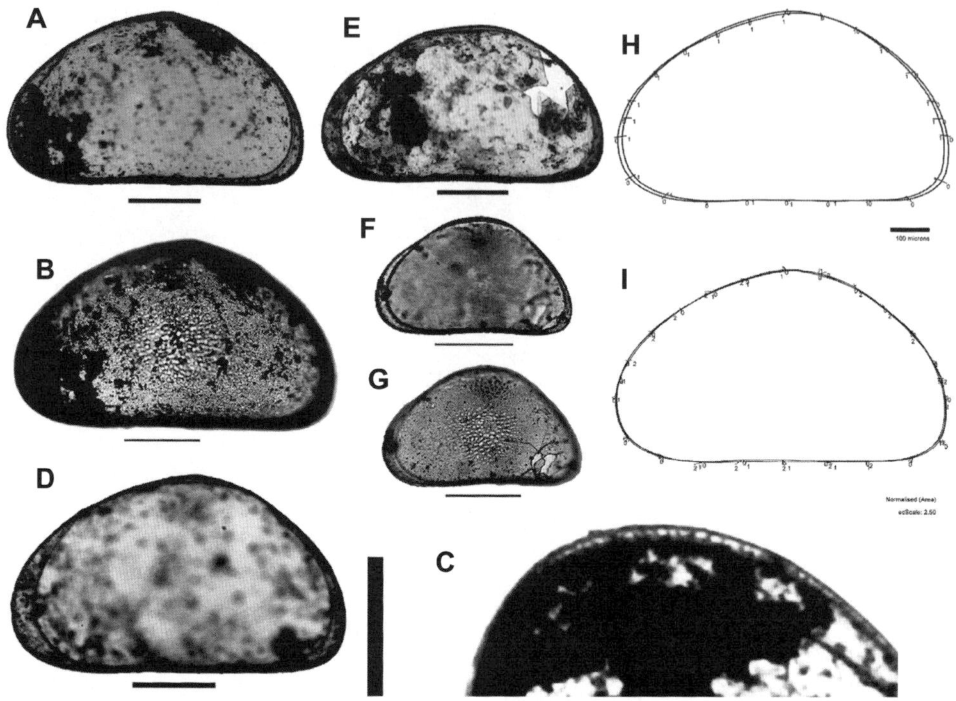

Fig. 4. *Typhlocypris eremita* (Vejdovský, 1882), specimens from Čečelice. A, Ad LV No. 8354 (new reference material) in outer lateral view; B, Ad LV No. 8354 as in A but with visible foveolae in the central surface; C, enlarged anterior end of LV No. 8354 (new reference material) with details on the fused zone and the short and straight marginal pore canals; D, Ad LV No. 8355 outer lateral view; E, Ad RV No. 8356, reversed outer lateral view; F, A-2 RV No. 8357, reversed outer lateral view; G, same as F but with visible foveolae in the central surface; H, superposition in non-standardised for equal surface mode of Ad LV-shapes of specimens No. 8354 (new reference material) = 0 and 8355 = 1; I, superposition in standardised for equal surface mode of LV No. 8354 (new reference material) = 0, LV no. 8355 = 1 and Absolon's (1978) LV from fig. 25 = 2. Scale bars: A, B, D-G, 0.2 mm, C and H, 0.1 mm. Delta vectors in H and I are 2.5× enlarged.

Typhlocypris eremita (*sensu lato*) from Hotarele and Tekovské Lužany (Nagysalló): their interest for redefinition of the genus *Typhlocypris*

Iepure et al. (2007) defined *T. eremita* (as *Pseudocandona eremita* in their paper) as a metaspecies, with the meaning of an entity which has a rather arbitrary definition for the moment and within which one can insert populations or even nominal taxa which are commonly difficult to characterise. Considering the parthenogenetic population from Hotarele in southern Romania studied first by Danielopol (1982) and, more recently, by Iepure et al. (2007), it was not possible to find clear morphological traits permitting to classify those ostracods as belonging to a new species different from *T. eremita* from Bohemia as characterised by Vávra (1891a) and Absolon (1978). Therefore, the population from Hotarele was

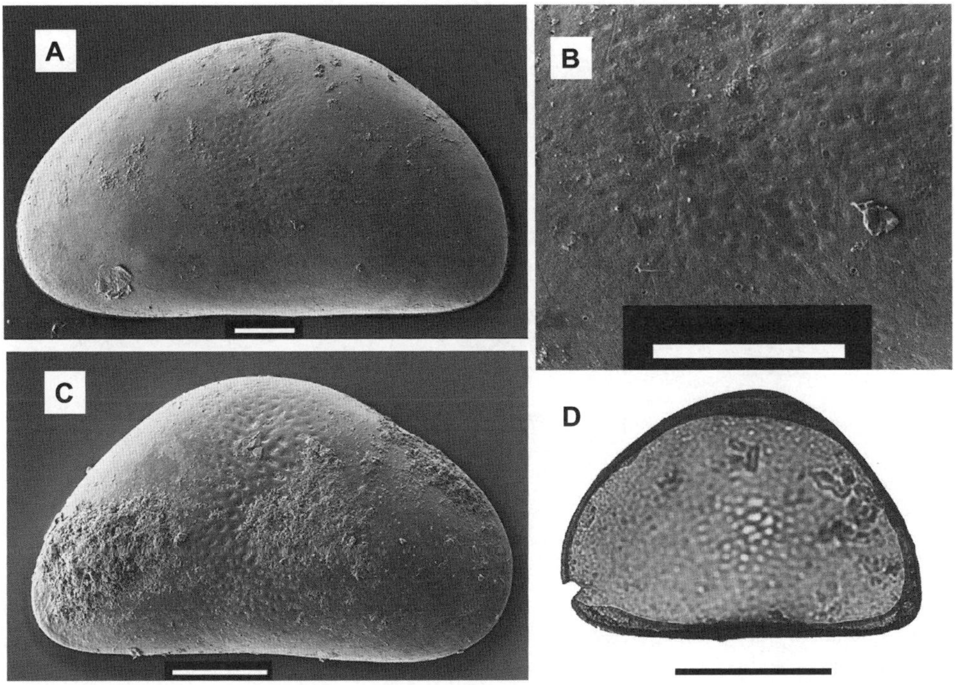

Fig. 5. A-C, *Typhlocypris eremita* (Vejdovský, 1882) specimens from Čečelice, SEM microphotographs. A, Ad LV No. 8354 (new reference material) outer lateral view; B, detail from A of central surface with enlarged foveolae; C, juv A-2 RV No. 8358 outer lateral view; D, *Typhlocypris* sp. juv ? A-3 LV No. 3915 outer lateral view from Hennersdorf near Vienna (Late Miocene, Pannonian E) in the layer Nα (cf. Danielopol et al., 2011, fig. 6). Scale bars: A-C, 0.1 mm, D, 0.2 mm.

considered to belong to a metaspecies termed *T. eremit*a (*sensu lato*), for which in addition to the existing morphological information, details on the life cycle traits are also available (Danielopol, 1980). Here we present information on the developmental stages of the valves starting from the A-3 stage to that of the adult. Fig. 6 displays images in TLM where several important aspects of the development trajectory of this species are seen: (1) the left valves (fig. 6A-G) change from a pronounced triangular shape and a well-developed umbo in the A-3 stage to an attenuated triangular outline and a slightly developed dorsal umbo in the adult. (2) The right valves (fig. 6H-K) show changes from a triangular shape with a round dorsal margin in the A-3 stage to a trapezoidal shape in the A-2 and A-1 stages. (3) The anterior fused zone with the marginal pore canals and the inner lamella are reduced and their extension does not change substantially from one stage to the next ones. (4) The number of the marginal pore canals increases from the A-3 stage towards the adult stage. (5) When the outlines of the A-3 left valves are superposed onto those of the subsequent stages (fig. 7A), the triangularity remains preserved. (6) However, if we look at the same valves superposed and standardised for equal

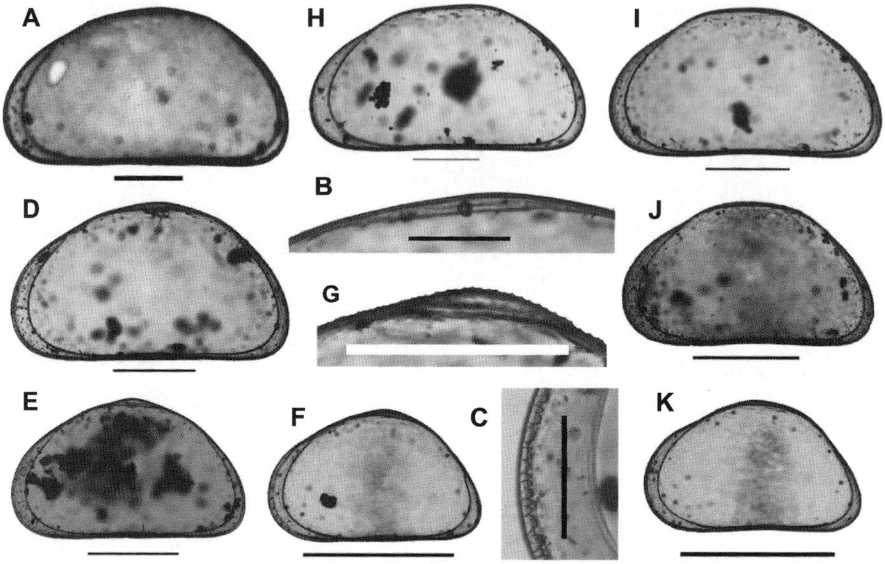

Fig. 6. *Typhlocypris eremita* (Vejdovský, 1882) (*sensu lato*), valves of living specimens from Hotarele, south of Bucharest. A-G, LVs outer lateral view; H-K, RVs reverse outer lateral view; A-C, female Ad valve No. 8299; A, general view; B, detail of the dorsal margin with slightly developed umbo; C, detail of the anterior margin with the short marginal pore canals; D, juv A-1 valve No. 8301, E, juv A-2 valve No. 8303; F, juv A-3 valve No. 8306; G, detail of the dorsal margin of juv A-3 valve No. 8305; H, Ad valve No. 8300; I, juv A-1 valve No. 8302; J, juv A-2 valve No. 8304; K, juv A-3 valve No. 8307. Scale bars: 0.2 mm, except B, C and G, 0.1 mm.

surface (fig. 7B) a gradual decrease of the triangular shape from the A-3 stage to the adult is seen. In the A-3 the maximum height is placed in the middle or slightly behind the middle of the valve (fig. 7B). This pattern is also seen in the adult right valve (fig. 7D).

When the left adult valve from Hotarele was superposed onto that from Čečelice and standardised for equal surface (fig. 7C), the Bohemian valve displays a greater height as compared to that from southern Romania. Both the left and right valves of *T. eremita* from Tekovské Lužany (Nagysalló) in Slovakia illustrated by Farkas (1958, fig. 51A-B) are higher than those from Čečelice and from Hotarele (figs. 3D and 7D). Similar differences were described by Iepure et al. (2007) for the closely related species *T. danubialis* Iepure, Namiotko & Danielopol, 2007 and *T. transylvanica* Iepure, Namiotko & Danielopol, 2007. Hence, a re-investigation of local populations originally identified as *Typhlocypris eremita* (or *Candona eremita* and *Pseudocandona eremita*) from Europe is needed in order to better understand the extent of intra- and/or interpopulational variability of valves, as well as microevolutionary aspects related to them. We have in mind *inter alia* the re-examination of the *Typhlocypris* material published by Klie (1936, 1940) and Löffler (1964).

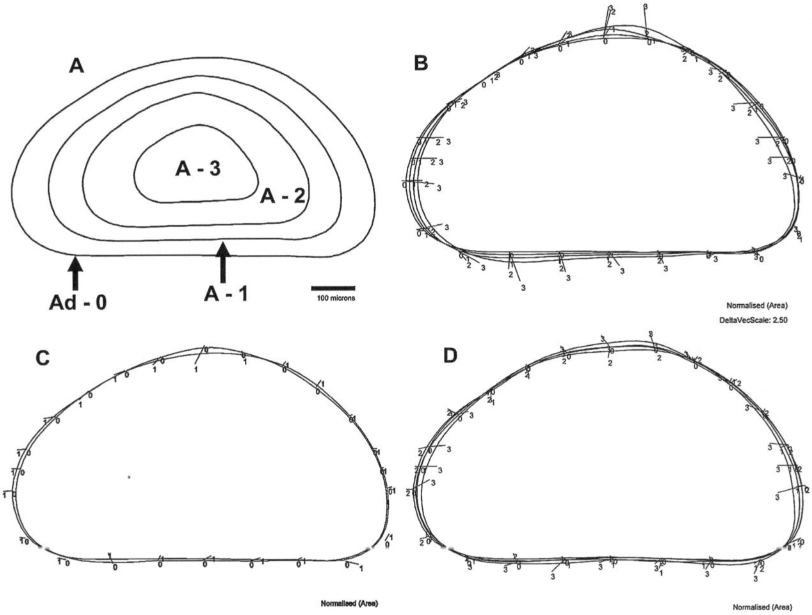

Fig. 7. *Typhlocypris eremita* (Vejdovský, 1882) (*sensu lato*). A and B, superposition of valve outlines of Ad female indicated as 0 and juv A-1 (=1), A-2 (=2) and A-3 (=3) stages of living individuals from Hotarele; A, in non-standardised for equal surface mode; B, in standardised for equal surface mode; C, superposition of Ad LV No. 8354 from Čečelice (0) and Ad LV No. 8299 from Hotarele in standardised for equal surface mode; D, superposition of Ad RV-shapes in standardised for equal surface mode of the specimens from Čečelice in Absolon (1978: fig. 25R) indicated as 0, our valve No. 8356 from Čečelice (=1), valve No. 8300 from Hotarele (2), and valve from Farkas (1958: fig. 51B) (3). Scale bar in A, 0.1 mm. Delta vectors in B-D are 2.5× enlarged.

Redefinition of the genus *Typhlocypris* Vejdovský, 1882

Class: OSTRACODA Latreille, 1802

Subclass: PODOCOPA G. W. Müller, 1894

Order: PODOCOPIDA Sars, 1866

Suborder: CYPRIDOCOPINA Baird, 1845

Superfamily: CYPRIDOIDEA Baird, 1845

Family: CANDONIDAE Kaufmann, 1900

Subfamily: CANDONINAE Kaufmann, 1900

Tribe: Candonini Kaufmann, 1900

Genus *Typhlocypris* Vejdovský, 1882

Diagnosis of the genus **Typhlocypris** Vejdovský, 1882

We use the diagnosis provided by Namiotko et al. (2012: 42), which was compiled from Danielopol (1982), Namiotko & Danielopol (2004), Namiotko et al. (2004) and also from additional information based on the new reference material from Čečelice.

Type species.— *Cypris eremita* Vejdovský, 1882.

New reference material (here designated).— One left adult valve from the Early Holocene deposits of an outcrop at Čečelice, 30 km north of Prague, Czech Republic (see Material and Methods for more details) registered under NHMW-25481.

Diagnosis.— Carapace of medium length (mostly $\leqslant$1.1 mm), approx. triangular in lateral view, with the greatest height at or just behind mid-length (at 50-60% of the length), and height to length ratio > 50%. Left valve dorsally with a hump (umbo) that overlaps the right valve. Valves thin, with a fine ornamentation consisting of shallow pits mostly located in the central valve area. Calcified inner lamella narrow, anteriorly usually $\leqslant$10% of the valve length and <2× as wide as posteriorly. Inner and outer margins more or less parallel. Valve shape, ornamentation and the relative width of the inner lamella remain almost unchanged throughout the last four stages of the postembryonic development. Sexual dimorphism in size and shape of the carapace weakly expressed; male carapace slightly larger than that of the female. A2 in both sexes with long apical claws: in females G1 $\geqslant$ 2.2× and GM $\geqslant$ 1.8× the length of the penultimate segment, in males G2 $\geqslant$ 2.0× and GM $\geqslant$ 1.6× the length of the 3rd and 4th endopodial segments combined (E II + III); male z1 claw relatively short, usually <3/4 of G2. Second segment of mandibular palp with 3 setae in the setal group, and with externo-distal (γ) seta smooth (not plumose). L5 with rudimentary exopodial branchial plate bearing two filaments. L7 with protopodite bearing three setae (d1, d2, dp), penultimate segment (E II + III) lacking the medial f seta and bearing long distal g seta, the terminal segment set with two long h2 and h3 setae and one short curved h1 seta. Female genital lobe with two fine folds separated by a flat depression or weakly developed and evenly rounded. Hemipenis: three lobes (a, b, h) well developed; M-process with a broad proximal plate, central part contracted and distal part weakly sclerotised, variously shaped and often crenulated; bursa copulatrix cornet-shaped. Zenker organ with five whorls of spines and two terminal cap-like structures.

Additional diagnostic valve differences between *Typhlocypris* Vejdovský and *Pseudocandona* Kaufmann

Namiotko & Danielopol (2004) offered a synthetic view on the diagnostic traits of the main living species of *Pseudocandona* subdivided at that time into six groups: *carribeana*, *compressa*, *eremita*, *prespica*, *rostrata* and *zschokkei*. The *eremita*-group is what we here define as the genus *Typhlocypris* Vejdovský. The *compressa*- and *rostrata*-groups form the genus *Pseudocandona* Kaufmann (*sensu stricto*). For the *zschokkei*-group Danielopol et al. (2012) proposed a new genus,

Marmocandona Danielopol, Namiotko & Meisch, 2012. However, we will not discuss here the affinities and differences of *Typhlocypris* and *Marmocandona* (this will be published elsewhere). Further, Namiotko & Danielopol (2004) pointed out that the *carribeana*- and *prespica*-groups need to be reviewed as their affinities with the other species-groups of *Pseudocandona* are still unclear. Therefore, we here tentatively consider *Pseudocandona* to comprise only the species of the *compressa*- and *rostrata*-groups.

Most of the European species of *Pseudocandona* display a more or less rectangular shape in the adult stage (Meisch, 2000). Absolon (1978), Kovalenko (1983) and Namiotko & Danielopol (2004) provided information on the carapace shape of the last three juvenile stages of the species of both the *rostrata*- and *compressa*-groups, showing that the juveniles down to the juvenile stage A-3 keep a more or less rectangular shape. Fig. 8, with superposed outlines of the A-3 shape of the left valve onto the adult valve shows clear differences between *Pseudocandona sarsi* (Hartwig, 1899) (fig. 8A), *Pseudocandona albicans* (Brady, 1864) (fig. 8B) and *Typhlocypris eremita* (*sensu lato*) (fig. 8C). It should be noticed

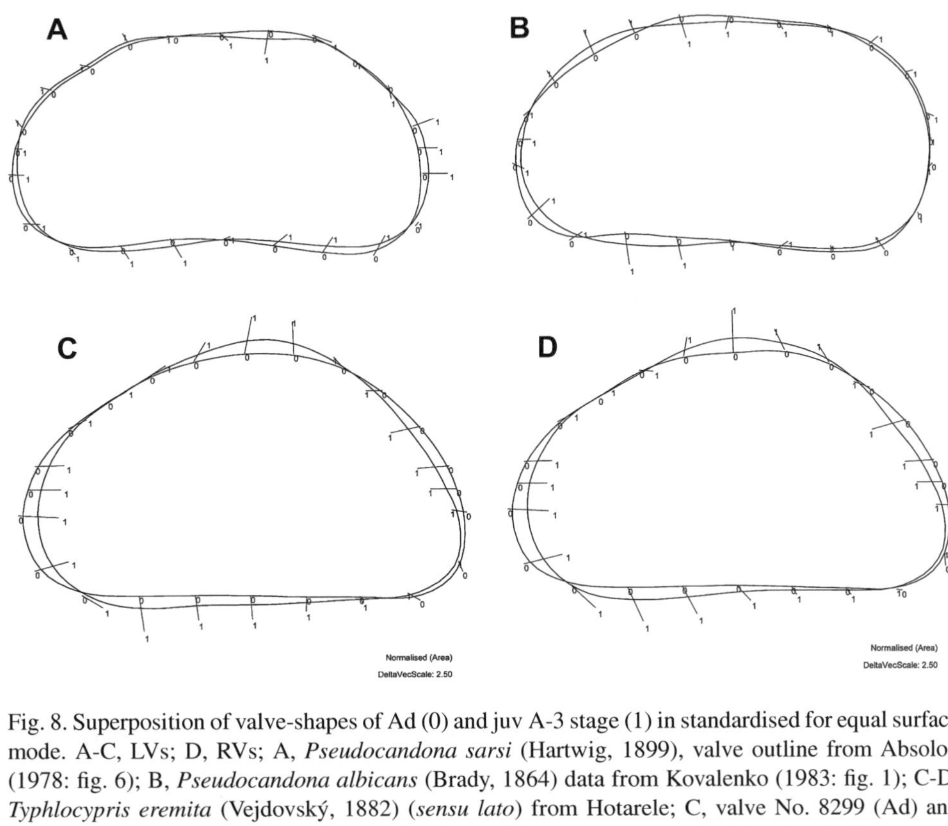

Fig. 8. Superposition of valve-shapes of Ad (0) and juv A-3 stage (1) in standardised for equal surface mode. A-C, LVs; D, RVs; A, *Pseudocandona sarsi* (Hartwig, 1899), valve outline from Absolon (1978: fig. 6); B, *Pseudocandona albicans* (Brady, 1864) data from Kovalenko (1983: fig. 1); C-D, *Typhlocypris eremita* (Vejdovský, 1882) (*sensu lato*) from Hotarele; C, valve No. 8299 (Ad) and valve No. 8306 (juv A-3); D, valve No. 8300 (Ad) and valve No. 8307 (juv A-3).

that in *Pseudocandona* the right valves of adults and A-3 juveniles are also more or less rectangular, while in *Typhlocypris eremita* (*sensu lato*) the A-3 valve is triangular to become quasi-rectangular in the adult stage (fig. 8D). We therefore would like to point out that details of the valve shapes of juvenile stages combined with those observed in the adults help to complement the differential diagnosis of *Typhlocypris* and *Pseudocandona*.

Notes on living and fossil species assigned to the genus *Typhlocypris* Vejdovský

There are currently 14 nominal living species of the genus *Typhlocypris* (see Danielopol, 1978; Namiotko & Danielopol, 2004; Namiotko et al., 2004; Karanovic, 2005; Iepure et al., 2007): *Typhlocypris cavicola* (Klie, 1935), *T. danubialis* (Iepure, Namiotko & Danielopol, 2007) comb. nov., *T. dispar* (Hartmann, 1964), *T. eremita* (Vejdovský, 1882), *T. marmonieri* (Danielopol & Namiotko, 2004) comb. nov., *T. pannonicola* (Löffler, 1960), *T. pretneri* (Danielopol, 1978) comb. nov., *T. puteana* (Klie, 1931), *T. serbani* (Danielopol, 1982) comb. nov., *T. skadari* Karanovic, 2005, *T. sywulai* (Namiotko, Danielopol & Rađa, 2004) comb. nov., *T. szoecsi* (Farkas, 1958), *T. transylvanica* (Iepure, Namiotko & Danielopol, 2007) comb. nov. and *T. trigonella* (Klie, 1931).

Typhlocypris eremita and the closely related species *T. danubialis*, *T. pannonicola*, *T. serbani*, *T. szoecsi* and *T. transylvanica* have left valves with a reduced umbo and a widely rounded postero-ventral margin. Their morphological similarity suggests close phylogenetic affinities. This observation led Meisch (2000) and Karanovic (2012) to suggest that the names of some of these species, mainly of *T. pannonicola* and *T. szoecsi* are synonyms of *T. eremita*.

Typhlocypris cavicola and *T. pretneri* are closely related species (Klie, 1935; Danielopol, 1982; Mori et al., 2011). Their left valves generally possess umbos larger than those seen in other species of the genus. Also, the pitted (foveolated) central valve area is larger. The taxonomic status of both species should be checked after a careful redescription. This is a current project in which two of us (C.M. and N.M.) are involved.

Typhlocypris trigonella (Klie) is so far only known from one single, 0.55-mm-long individual collected from the Postojna cave in Slovenia (Klie, 1931). The species was not found in the samples taken from that cave or from any other locality in Slovenia since 1931. *T. trigonella* has the typical morphological traits of *Typhlocypris*: a triangular left valve with a strongly developed dorsal umbo and foveolae in the central valve area (Klie, 1931: figs. 9 and 10). However, the distal segment of L7 displays a long h1 seta, which does not conform to the short, hook-like seta that characterises *Typhlocypris*. Note that also the setal equipment of the basal segment of that limb, which is important for the generic assignment (Klie,

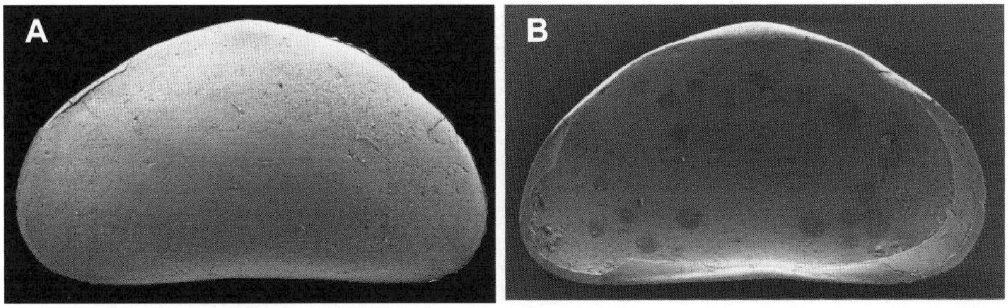

Fig. 9. *Typhlocypris incertae sedis*, *Candona* (*Typhlocypris*) aff. *eremita* (Vejdovský, 1882) in Gross (2004); A, LV outer view from Gross (2004: Plate 3, fig. 13); B, same specimen as in A but inner view from Gross (2004: Plate 4, fig. 1). Length 0.8 mm.

1931) remains unknown. Therefore, the exact taxonomic status of *T. trigonella* remains to be elucidated.

Typhlocypris puteana displays the left valve dorsally less prominent and more elongated, resembling that of the species of the genus *Cryptocandona* Kaufmann, 1900 (Danielopol, 1982, fig. 14F) and the central valve area displays only a faint

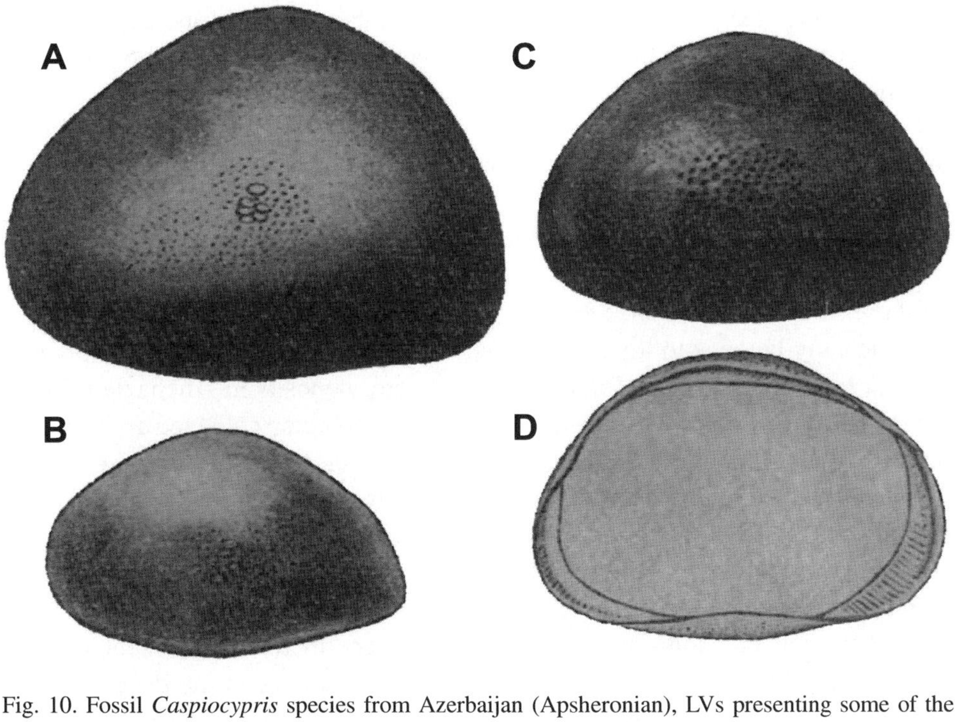

Fig. 10. Fossil *Caspiocypris* species from Azerbaijan (Apsheronian), LVs presenting some of the typical traits of *Typhlocypris*. A-B, *Caspiocypris schneiderae* Livental, 1961; A, Ad outer view; B, juv outer view; C-D, *Caspiocypris ola* Markova, 1961 Ad valves; C, outer view; D, inner view. From Agalarova et al. (1961: Plate 6, figs. 1, 2a, 2b, 3).

ornamentation (Namiotko, unpubl.). The endopodite of L7 was figured by Klie (1931) as having 4 segments. As pointed out above in the Material and Methods section we consider that the endopodite E II + III displays a well expressed annulation. In our opinion this species may belong to a separate lineage within the genus *Typhlocypris*.

A number of fossil species closely resemble the living stygobitic species of *Typhlocypris*. This is the case of *T. eremita* illustrated by Sokač (1978) from Quaternary deposits at Vinkovci in Slavonia, Croatia and *Typhlocypris* cf. *serbani* described by Krstić (1993) from Quaternary deposits at Bačka in Serbia. The latter species resembles *T. serbani* described from the groundwater habitats in the Olt Valley in Romania. More recently, Pipík and Bodergat (2008) mentioned *Pseudocandona* aff. *eremita* from Miocene deposits of the palaeolake Turiec in Slovakia. The latter species is clearly similar to *T. eremita* from Bohemia. *Candona* (*Typhlocypris*) *eremita* in Freels (1980), from limnic Pliocene deposits in Turkey, clearly also belongs to *Typhlocypris* and closely resembles extant *T. eremita*. All these fossil species were found in sediments deposited in freshwater habitats. They all have a very small umbo of about 2.5% of the maximal height of the left valve; the dorsal angle of that valve is about 120° for both the Recent living species and the fossil ones. They all come from central and south-eastern Europe and Anatolia. We therefore hypothesise that the extant hypogean species are related to the fossil taxa mentioned above.

Krstić & Guan (2000) noted that the tribe Advenocyprini, which they introduced as new, comprises two groups of species: (1) species with very high and triangular carapaces, grouped in the genus *Advenocypris* (type species *A. alpherovi* Schneider, 1956), and (2) species with more roundly shaped carapaces belonging to the *eremita*-group. The valves of *Advenocypris schneiderae* Vekua, 1975 resemble those of the living species *Typhlocypris dispar* and in our opinion *A. schneiderae* actually belongs to the genus *Typhlocypris* as defined herein. *Typhlocypris schneiderae* is known from Quaternary freshwater deposits in Abkhazia (Vekua, 1975). The living species *T. dispar* is a stygobitic species collected from a cave near Erekli in Anatolia (Hartmann, 1964). Both species display highly triangular left valves with an umbo of about 7.5-10% of the maximum valve height. The dorsal angle of the left valve varies from 85° (the fossil *T. schneiderae*) to 110° in the cavernicolous *T. dispar*. We therefore hypothesise that both these species known from the same geographic area are phylogentically linked.

We point out the similarity in the carapace shape of the Recent subterranean species *T. marmonieri* from the karstic Massif des Causses in the department of Tarn, southern France (Namiotko & Danielopol, 2004) and the fossil Miocene species *Cavernocandona roaixensis* Carbonnel, 1969 recorded from the department of Vaucluse, also in southern France. The carapaces of both species are very high, with a height reaching about 70% of the length, and we therefore suggest a close phylogenetic relationship between the two species. *T. marmonieri* also appears similar to *Typhlocypris pechelbronnensis* (Stchepinsky, 1960). The latter species is geologically the most ancient species of the genus. It was originally assigned to the subgenus *Kochia* of the genus *Candona*, then redescribed and placed in the subgenus *Typhlocypris* by Triebel (1963), and in the genus *Typhlocypris* by Carbonnel (1969). The species was found in non-marine Oligocene/Early Miocene deposits at Keffenach and Entzheim (France) and Nieder-Ingelheim (Germany). Additionally to the triangular carapace shape and the presence of an umbo on the left valve, the species presents foveolae in the central valve area as well as short anterior marginal pore canals (Triebel, 1963: Plate 26, figs. 12a, 14).

The fossil *Typhlocypris ratisbonensis* (Lutz, 1965) occurred in Central Europe. First identified by Lutz (1965) in a sample of Middle Miocene age (Badenian, Upper Freshwater Molasse) from near Regensburg in Bavaria, it was rediscovered by Pokorný (1988) in Bohemia, at Tuchořice, in a species rich sample of Early Miocene age. The latter author provided for the first time a detailed description of *T. ratisbonensis*, including details of the left and right valves belonging to both adults and the last four juvenile stages. The juvenile left valves display the typical triangular shape and the central ornamentation with foveolae. The adult left valve is approximately triangular, but instead of an umbo, it displays an antero-dorsal plication. Because of the latter character, *T. ratisbonensis* occupies an isolated position within the genus *Typhlocypris*.

There are several taxa commonly assigned to *Typhlocypris*, but which might not belong there. (1) *Candona* (*Typhlocypris*) aff. *eremita*, described by Gross (2004) from Mataschen, located in the area of (palaeo)Lake Pannon in the Styrian Basin (Late Miocene, Pannonian B). The left valve is triangular in shape and displays an umbo (fig. 9A and B). However, the surface of the valves is smooth. It is not clear if this morphotype belongs to *Typhlocypris* or if it displays a homeomorphic shape evolved through convergence (see next section). For such unresolved taxonomic cases we suggest to use the generic name "*Typhlocypris*" *incertae sedis* until new information will become available. (2) *Caspiocypris schneiderae* Livental, 1961 and *Caspiocypris ola* Markova, 1961 described in the monography of Agalarova et al. (1961) dealing with Pliocene and Quaternary ostracods from Azerbaijan. Both species possess triangular adult left valves with a dorsal umbo and the central valve area is covered with foveolae (fig. 10A and C). Juveniles of *C. schneiderae* are also triangular (fig. 10B). However, the carapace of both species is conspicuously inflated and the ventral margin is convex, traits that are not seen in the well established *Typhlocypris* taxa. Moreover, *Caspiocypris ola* seems to display long anterior marginal pore canals (fig. 10D), a trait atypical for *Typhlocypris*. We propose to consider these species as "*Typhlocypris*" *incertae sedis*, hoping that new data allowing one to resolve the question of their generic assignment will soon be available.

Hou et al. (2002) mentioned the following fossil *Typhlocypris* species from China: *Typhlocypris acuminata* Ho, 1978, *T. arrecta* (Chen & Ho, 1982), *T. brevialis* (Shi & Ho, 1963), *T. camerata* (Bojie, 1978), *T. definita* (Tsao, 1959), *T. dorsicurta* (Bojie, 1978), *T. gibbosa* (Chen & Ho, 1982), *T. sinensis* (Bojie, 1978) and *T. truncata* (Ho, 1978). All the above mentioned species do not belong to *Typhlocypris* as defined here and are therefore not discussed here.

Brief review of the ecological and geographical distribution of the species of *Typhlocypris*

Living species of *Typhlocypris* mainly occur in the interstitial habitats of alluvial aquifers, in the hyporheic zone along rivers and in cavernicolous habitats in central and south-eastern Europe (*Pseudocandona eremita* included in the checklist of Recent non-marine ostracod species of China by You et al., 2002 most probably constitutes an erroneous record that needs verification). They were occasionally collected in surface waters of irrigation ditches fed by underground waters (Namiotko, 1990; Namiotko & Sywula, 1993). In Slovenia, *T. pretneri* and *T. cavicola* were found in karstic outflows, and the latter species also in caves and in the hyporheic zone (Danielopol, 1982; Mori et al., 2011; Mori & Meisch, 2012). At present it is unclear if they are capable of maintaining stable populations in those springs.

Absolon (1973) reported *T. eremita* from Holocene marl sediments, rich in calcium carbonate, which were either deposited close to springs forming travertine substrates (e.g., Hurichův Dolec in Czech Republic) or in shallow bogs (e.g., Čečelice, also Czech Republic).

Fossil species now assigned to *Typhlocypris* are mentioned from lacustrine sediments (e.g., Lutz, 1965; Carbonnel, 1969; Vekua, 1975; Sokač, 1978; Freels, 1980; Pokorný, 1988; Krstić, 2006; Pipik & Bodergat, 2008). Species of *Typhlocypris* are also known from brackish water environments in basins linked to the Paratethys domain. For instance, Krstić (1972) mentioned *T. beogradica* (Krstić, 1972) from sediments of Lake Pannon dated from the Middle Pannonian (Late Miocene). Gross (2004) recorded *Typhlocypris* aff. *eremita* from the Styria Basin during the Early Pannonian, Late Miocene. Minati and Danielopol (unpubl.) found one single juvenile of *Typhlocypris* sp. at Hennersdorf in Austria (Pannonian E, Late Miocene) (fig. 5D). In our opinion, the latter specimen had been drifted from the nearby non-saline, limnic habitats into the brackish Lake Pannon.

Tunoglu (2003) mentions two *Typhlocypris* species (left in open nomenclature) sampled from a Pontian (Early Pliocene) oligohaline environment of the (palaeo) Euxinic Basin, a part of the Eastern Paratethys.

The problem of homeomorphism in the triangular Candoninae — comparative data related to living species of *Typhlocypris*

Danielopol (1978) showed in a diagram that living Candoninae with triangular carapace shapes are relatively rare in the Palaearctic, with a higher number of species living in hypogean aquatic habitats as compared to those known from surface water systems. In a number of species with triangular carapaces, the generic characters are easily recognised. For instance, in *Mixtacandona tabacarui* Danielopol & Cvetkov, 1979 the distal setae of the L7 and the structure of the hemipenis unambiguously point towards *Mixtacandona* (fig. 11); the valves are also easily separated from those characteristic of *Typhlocypris* because of the wide calcified inner lamellae, which form the anterior and posterior vestibulae. Two

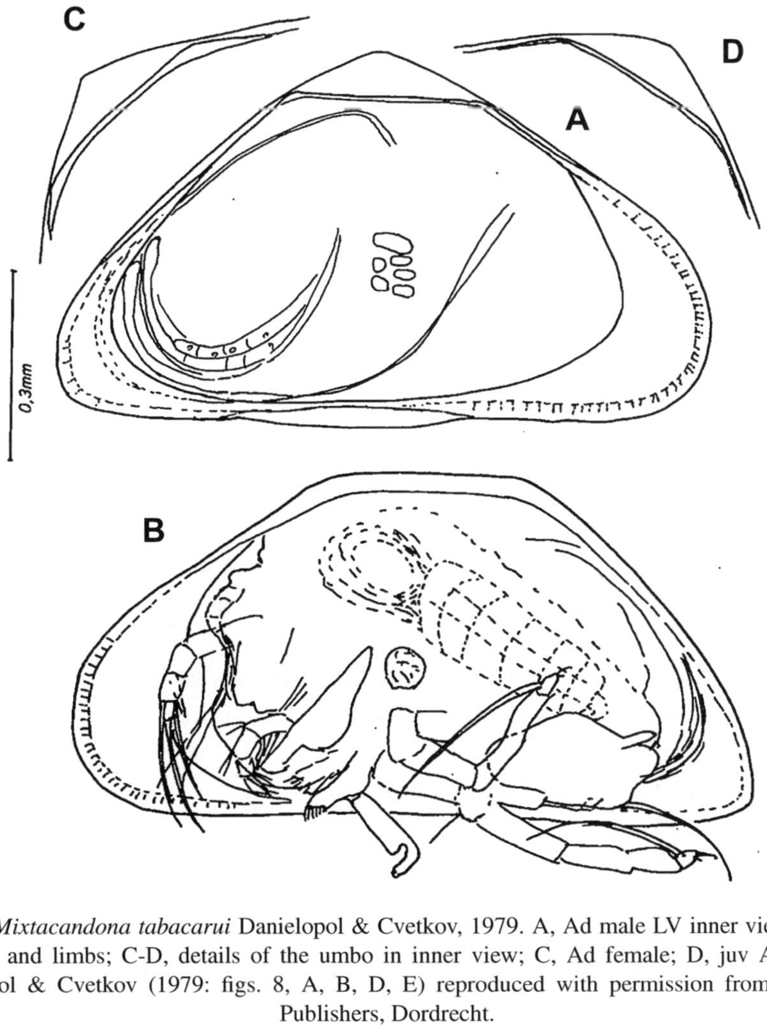

Fig. 11. *Mixtacandona tabacarui* Danielopol & Cvetkov, 1979. A, Ad male LV inner view; B, Ad male RV and limbs; C-D, details of the umbo in inner view; C, Ad female; D, juv A-1. From Danielopol & Cvetkov (1979: figs. 8, A, B, D, E) reproduced with permission from Springer Publishers, Dordrecht.

other European species with triangular valves according to characters of both the soft parts and carapace were documented to belong to other genera than *Typhlocypris*: *Fabaeformiscandona aemonae* (Klie, 1935) or *Schellencandona triquetra* (Klie, 1936) (see Meisch, 2000). Danielopol (1982) showed that *Pseudocandona inaequivalvis baikalensis* (Bronstein, 1930) displays limbs and the Zenker Organ, which differ from those of *Typhlocypris*. The valves of that species are asymmetric and differ from those of *T. eremita* and the related species in fine details, e.g., the smooth valve surface (Danielopol, unpubl.). Detailed descriptions of the valves of the triangular species belonging to the *neglecta*-group of *Candona* living in Lake Ohrid (see Mikulić, 1961) are so far missing. This is also true for *Candona punctata* Furtos, 1933, the carapace of which is triangular in shape, the left valve being dorsally more developed than the right one. Furtos (1933) showed that the entire surface of the right valve is strongly pitted. No further details of the valve structure were provided by the authors who studied that species (Furtos, 1933; Staplin, 1963; Karanovic, 2006). However, from the original description by Furtos (op. cit.) and the more recent one by Karanovic (2006) it appears that the L5 male endopodite and also the hemipenis differ markedly from those of *Typhlocypris*. Also, the distal h1 seta of L7 in the redescription of Karanovic (op. cit.) is long and flexible, unlike that seen in the European species of *Typhlocypris*, where that seta is curved and conspicuously hook-like. In our opinion, *C. punctata* therefore does not belong to the genus *Typhlocypris* as redefined here.

CONCLUSION

1. In addition to the morphological traits already discussed by Meisch (1996, 2000), Namiotko & Danielopol (2004) and Iepure et al. (2007), the traits presented here, in particular those of the developmental trajectory of the valve shapes, provide robust evidence for considering *Typhlocypris* Vejdovský, 1882 a valid genus.
2. The generic redefinition presented here permits to allocate a number of Neogene and Quaternary fossil species from the Palaeartic to the genus *Typhlocypris*.
3. The genus *Typhlocypris* is an evolutionary entity with its own history and biogeography.
4. *Typhlocypris* presently appears to form a bundle of evolutionary lineages (species) with slightly diverging trajectories.
5. The origin of the living stygobitic species of *Typhlocypris* has to be sought in limnic ostracods which were spread over the western Palaearctic since the Late Palaeogene.

6. Investigations should be undertaken on living species using geometric morphometrics combined with classic morphological analysis of traits and, if possible, molecular techniques, in order to clarify the taxonomic status of various populations recognised as *Typhlocypris incertae sedis*.
7. Fossil Candoninae species presently classified in other genera but presenting similarities to *Typhlocypris* should be re-examined with both SEM and TLM techniques. Future investigators should use a large number of valves in order to treat their morphology with geometric morphometrics and multivariate statistic techniques.

ACKNOWLEDGEMENTS

Professor Eugen Kempf (Cologne) and his Cologne Database are here gratefully acknowledged. We acknowledge the cooperation and support with material, literature and/or unpublished information offered by A. Absolon (Prague), A. Baltanás (Madrid), G. Carbonnel (Lyon), P. Dolejš (Prague), P. Dworschak (Vienna), S. Filipescu (Cluj-Napoca), Z. Gidó (Szentlörinc), E. Gliozzi (Rome), S. Iepure (Cluj), V. Hajek-Tadesse (Zagreb), D. J. Horne (London), E. Kempf (Cologne), the late V. Kulhavy (Prague), A. Lord (Frankfurt/Main), R. Matzke-Karasz (Munich), R. Pipik (Banská Bystrica), P. Opreanu (Constanţa), the late V. Pokorný (Prague), L. Silye (Cluj-Napoca), B. Sket (Ljubljana), A. Smith (Kent), M. Stoica (Bucharest), C. Wrozyna (Graz) and an anonymous reviewer. An inquiry and comments of K. Martens (Brussels) stimulated the development of a part of this project and D. J. Horne (London) drew our attention to ambiguous descriptions of the cleaning leg (L7 limb) of the subfamily Candoninae existing in the literature. G. Boxshall (London) and R. Smith (Shiga) helped in clarifying the question of the structure and nomenclature we adopted here for the endopodial segments of L7 limb of the Candoninae. W. Piller (Graz) and the Austrian Academy of Sciences through its Commission for the Stratigraphical and Palaeontological Research of Austria supported logistically and financially the completion of the final part of the project and also the participation of D. L. D. to the Cologne Meeting in October, 2012. The following colleagues helped with the technical work: J. Knoblechner (Mondsee) and K. Minati (Graz). The Austrian Science Fund financially supported part of the work of D. L. D. and M. G. through the grants P17738-B03 and P. 21748-N21.

REFERENCES

ABSOLON, A., 1973. Ostracoden aus einigen Profilen spät- und postglazialer Karbonatablagerungen in Mitteleuropa. Mitteilungen der Bayerischen Staatssammlung für Paläontologie und historische Geologie, **13**(1): 47-94.

— —, 1978. Die Gattung *Candona* (Ostracoda) im Quartär von Europa. Rozpravy Československé Akademie Věd, Řada Matematických a Přírodních Věd, **88**(5): 1-76.

AGALAROVA, D. A., Z. K. KADYROVA & S. A. KULIEVA, 1961. Ostracoda from Pliocene and Postpliocene deposits of Azerbaijan: 1-420. (Azerbaijan State Publishers, Baku). [In Russian.]

ALM, G., 1915. Monographie der schwedischen Süsswasser-Ostracoden nebst systematischen Besprechungen der Tribus Podocopa. Zoologiska Bidrag fran Uppsala, **4**(1): 1-248.

BRONSTEIN, Z. S., 1947. Ostracodes des eaux douces. In: E. N. PAVLOVSKY & A. A. STACKELBERG (eds.), Faune de l'URSS. Crustacés, Vol. 2, n° 1, **31**: 1-339. (Institut Zoologique de l'Académie des Sciences de l'URSS, N.S, Moscou/Leningrad). [In Russian.]

BROODBAKKER, N. & D. L. DANIELOPOL, 1982. The chaetotaxy of Cypridacea (Crustacea, Ostracoda) limbs: proposals for a descriptive model. Bijdragen tot de Dierkunde, **52**(2): 103-120.

CARBONNEL, G., 1969. Les ostracodes du Miocene Rhodanien: Systematique, biostratigraphie, ecologique, paleobiologie. Documents des Laboratoires de Geologie de la Faculte des Sciences de Lyon, **32**(1-2): 1-469.

CHU, G.-L. & S. C. SANDERSON, 2008. The genus *Kochia* (Chenopodiaceae) in North America. Madroño, **55**: 251-256.

DANIELOPOL, D. L., 1971. Quelques remarques sur le peuplement ostracodologique des eaux douces souterraines d'Europe. Bulletin du Centre de Recherches Pau-SNPA, **5**(Suppl.): 179-190.

— —, 1978. Über Herkunft und Morphologie der Süßwasser-hypogäischen Candoninae (Crustacea, Ostracoda). Sitzungsberichte der Österreichischen Akademie der Wissenschaften, Mathematisch-naturwissenschaftliche Klasse, Abt. 1, **187**(1/5): 1-162.

— —, 1980. Sur la biologie de quelques ostracodes Candoninae épigés et hypogés d'Europe. Bulletin du Musée National d'Histoire Naturelle de Paris, ser. 4, sect. A, **2**(2): 471-506.

— —, 1982. Nouvelles données sur les Candoninae (Ostracoda) hypogés de Roumanie et Yougoslavie. Bulletin du Musée National d'Histoire Naturelle de Paris, ser. 4, sect. A, **4**(3-4): 369-396.

DANIELOPOL, D. L., A. BALTANÁS, U. MOROCUTTI & F. ÖSTERREICHER, 2011. On the need to renew the taxonomic system of the Candoninae (non-marine Ostracoda, Crustacea). Reflexions from an analysis of data using the Yule Process. Geo-Eco-Marina, **17**: 5-20.

DANIELOPOL, D. L. & L. CVETKOV, 1979. Trois nouvelles espèces du genre *Mixtacandona* (Ostracoda, Cyprididae, Candoninae). Hydrobiologia, **67**(3): 249-266.

DANIELOPOL, D. L., T. NAMIOTKO & C. MEISCH, 2012. *Marmocandona* nov. gen. (Ostracoda, Candoninae), with comments on the contribution of stygobitic organisms to micropalaeontological studies. Kölner Forum für Geologie und Paläontologie, **21**: 13-16.

FARKAS, H., 1958. Kagylosrakok — Ostracoda. Fauna Hungariae, Magyarorszag Allatvilaga, **39**, 4(3): 1-68. [In Hungarian.]

FREELS, D., 1980. Limnische Ostrakoden aus Jungtertiär und Quartär der Türkei. Geologisches Jahrbuch, B, **39**: 1-169.

FURTOS, N., 1933. The Ostracoda of Ohio. Bulletin of the Ohio Biological Survey, **29**: 411-524.

GROSS, M., 2004. Zur Ostracodenfauna (Crustacea), Paläoökologie und Stratigraphie der Tongrube Mataschen (Unter-Pannonium, Steirisches Becken, Österreich). Joannea Geologie Paläontologie, **5**: 49-129.

HARTMANN, G., 1964. Asiatische Ostracoden. Systematische und zoogeographische Untersuchungen. Internationale Revue der gesamten Hydrobiologie, Systematische Beihefte, **3**: 1-155.

HARTWIG, W., 1901. Die Arten der Ostracoden-Unterfamilie Candoninae der Provinz Brandenburg. Sitzungs-Bericht der Gesellschaft naturforschender Freunde zu Berlin, **1901**(4): 87-129.

HÉJJAS, E., 1894. Neue Beiträge zur fossilen Ostracoden-Fauna Siebenbürgens. Ertesitö az Erdelyi Muzeum-Egylet Orvos-Termeszettudomanyi Szakosztalyabol, Ser. 2, **19**(1): 99-112.

HORNE, D. J., A. COHEN & K. MARTENS, 2002. Taxonomy, morphology and biology of Quaternary and living Ostracoda. Geophysical Monograph, **131**: 5-36.

HOU, Y., Y. GUO & D. CHEN, 2002. Fossil Ostracoda of China, (Vol. 1) Superfamilies Cypridacea and Darwinulidacea: 1-1090. (Science Press, Beijing). [In Chinese with Engl. abstract.]

ICZN (INTERNATIONAL COMMISSION ON ZOOLOGICAL NOMENCLATURE), 1999. International Code of Zoological Nomenclature (4th ed.): 1-126. (Tipografia La Garangola, Padova).

IEPURE, S., T. NAMIOTKO & D. L. DANIELOPOL, 2007. Evolutionary and taxonomic aspects within the species group *Pseudocandona eremita* Vejdovský (Ostracoda, Candonidae). Hydrobiologia, **585**: 159-180.

JANČAŘÍK, A., 1952. Nový nález lasturnatky *Candona eremita* Vejdovský 1882. Sborník Klubu Přírodovědeckého v Brně, **29**(1951): 232-235. [In Czech.]

KARANOVIC, I., 2005. On the genus *Typhlocypris* Vejdovský, 1892 (Crustacea: Ostracoda: Candoninae), with description of two new species. Systematics and Biodiversity, **3**(4): 375-406.

— —, 2006. Recent Candoninae (Crustacea, Ostracoda) of North America. Records of the Western Australian Museum, Suppl., **71**: 1-75.

— —, 2012. Recent freshwater ostracods of the World. Crustacea, Ostracoda, Podocopida: 1-608. (Springer-Verlag, Berlin, Heidelberg).

KAUFMANN, A., 1900. Cypriden und Darwinuliden der Schweiz. Revue Suisse de Zoologie, **8**: 209-423.

KEMPF, E. K., 1980a. Index and bibliography of nonmarine Ostracoda. Part 1, Index A. Sonderveröffentlichungen des Geologischen Instituts der Universität zu Köln, **35**: 1-188.

— —, 1980b. Index and bibliography of nonmarine Ostracoda. Part 2, Index B. Sonderveröffentlichungen des Geologischen Instituts der Universität zu Köln, **36** (Part 2: Index B): 1-180; **37** (Part 3: Index C): 1-204; **38** (Part 4: Bibliography A): 1-186.

— —, 1980c. Index and bibliography of nonmarine Ostracoda. Part 3, Index C. Sonderveröffentlichungen des Geologischen Instituts der Universität zu Köln, **37**: 1-204.

— —, 1980d. Index and bibliography of nonmarine Ostracoda. Part 4, Bibliography A. Sonderveröffentlichungen des Geologischen Instituts der Universität zu Köln, **38**: 1-186.

— —, 1997a. Index and bibliography of nonmarine Ostracoda. Part 6, Index A, Supplement 1. Sonderveröffentlichungen des Geologischen Instituts der Universität zu Köln, **109**: 1-142.

— —, 1997b. Index and bibliography of nonmarine Ostracoda. Part 7, Index B, Supplement 1. Sonderveröffentlichungen des Geologischen Instituts der Universität zu Köln, **110**: 1-134.

— —, 1997c. Index and bibliography of nonmarine Ostracoda. Part 8, Index C, Supplement 1. Sonderveröffentlichungen des Geologischen Instituts der Universität zu Köln, **111**: 1-152.

— —, 1997d. Index and bibliography of nonmarine Ostracoda. Part 9, Bibliography C. Sonderveröffentlichungen des Geologischen Instituts der Universität zu Köln, **112**: 1-144.

KLIE, W., 1931. Zwei neue Arten der Ostracoden-Gattung *Candona* aus unterirdischen Gewässern im südöstlichen Europa. Zoologischer Anzeiger, **96**(7/8): 161-168.

— —, 1935. Drei neue Höhlenostracoden aus der Umgebung von Laibach. Zoologischer Anzeiger, **111**(7/8): 189-198.

— —, 1936. Entomostraken aus der bulgarischen Höhle "Lakatnik". Mitteilungen aus den Königlichen. Naturwissenschaftlichen Instituten in Sofia — Bulgarien, **9**: 108.

— —, 1938. Ostracoda, Muschelkrebse. In: Die Tierwelt Deutschlands und der angrenzenden Meeresteile nach ihren Merkmalen und nach ihrer Lebensweise, **34**. Teil: Krebstiere oder Crustacea: 1-230. (Gustav Fischer Verlag, Jena).

— —, 1940. Zur Kenntnis der Ostracodenarten *Candona eremita* (Vejdovsky) und *Candona reducta* Alm. Mitteilungen über Höhlen- und Karstforschung, **1940**(1): 24-29.

KOVALENKO, A. L., 1983. Postembryonic evolution of *Candona parallela* G.W. Müller (Crustacea, Ostracoda). Izvestiya Akademii Nauk Moldavskoy SSR, Seriya Biologicheskich i Kimicheskich Nauk, **1983**(1): 33-38.

KRSTIĆ, N., 1972. Genus *Candona* (Ostracoda) from *Congeria* beds of southern Pannonian Basin. Serbian Academy of Sciences and Arts, Monographs, **140**, Section of Natural and Mathematical Sciences, **39**: 1-145.

— —, 1993. Several rare ostracode species of Bačka Quaternary. Proceedings of Geoinstitute Belgrade, **28**: 173-185.

— —, 2000. Some ostracodes from the Dinaric Alps Neogene and its paleoecology. In: S. KARAMATA & S. JANKOVIĆ (eds.), Geology and metalogeny of the Dinarides and the Vardar Zone. Academy of Sciences and Arts of the Republika Srpska, Monographs, **1**: 207-214.

— —, 2006. Pliocene ostracodes of the Paludinian Beds in Pannonian plain, Serbian Part. Herald of the Nature History Museum Belgrade, Spec. Publ.: 1-409.

KRSTIĆ, N. & S. GUAN, 2000. A proposal for the systematics of the subfamily Candoninae (Ostracodes) with the description of the *Macedocandona*, new genus. Geologica Macedonica, **14**: 25-48.

LINHART, J., W. BRAUNEIS & W. NEUBAUER, 2007. Morphomatica, version 1.6.0.1, available online at http://palstrat.uni-graz.at/methods in ostracodology.htm. Graz.

LÖFFLER, H., 1964. 3. Beitrag zur Kenntnis der Entomostrakenfauna burgenländischer Brunnen und Quellen (Südliches Burgenland). Wissenschaftliche Arbeiten aus dem Burgenland, **31**: 156-169.

LUBIMOVA, P. S., M. I. MANDELSTAM & G. F. SCHNEIDER, 1960. Ostracoda: Nadsemeystwo Cypracea. In: J. A. ORLOV (ed.), Osnovy Paleontologii, **8**: 264-421. (Nauka, Moskva). [In Russian.]

LUTZ, A. K., 1965. Jungtertiäre Süsswasser-Ostracoden aus Süddeutschland. Geologisches Jahrbuch, **82**: 271-329.

MANDELSTAM, M. I. & G. F. SCHNEIDER, 1956. Materialien zur Paläontologie: Neue Familien und Gattungen. Trudy Vsesojuznogo Nauchno-issledovatelskogo Geologicheskogo Instituta (VSEGEI) Ministerstva Geologii i Ochrany Nedr SSSR, NS, **12**: 87-144. [In Russian.]

— — & — —, 1963. Fossil Ostracoda of the USSR: family Cyprididae. Trudy Vsesoyuznogo Nauchno-Issledovatelskogo Geologo-Razvedochnogo Neftyanogo Instituta (VNIGRI), NS, **203**: 1-331. [In Russian, English translation 1968 by National Lending Library of Science and Technology, Boston Spa.]

MARMONIER, P., 1988. Biocénoses interstitielles et circulation des eaux dans le sous-écoulement d'un chenal aménagé du Haut-Rhône français: 1-317. (Ph.D. Thesis, University Claude Bernard Lyon I, Lyon).

MARTENS, K., 1984. Annotated check-list of non-marine ostracods (Crustacea, Ostracoda) from African inland waters. Musée Royal de l'Afrique Centrale Tervuren, Documentation zoologique, **20**: 1-51.

— —, 1987. Homology and functional morphology of the sexual dimorphism in the antenna of *Sclerocypris* Sars, 1924 (Crustacea, Ostracoda, Megalocypridinae). Bijdragen tot de Dierkunde, **57**(2): 183-190.

MARTENS, K. & S. SAVATENALINTON, 2011. A subjective checklist of the Recent, free-living, non-marine Ostracoda (Crustacea). Zootaxa, **2855**: 1-79.

MARTENS, K., I. SCHÖN, C. MEISCH & D. J. HORNE, 2008. Global diversity of ostracods (Ostracoda, Crustacea) in freshwater. Hydrobiologia, **595**: 185-193.

MEISCH, C., 1996. Contribution to the taxonomy of *Pseudocandona* and four related genera, with the description of *Schellencandona* nov. gen., a list of the Candoninae genera, and a key to the European genera of the subfamily (Crustacea, Ostracoda). Bulletin de la Société des naturalistes luxembourgeois, **97**: 211-237.

— —, 2000. Freshwater Ostracoda of Western and Central Europe. In: J. SCHWOERBEL & P. ZWICK (eds.), Die Süßwasserfauna von Mitteleuropa, **8/3**: 1-522. (Spektrum Akademischer Verlag, Berlin, Heidelberg).

MIKULIĆ, F., 1961. Neue Arten Candona aus dem Ohridsee. Bulletin du Muséum d'Histoire Naturelle Belgrade, Ser. B, **17**: 87-107. [In Serbo-Croatian with German abstract.]

MORI, N. & C. MEISCH, 2012. Contribution to the knowledge on the distribution of Recent free-living freshwater ostracods (Podocopida, Ostracoda, Crustacea) in Slovenia. Natura Sloveniae, **14**: 5-22.

MORI, N., C. MEISCH & A. BRANCELJ, 2011. Morphology, ecology and distribution of *Typhlocypris cavicola* (Klie, 1935) and *Fabaeformiscandona aemonae* (Klie, 1935), two triangulary shaped ostracods from Slovenia. Joannea Geologie und Paläontologie, **11**: 144-146.

MÜLLER, G. W., 1912. Crustacea: Ostracoda. Das Tierreich, **31**: 1-434.

NAMIOTKO, T., 1990. Freshwater Ostracoda (Crustacea) of Żuławy Wiślane (Vistula Fen Country, Northern Poland). Acta Zoologia Cracoviensia, **33**(19): 459-484.

NAMIOTKO, T. & D. L. DANIELOPOL, 2004. Review of the *eremita* species-group of the genus *Pseudocandona* Kaufmann (Ostracoda, Crustacea), with the description of a new species. Revista Española de Micropaleontologia, **36**: 117-134.

NAMIOTKO, T., D. L. DANIELOPOL & T. RAĐA, 2004. *Pseudocandona sywulai* sp. nov., a new stygobitic ostracode (Ostracoda, Candonidae) from Croatia. Crustaceana, **77**(3): 311-331.

NAMIOTKO, T., D. L. DANIELOPOL, C. MEISCH, M. GROSS & N. MORI, 2012. Reassessment of the genus *Typhlocypris* Vejdovský (Ostracoda, Candoninae), with a brief reflexion on the social role of the "Kempf Database". Kölner Forum für Geologie und Paläontologie, **21**: 48-50.

NAMIOTKO, T. & T. SYWULA, 1993. Crustacean assemblages from the irrigation ditches near Szymankowo (Vistula Delta). Zeszyty Naukowe Uniwersytetu Gdańskiego, Biologia, **10**: 159-162. [In Polish with Engl. abstract.]

PETKOVSKI, T. K., 1969a. Einige neue und bemerkenswerte Candoninae aus dem Ohrid-See und einigen anderen Fundorten in Europa (Crustacea, Ostracoda). Acta Musei Macedonici Scientium Naturalium, **11**(5): 81-111.

— —, 1969b. Über die Notwendigkeit einer Revision der Süßwasser-Ostracoden Europas. In: J. W. NEALE (ed.), The taxonomy, morphology and ecology of Recent Ostracoda: 76-81. (Oliver & Boyd, Edinburgh).

PIPÍK, R. & A.-M. BODERGAT, 2008. Nouvelle faune d'ostracodes d'eau douce du Miocène supérieur de la région Carpatho-Pannonienne et description d'espèces appartenant aux genres *Pseudocandona*, *Fabaeformiscandona* et *Candonopsis*. Palaeontographica Abt. A, **286**(4-6): 89-121.

POKORNÝ, V., 1988. Freshwater ostracodes from the Lower Miocene of Tuchořice (Bohemia, Czechoslovakia). Acta Universitatis Carolinae, Geologica, **3**(1986): 281-315.

ŘEHÁČKOVÁ, V., 1956. Hydrobiology of 4 wells in the city of Prague. Acta Societatis Zoologicae Bohemoslovenicae, **20**(2): 162-176. [In Czech with Engl. abstract.]

ROHLF, F. J., 2003. Tps-dig, Program, Version 1.37, available online at http://Life.Bio.SUNYSB.edu/morph/morph/html. (Department of Ecology and Evolution, State University of New York, Stony Brook, NY).

SOKAČ, A., 1978. Pleistocene ostracode fauna of the Pannonian Basin in Croatia. Palaeontologia Jugoslavica, **20**: 1-51.

ŠOŠTARIĆ, D., 1888. Prilog poznavanju faune slatkovodnih korepnjaka hrvatske. Rad Jugoslavenske Akademije znanosti i umjetnosti, **92** (Matematicko-prirodoslovni razred 9): 103-214. (Ostracoda: 103-112, 144-153, 207-214).

STAPLIN, F. L., 1963. Pleistocene Ostracoda of Illinois. Part 1. Subfamilies Candoninae, Cyprinae, general ecology, morphology. Journal of Paleontology, **37**: 757-797.

SYWULA, T., 1973. Notes on Ostracoda. 13. General plan of structure of the penis in the subgenera *Candona* Baird, *Eucandona* Daday and *Typhlocypris* Vejd. of the genus *Candona* Baird. Bulletin de l'Academie Polonaise des Science, Série des Sciences Biologiques, Cl. 2, **21**: 123-125.

— —, 1974. Małżoraczki (Ostracoda). Fauna Słodkowodna Polski, **24**: 1-315.

TER BORGH, M., I. VASILIEV, M. STOICA, S. KNEŽEVIĆ, L. MATENCO, W. KRIJGSMAN, L. RUNDIĆ & S. CLOETINGH, 2013. The isolation of the Pannonian Basin (Central Paratethys): new constraints from magnetostratigraphy and biostratigraphy. Global and Planetary Change, **103**: 99-118.

TRIEBEL, E., 1963. Ostracoden aus dem Sannois und jüngeren Schichten des Meinzer Beckens: I. Cyprididae. Senckenbergiana lethaea, **44**(3): 157-207.

TUNOGLU, C., 2003. Systematics and biostratigraphy of the Pontian Candonidae (Ostracoda) from the eastern Black Sea region (Northern Turkey). Geologica Carpathica, **54**(1): 21-40.

VÁVRA, W., 1891a. Monographie der Ostracoden Böhmens. Archiv der naturwissenschaflichen Landesdurchforschung von Böhmen, **8**(3): 1-116.

— —, 1891b. Kritisches Verzeichniss der Ostracoden Böhmens. Sitzungsberichte der Königlichen Böhmischen Gesellschaft der Wissenschaften, **1891**: 159-168.

VEJDOVSKÝ, F., 1880. Über den Ursprung der Brunnenfauna. Jahresbericht der königlich-böhmischen Gesellschaft der Wissenschaften, **1880**: 49-56. [In Czech.]

— —, 1882. Thierische Organismen der Brunnenwässer von Prag: 1-70 (Ostracoda: 64-65, table 7).

VEKUA, M. L., 1975. The ostracods of the Kimmerian and Kuyalnikian deposits of Abhazia and their stratigraphic significance: 1-137. (Akademia Nauk Gruzinskoy SSR, Tbilisi). [In Russian.]

YU, N., Q. ZHAO, E. LI, S. CHEN & L. CHEN, 2009. An updated and annotated checklist of recent nonmarine ostracods from China. Zootaxa, **2067**: 29-50.

Please see also the Supplementary Material provided by the authors of the above article in an Addition, that can be accessed in the Online Edition of Crustaceana.

First received 4 February 2013.
Final version accepted 29 July 2013.

[When citing this chapter, refer to Crustaceana 87 (2014) 985–1026]

This paper is dedicated to Prof. Dr. Eugen Kempf on the occasion of his 80th birthday (16 April 2012)

ON SOME RECENT NON-MARINE OSTRACODS FROM NORTHERN CHINA, WITH DESCRIPTION OF ONE NEW SPECIES

BY

DAYOU ZHAI[1,4]) and WANHE ZHAO[2,3])

[1]) Key Laboratory of Cenozoic Geology and Environment, Institute of Geology and Geophysics, Chinese Academy of Sciences, 19 Beitucheng West Road, Chaoyang District, Beijing 100029, P.R. China

[2]) Scanning Electron Microscope Laboratory, School of Materials Science and Engineering, University of Science and Technology Beijing, 30 Xueyuan Road, Haidian District, Beijing 100083, P.R. China

[3]) The High School Affiliated to Renmin University of China, 37 Zhongguancun Street, Haidian District, Beijing 100080, P.R. China

ABSTRACT

Preliminary investigations of urban waters in Beijing City and small water-bodies in Inner Mongolia yielded a total of 12 living ostracod species. Five species are new records for China: *Candona quasiakaina* Karanovic & Lee, 2012, *Fabaeformiscandona myllaina* Smith & Kamiya, 2007, *Heterocypris auricularis* sp. nov. (previously known as *Heterocypris sanukiensis* Okubo, 2004 nomen nudum), *Ilyocypris mongolica* Martens, 1991 and *Plesiocypridopsis newtoni* (Brady & Robertson, 1870). *Bradleycypris vittata* (Sars, 1903), *Ilyocypris angulata* Sars, 1903, *Limnocythere stationis* Vávra, 1891 and *Physocypria kraepelini* G. W. Müller, 1903 are the first records for Palaearctic China. *Cypridopsis vidua* (O. F. Müller, 1776), *Heterocypris incongruens* (Ramdohr, 1808) and *Limnocythere inopinata* (Baird, 1843) are also reported in this paper, although they had already been found in China before. Of the above species, *B. vittata*, *I. angulata* and *I. mongolica* are re-described. The present study expands our knowledge on the taxonomy and geographical distribution of ostracod species.

Key words. — Freshwater ostracods, taxonomy, redescription, zoogeography, China

ZUSAMMENFASSUNG

In ersten Untersuchungen von städtischen Gewässer von Peking und Kleingewässern in der Inneren Mongolei konnten insgesamt 12 lebende Ostrakodenarten nachgewiesen werden. Davon konnten fünf Arten für China erstmalig nachgewiesen werden, namentlich *Candona quasiakaina* Karanovic & Lee, 2012, *Fabaeformiscandona myllaina* Smith & Kamiya, 2007 *Heterocypris*

[4]) Corresponding author; e-mail: dyzhai@mail.iggcas.ac.cn

auricularis sp. nov. (bisher beschrieben als *Heterocypris sanukiensis* Okubo, 2004 nomen nudum), *Ilyocypris mongolica* Martens, 1991 und *Plesiocypridopsis newtoni* (Brady & Robertson, 1870). Ferner sind folgende Arten Erstnachweise für die Paläarktis in China, *Bradleycypris vittata* (Sars, 1903), *Ilyocypris angulata* Sars, 1903, *Limnocythere stationis* Vávra, 1891 und *Physocypria kraepelini* G.W. Müller, 1903. Weitere Arten, die bereits für China bekannt sind, werden in der vorliegenden Arbeit aufgeführt *Cypridopsis vidua* (O.F. Müller, 1776), *Heterocypris incongruens* (Ramdohr, 1808) und *Limnocythere inopinata* (Baird, 1843). Von den oben genannten Arten werden *B. vittata*, *I. angulata* und *I. mongolica* neuerlich beschrieben. Die vorliegende Studie erweitert die Erkenntnis über die Taxonomie und geographische Verteilung der genannten Ostrakoden-Arten.

Schlüsselworte. — Süßwasserostracoden, Taxonomie, Neubeschreibung, Zoogeographie, China

INTRODUCTION

There are few taxonomic data on the Recent non-marine ostracods of China. More than 90 years after the pioneering work of Sars (1903), only 47 living species were recorded from China (Yin & Martens, 1997). A total of 154 species were included in the checklist of Yu et al. (2009), of which only 94 were considered to be "living" (Martens & Savatenalinton, 2011). Recent works, such as Wrozyna et al. (2009), Mischke et al. (2010), and Yu et al. (2010), recorded more species (mostly as subfossil material) that are new for China. In general, however, the living ostracod fauna of China still remains poorly known (Horne et al., 2011). Due to the lack of knowledge on the Recent species, many ostracodologists in this country have to rely on taxonomic guides to fossil species (Hou & Gou, 2002; Hou et al., 2007) for identification. Meanwhile, most Chinese ostracodologists describe species based on shell morphology, and little attention has been paid to the soft parts (Chen, 1990; Neale & Zhao, 1991; Yin & Martens, 1997; Yu et al., 2009, 2010). Species identification based on carapace alone can potentially be problematic (especially for Candonidae, *Ilyocypris* and *Heterocypris*), and can lead to wrong biodiversity estimations, because many closely related species may not be detected using this method alone.

Because of a potentially high biodiversity in China, there is a need for a systematic study of ostracods based on careful examinations of both carapace and soft parts. In the present study, several living ostracod species from Beijing and Inner Mongolia in northern China are recorded, with their diagnostic morphology and geographical distribution described. It is only a small account of the ostracods of China, but may be a useful reference for future studies in this country.

MATERIAL AND METHODS

The investigated water-bodies can be divided into two parts: those in the City of Beijing and those in the eastern Inner Mongolia Province (fig. 1). The sites in

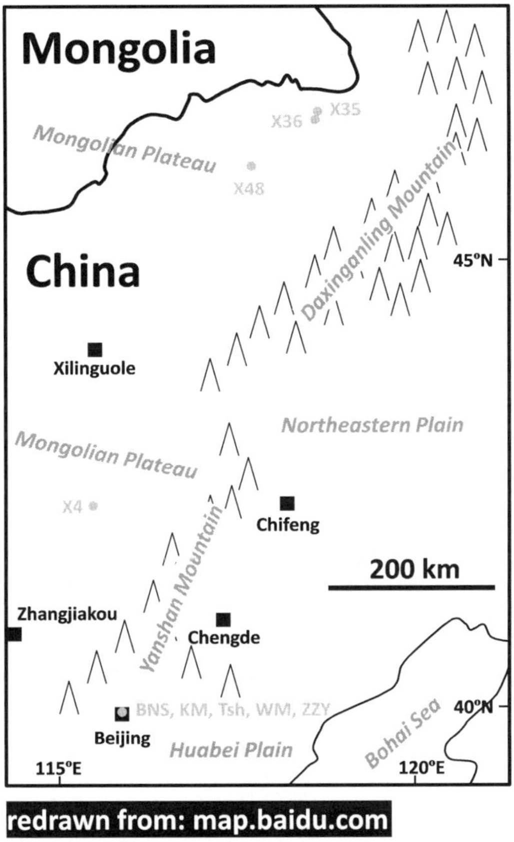

Fig. 1. Map of the study area (redrawn from: map.baidu.com), showing the main geomorphological settings, locations of the sampling sites (solid blue circles) and the major cities (solid black squares). Also see table I. This figure is published in colour in the online edition of this journal, which can be accessed via http://booksandjournals.brillonline.com/content/journals/15685403.

Beijing City are artificial lakes, and are coded with abbreviations of the locality names. The sites in Inner Mongolian grassland consist of small ponds and small lakes near the paved road, and are coded by the character 'X' followed by a number, e.g., X35. Table I gives a summary of the coordinates and brief descriptions of the sites.

Some samples were collected by sucking the surface sediment and the bottom water with a pipette. A few were collected with a bottle or a spoon moving above the substrate. At the sites X35, X36 and X48 the ostracods were very abundant and were captured directly with a pipette.

For the samples from Beijing, animals were picked alive under an XTD-3 stereomicroscope, killed with 15% ethanol and stored in 70% ethanol. Some recently dead specimens were also picked. The samples from Inner Mongolia were

TABLE I
Information on the habitat types, geographical coordinates, sampling time and species records of the study sites

Site	Habitat description	Time of sampling	Ostracods (adult records)
BNS	Pond, 39.988°N 116.389°E	10/2012	Bvi, Cvi, Phk
KM	Lake, 39.992°N 116.262°E	07/2011	Cvi, Phk
KM	Lake, 39.988°N 116.266°E	09/2011	Lst, Phk
Tsh	Artificial lake, 40.001°N 116.313°E	09/2011	Cqu, Fmy, Ian, Phk
Tsh	Artificial lake, 40.000°N 116.314°E	03/2012	Fmy, Ian, Phk
Tsh	Artificial lake, 40.000°N 116.314°E	11/2012	Cqu, Fmy, Ian, Lin
Tsh	Artificial lake, 40.000°N 116.314°E	12/2012	Bvi, Cqu, Fmy, Ian, Phk
WM	Artificial lake, 39.993°N 116.302°E	07/2011	Phk
WM	Artificial lake, 39.992°N 116.304°E	09/2011	Phk
ZZY	Artificial lake, 39.940°N 116.310°E	05/2011	Phk
X4	Dead arm of a river, 42.236°N 115.625°E	07/2012	Hau, Imo, Lin, Pln
X35	Small puddle, 45.516°N 118.626°E	07/2012	Hin
X36	Small puddle, 46.514°N 118.608°E	07/2012	Imo
X48	Small pond, 46.008°N 117.756°E	07/2012	Hin

Abbreviations: Bvi, *Bradleycypris vittata* (Sars); Cqu, *Candona quasiakaina* Karanovic & Lee; Cvi, *Cypridopsis vidua* (O. F. Müller); Fmy, *Fabaeformiscandona myllaina* Smith & Kamiya; Hau, *Heterocypris* auricularis sp. nov.; Hin, *Heterocypris incongruens* (Ramdohr); Ian, *Ilyocypris angulata* Sars; Imo, *Ilyocypris mongolica* Martens; Lin, *Limnocythere inopinata* (Baird); Lst, *Limnocythere stationis* Vávra; Phk, *Physocypria kraepelini* G. W. Müller; Pln, *Plesiocypridopsis newtoni* (Brady & Robertson).

transferred to 70% ethanol within 5 days after sampling and the animals were picked dead.

Some animals were dissected in Hydro-Matrix (Micro-Tech-Lab, Graz, Austria). Others were dehydrated in ethanol before being dissected in a solution containing 60% fresh pine resin and 40% ethanol. Soft parts were sealed and examined with a BK3300 light microscope. Valves were stored dry in micropalaeontology slides and examined with an XTD-3 stereomicroscope or an LEO 1450VP scanning electron microscope. Line drawings were produced with the aid of microscopic photos. All specimens are housed in the Institute of Geology and Geophysics, Chinese Academy of Sciences.

Abbreviations.— LV, left valve; RV, right valve; lat., lateral view; int., internal view; ext., external view; dor., dorsal view; ven., ventral view; fron., frontal view; A1, antennule; A2, antenna; Md, mandible; Mx, maxillula; L5, fifth limb; L6, sixth limb; L7, seventh limb; Ur, uropod; Hp, hemipenis; ZO, Zenker organ. Terminology of the appendage chaetotaxy follows Martens (1987), Meisch (2000) and Karanovic (2012).

RESULTS

During this survey, a total of 12 species were found, which are listed below, with a description of each species' morphological features and known geographical distribution. All the referred specimens were adults unless otherwise noted.

Superfamily CYPRIDOIDEA Baird, 1845
Family CANDONIDAE Kaufmann, 1900
Subfamily CANDONINAE Kaufmann, 1900
Genus *Candona* Baird, 1845
Candona quasiakaina Karanovic & Lee, 2012

2012 *Candona quasiakaina* n. sp. Karanovic & Lee: 3368; figs. 1-4

Material examined.— Two dissected males (dyzoc368, dyzoc393) and one dissected female (dyzoc394) from a small lake (Tsh, fig. 1, table I) in Tsinghua University. Two dissected females (dyzoc195, dyzoc236) from the same site with the soft parts deteriorated to different extents.

Dimensions.— Male (LV, $n = 2$) length 1440-1520 μm, H/L ratio 0.53-0.54. Female (LV, $n = 3$) length 1420-1480 μm, H/L ratio 0.49-0.51.

Brief description.— Carapace large for the genus. Sub-triangular in lat. Dorsal margin strongly arched with the highest point situated behind mid-length. Male dorsum more strongly arched than female, hump-like (fig. 2). Lobe a (i.e., outer lobe) of Hp with a spine-like extremity. Male L5 palps not equal, with right palp larger and slightly inflated at the middle. Both male L5 palps with a small lobe-like expansion at the ventro-subapical part of the finger (fig. 3). Female genital lobe with large, inflated basal part and narrow, rounded posterior end. Sp seta of Ur extending significantly beyond base of claws.

Comparisons with *C. quasiakaina* from South Korea (Karanovic & Lee, 2012).— Some morphological differences are found between the Chinese specimens and the South Korean specimens. For example, the male t2 of A2 of the Chinese specimens has a sharp and pointed end, while that of the South Korean specimens has a rounded end. The shortest seta on 1st A2 protopodal segment and the α seta of Md palp are longer and thinner in the Chinese specimens than those in the South Korean specimens. Despite the observed differences, the two populations resemble each other in the morphology of male and female reproductive organs and male prehensile palps, and most of the details of chaetotaxy. They are here considered to be conspecific.

Comparisons with other Candoninae species.— Within the subfamily, some other species also have a spine-like extremity on the outer lobe a of Hp. Nonetheless, they can be distinguished from *C. quasiakaina*. (1) Shell of *Fabaeformiscandona akaina* Smith & Janz, 2008 has a very narrow, nearly pointed posterior end in lat. (Smith & Janz, 2008). Sp of Ur extends only slightly beyond base of claws.

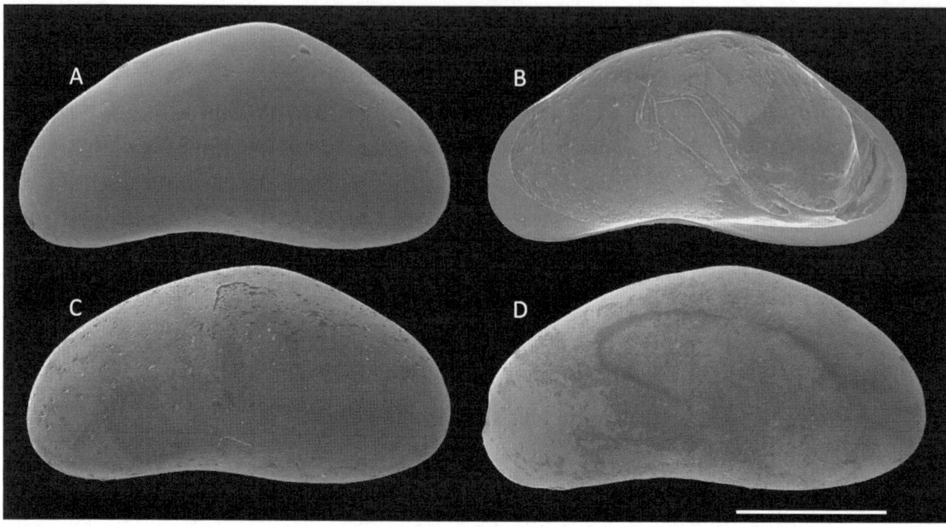

Fig. 2. *Candona quasiakaina* Karanovic & Lee, 2012, male and female. A, male LV, ext. (dyzoc368); B, male RV, int. (dyzoc368); C, female LV, ext. (dyzoc195); D, female LV, ext. (dyzoc236). Scale bar = 500 μm.

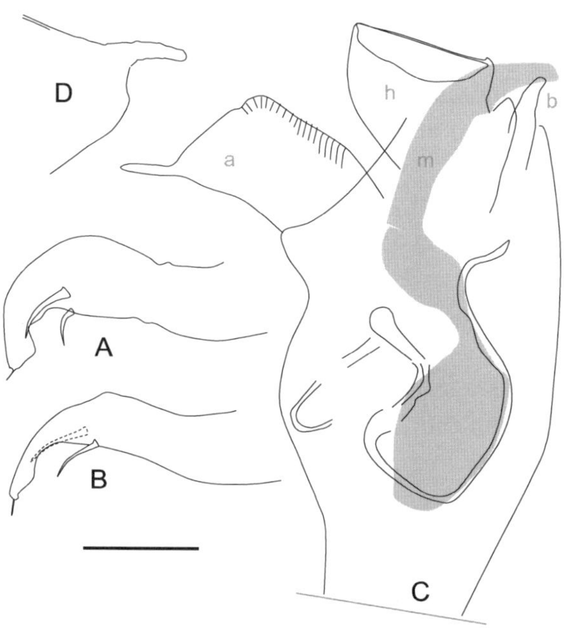

Fig. 3. *Candona quasiakaina* Karanovic & Lee, 2012, male and female. A, male right L5 palp (dyzoc368); B, male left palp (dyzoc368); C, male Hp (dyzoc368); D, posterior part of female genital lobe (dyzoc394). Scale bar = 100 μm. This figure is published in colour in the online edition of this journal, which can be accessed via http://booksandjournals.brillonline.com/content/journals/15685403.

Female genital lobe becomes gradually narrower posteriorly, rather than contracting suddenly at the middle part as in *C. quasiakaina*. (2) Male right L5 palp of *F. danielopoli* Yin & Martens, 1997 has a sinuated dorsal margin, and the Hp has a smaller lobe a with a stronger spine-like projection (Yin & Martens, 1997). (3) Hp of *F. hyalina* (Brady & Robertson, 1870) has a much narrower lobe a, and the basal part of Ur ramus is less swollen (Martens & Dumont, 1984). (4) Lobe a of Hp of *F. nishinoae* Smith & Janz, 2008 has a progressively narrower distal half (Smith & Janz, 2008), compared with the quadrate one in *C. quasiakaina*. Apical setae on the trunk of the prehensile palp are much shorter in *F. nishinoae*.

Distribution.— This species was originally described from a freshwater stream in South Korea (Karanovic & Lee, 2012) and this is the first record outside of the type locality.

Remarks.— (1) *Candona quasiakaina* has a smooth γ seta on Md palp, and should be placed in the genus *Fabaeformiscandona* according to the criterion of Meisch (2000). However, Karanovic & Lee (2012) assigned their specimens to genus *Candona* by arguing that the genus *Fabaeformiscandona* should be much narrower (Karanovic, 2006) and that some of the species included in this genus at the present should be given new generic positions. The allocation of this species to *Candona* is adopted here. (2) In fig. 3A of Karanovic & Lee (2012) a walking leg with 5-segmented endopod was drawn for *C. quasiakaina*. According to the description in the text and personal communication to one of those authors (Ivana Karanovic), this leg of the holotype specimen is an anomaly, and that the normal L6 endopod is 4-segment. (3) The lobe h (both sides of the Hp) of specimen dyzoc368 seems to be folded and shows a different shape. They are expected to be rounded in normal condition. This folding may indicate a gap of the internal structure in lobe h, and thus was retained in fig. 3.

Genus *Fabaeformiscandona* Krstić, 1972

Fabaeformiscandona myllaina Smith & Kamiya, 2007

2004 *Fabaeformiscandona* cf. *subvomerina* — Okubo: 19; fig. 5
2007 *Fabaeformiscandona myllaina* n. sp. Smith & Kamiya: 228; figs. 1, 2, 4, 6-8

Material examined.— Two dissected males (dyzoc203, dyzoc375) and two dissected females (dyzoc205, dyzoc223) from the site Tsh (fig. 1, table I).

Dimensions.— Male (LV, $n = 2$) length 1310-1420 μm, H/L ratio 0.51. Female (LV, $n = 2$) length 1260 μm, H/L ratio 0.48-0.49.

Brief description.— Carapace sub-reniform and elongated in lat., greatest height situated well behind mid-length. LV slightly overlapping RV posteriorly. Female carapace with truncated posterior end. Male with inflated, rounded posterior end and a more concave ventral margin (fig. 4). Male L5 palps unequal, with right palp

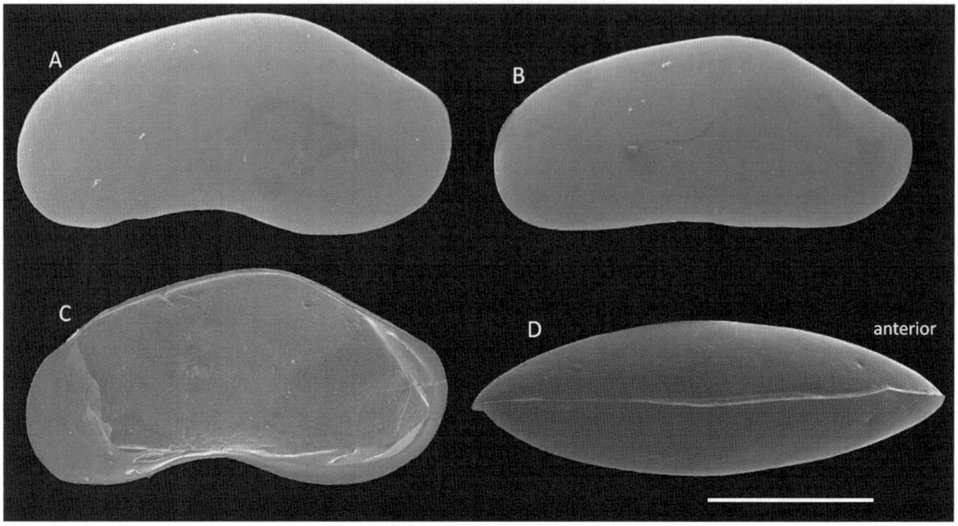

Fig. 4. *Fabaeformiscandona myllaina* Smith & Kamiya, 2007, male and female. A, male LV, ext. (dyzoc203); B, female LV, ext. (dyzoc205); C, male RV, int. (dyzoc203); D, male carapace, dor. (dyzoc375). Scale bar = 500 μm.

significantly inflated. Hp with S-shaped lobe a (i.e., outer lobe) directed nearly vertically away, m process (an elongated, sclerotized structure surrounding the bursa copulatrix) slightly inflated at sub-distal part. Female with large genital lobes (fig. 5).

Comparisons with the Japanese specimens.— In chaetotaxy, the Chinese specimens are identical to those described by Smith & Kamiya (2007), except that the dorso-apical seta on the 3rd segment of A1 is not smooth but setulose. This minor difference is here considered as within-species variation.

Distribution.— This species has been reported from irrigation ponds and lotus fields of Japan (as *Fabaeformiscandona* cf. *subvomerina* Liu & Lu, 1999) (Okubo, 2004). It was formally named by Smith & Kamiya (2007) who collected the specimens from an artificial pond in Japan. The species was also found in Lake Biwa (Smith & Janz, 2008). This is its first report from China.

Subfamily CYCLOCYPRIDINAE Kaufmann, 1900

Genus *Physocypria* Vávra, 1897

Physocypria kraepelini G. W. Müller, 1903

1903 *Physocypria kraepelini* n. sp. G. W. Müller: 165; figs. 1-7
1927 *Cypria* (*Physocypria*) *fadeewi* — Dubowsky: 84; figs. 10-18
1934 *Physocypria kliei* — Schäfer: 559; figs. 1-6
1941 *Cypria inversa* — Klie: 236; figs. 1-9

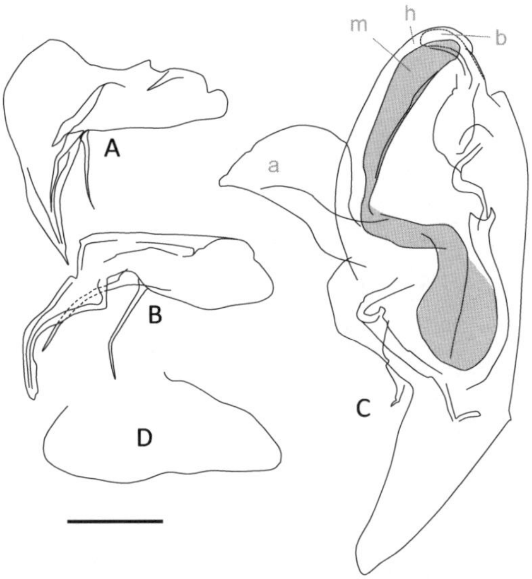

Fig. 5. *Fabaeformiscandona myllaina* Smith & Kamiya, 2007, male and female. A, male right L5 palp (dyzoc203); B, male left L5 palp (dyzoc203); C, male Hp (dyzoc203); D, female genital lobe (dyzoc205). Scale bar = 100 μm. This figure is published in colour in the online edition of this journal, which can be accessed via http://booksandjournals.brillonline.com/content/journals/15685403.

1943 *Physocypria perlata* — Rome: 1; fig. 1

Material examined.— Six dissected males (dyzoc2, dyzoc229, dyzoc237, dyzoc242, dyzoc243, dyzoc248) and nine dissected females (dyzoc1, dyzoc96, dyzoc97, dyzoc99, dyzoc204, dyzoc211-213, dyzoc359) from several sites in Beijing (table I).

Dimensions.— Male (LV, $n = 5$) length 540-560 μm, H/L ratio 0.67-0.75. Female (LV) length 585-660 μm ($n = 7$), H/L ratio 0.67-0.72 ($n = 4$).

Brief description.— Carapace very stout in lat. with a more or less arched dorsal margin (fig. 6). RV overlaps LV with a dorsal hump in fron. RV with small pustules on both the antero- and postero-ventral margins. These pustules are sometimes hard to observe. Colour brownish when alive, becoming whitish several days after death. A1 with very long swimming setae, some of which feathery. A2 with long swimming setae extending beyond the terminal claws with ca. 2/3 of their length. L6 without d1 and d2 setae, and with very long e seta reaching beyond the terminal segment. L7 with long h3 seta and short, sub-equally long h1 and h2. Ur with Gp shorter than Ga. Male L5 palps asymmetric with right palp stronger (fig. 7). Hp with slender, distally pointed lobes a (i.e., outer lobe) and b (i.e., inner lobe); b shorter than a, more or less curved.

Distribution.— This species is common in Europe (Meisch, 2000). *Physocypria nipponica* Okubo, 1990 reported from Japan may be conspecific with *P. kraepelini* (Smith & Janz, 2008). In China, *P. kraepelini* has been reported from Taihu Lake

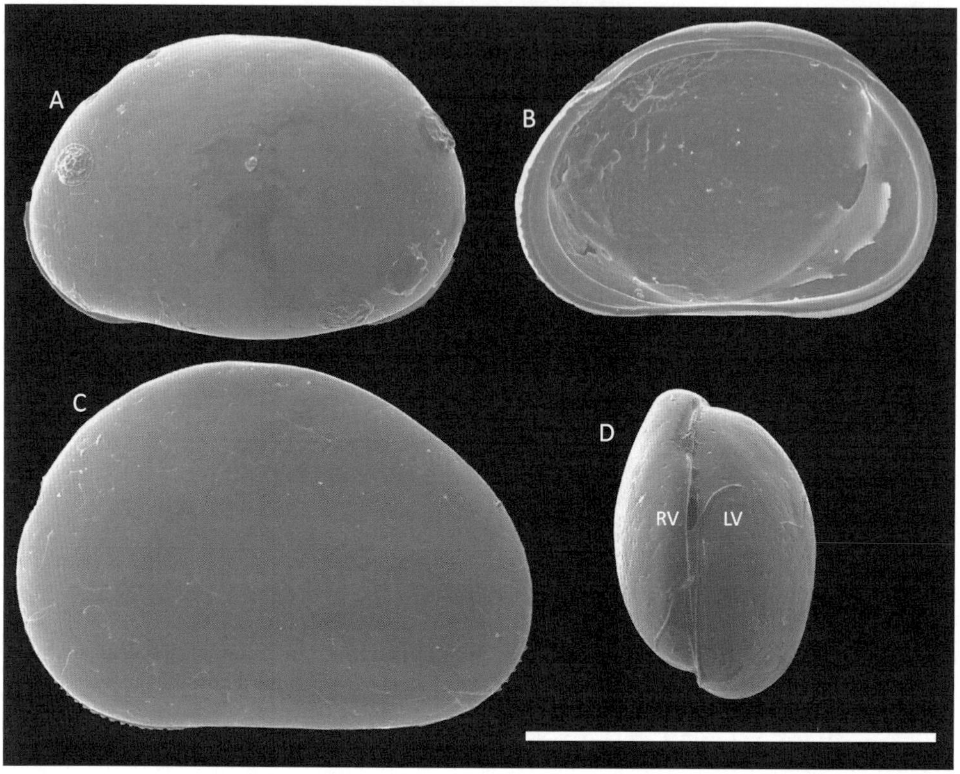

Fig. 6. *Physocypria kraepelini* G. W. Müller, 1903, male and female. A, male LV, ext. (dyzoc243); B, male RV, int. (dyzoc243); C, female RV, ext. (dyzoc359); D, male carapace, fron. (dyzoc248). Scale bar = 500 μm.

in Jiangsu Province (Yu et al., 2005), Dianchi Lake in Yunnan Province (Yu et al., 2010) and from Taiwan Island (Hsiao & Young, 2001). In Beijing City, *P. kraepelini* was found in all of the 5 sites investigated, and seemed to be present at least throughout the ice-free months (March to December; table I), being the most common ostracod in these habitats.

Family ILYOCYPRIDIDAE Kaufmann, 1900
Subfamily ILYOCYPRIDINAE Kaufmann, 1900
Genus *Ilyocypris* Brady & Norman, 1889
Ilyocypris angulata Sars, 1903

1903 *Ilyocypris angulata* n. sp. Sars: 22; pl. IV, fig. 8
1981a *Ilyocypris dentifera* form *angulata* — Victor & Fernando: 1106; figs. 23-38

Material examined.— Five dissected males (dyzoc383, dyzoc389, dyzoc401-403) and six dissected females (dyzoc214, dyzoc241, dyzoc373, dyzoc385, dyzoc390, dyzoc392) from a small lake in Tsinghua University of Beijing (Tsh, fig. 1, table I). Soft parts of dyzoc241, dyzoc373, dyzoc392

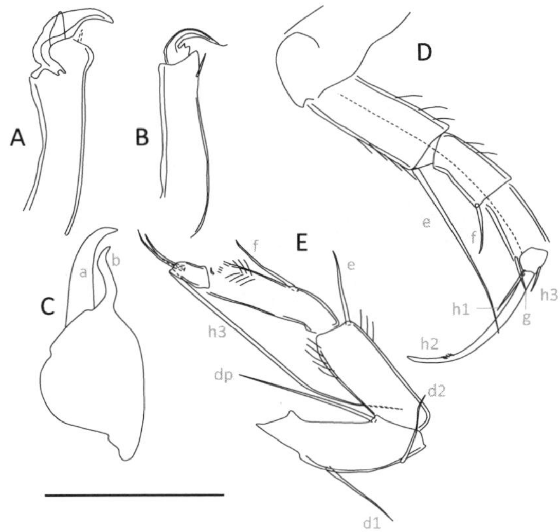

Fig. 7. *Physocypria kraepelini* G. W. Müller, 1903, male. A, right L5 palp (dyzoc243); B, left L5 palp (dyzoc243); C, Hp (dyzoc2); D, L6 (dyzoc243); E, L7 (dyzoc242). Scale bar = 100 μm. This figure is published in colour in the online edition of this journal, which can be accessed via http://booksandjournals.brillonline.com/content/journals/15685403.

and dyzoc401-403 are deteriorated to different extents, yet readily identifiable. Carapaces of dyzoc392, dyzoc401 and dyzoc402 are deteriorated to some extent.

Dimensions.— Male (LV, $n = 5$) length 850-900 μm, H/L ratio 0.51-0.54. Female (LV, $n = 5$) length 860-1020 μm, H/L ratio 0.51-0.54.

Re-description.— Carapace (fig. 8) sub-reniform in lat. Dorsal margin straight, greatest height situated at ca. first 1/5. Anterior and posterior margins broadly rounded. Ventral margin concave. Shell surface strongly pitted, with four prominent nodes, the one near the postero-dorsum usually being the largest and usually pointing posteriorly. Anterior margin of the valve with ca. 12 short spines. Posterior margin of valve with ca. 8 longer spines. Spines also developed on surface of valve. Number, position and size of the spines relatively constant. Anterior part of carapace moderately compressed in dor. Nodes and spines protruding beyond valve outline. In int., under the SEM, medial part of postero-ventral margin of LV with 2-4 conspicuous ripplets and numerous small striations. Color of carapace pearly white when alive, dirty white some time after deadth.

A1 (fig. 9) with 8 segments. First two segments with short spinules on the internal side. Second segment with 1 dorsal seta and 2 long ventral setae. Third segment quadrate or slightly elongated, with 1 short dorso-apical seta. Fourth segment with 1 short dorso-apical seta and 1 short ventro-apical seta. Fifth segment with 2 long dorso-apical setae and 2 unequal ventro-apical setae. Sixth segment with 2 long dorso-apical setae and 2 intermediately long ventro-apical setae, one

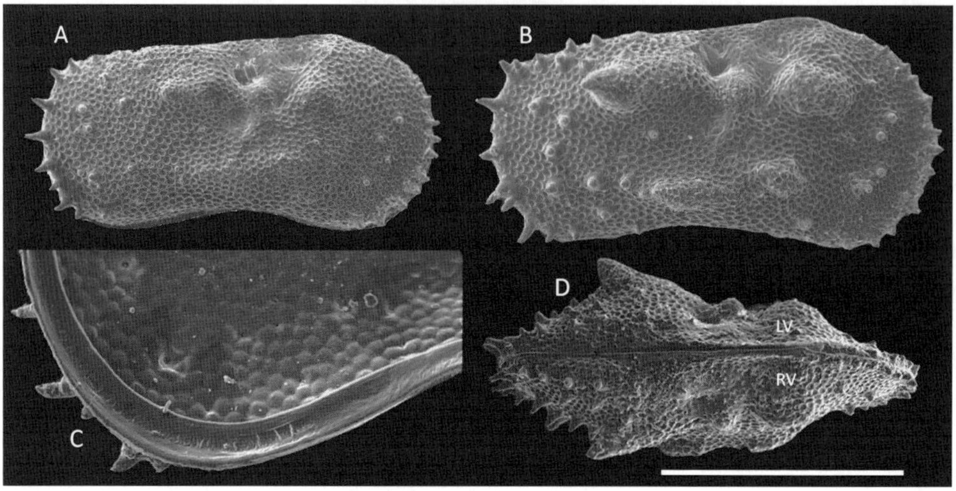

Fig. 8. *Ilyocypris angulata* Sars, 1903, male and female. A, male RV, ext. (dyzoc383); B, female RV, ext. (dyzoc385); C, male LV, showing ripplets on the postero-ventral margin (dyzoc383); D, female carapace, oblique-dor. (dyzoc214). Scale bar = 100 μm for A, B and D, 200 μm for C.

of which claw-like. Seventh segment with 4 long apical setae besides the α seta. α seta extending somewhat beyond the terminal segment. Eighth segment with 4 apical setae. Ventral-most seta claw-like, conspicuously longer than y_a.

A2 (fig. 9) with 5 segments, penultimate segment undivided in both sexes. One of the ventral setae of first segment strongly plumose. Second, third and penultimate segments with well developed pseudochaetae and spinules most of which situated on the internal side of the limb. Arrangements of spinules on iv-b segment very characteristic. Swimming setae sub-equal, extending beyond the longest claw with ca. 25% of their length. Sixth seta only extending to middle of claws. t1 strong, reaching beyond the terminal segment. t2-t4 progressively shorter. Female with strong G1, G2, G3 and GM claws and untransformed z setae. Male with strong G1, G2 and GM claws, and claw-like z1. y3 ca. 1/3 the length of Gm. All claws smooth.

Md palp (fig. 9) with few specific features. Unnamed seta of first segment of palp clearly branchial at the distal part. S1 and S2 plumose. α seta short and thin. Grouped setae on second segment sparsely plumose, extending beyond the terminal claws. The fourth seta much shorter, smooth. β seta short, slender, strongly plumose. Third segment with a total of 4 sub-apical and 4 apical setae one of which plumose. Fourth segment with 1 medio-lateral seta and 4 terminal claws/setae.

Mx (not shown) typical of the genus, with few specific features.

Male L5 palps (fig. 10) sub-equal in shape, but slightly different in chaetotaxy. Trunks of both sides with unequal apical setae. Finger of right side with one

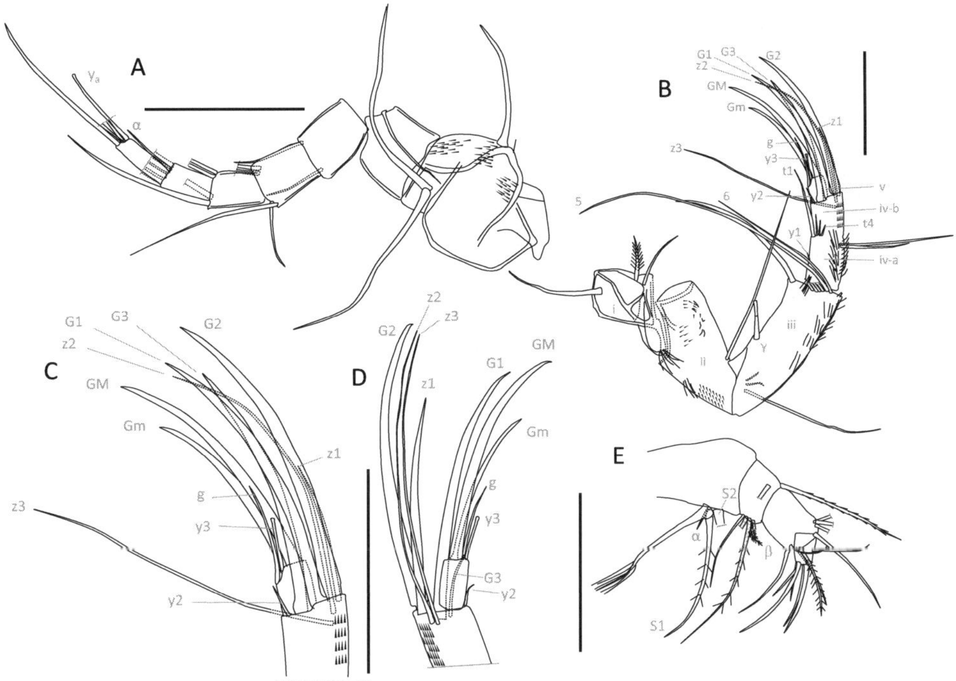

Fig. 9. *Ilyocypris angulata* Sars, 1903, male and female. A, male A1 (dyzoc383); B, female A2 (dyzoc385); C, distal part of female A2 (dyzoc385); D, distal part of male A2 (dyzoc383); E, palp of male Md (dyzoc389). Scale bars = 100 μm. This figure is published in colour in the online edition of this journal, which can be accessed via http://booksandjournals.brillonline.com/content/journals/15685403.

prominent apical seta and two very thin setae. Finger of left side with one very long apical seta and one short sub-apical seta. Female L5 palp 2-segmented (not shown).

L6 (fig. 10) penultimate segment undivided. h1 conspicuously longer than h3. h2 extraordinarily long, sub-equal to total length of the endopodal segments, slightly serrated at the middle part. First endopodal segment with well-developed, long pseudochaetae.

L7 (fig. 10) penultimate segment with two long setae (f and g). Distal end of penultimate segment without any spinules or pseudochaetae. h2 slightly longer than h3. h1 much shorter but stronger. All setae smooth. Posterior margin of first endopodal segment with several groups of stiff pseudochaetae.

Ur (fig. 10) with arched ramus. Ga and Gp sub-equal, bare. Sp reaching about base of claws in male, extending slightly beyond ramus in female. Sa short. Distal half of ramus with long pseudochaetae.

Hp (fig. 10) with short lobe a. Lobe b reaching slightly beyond mid-way of lobe h, with curved and pointed distal end. Lobe h with strong neck and inflated distal

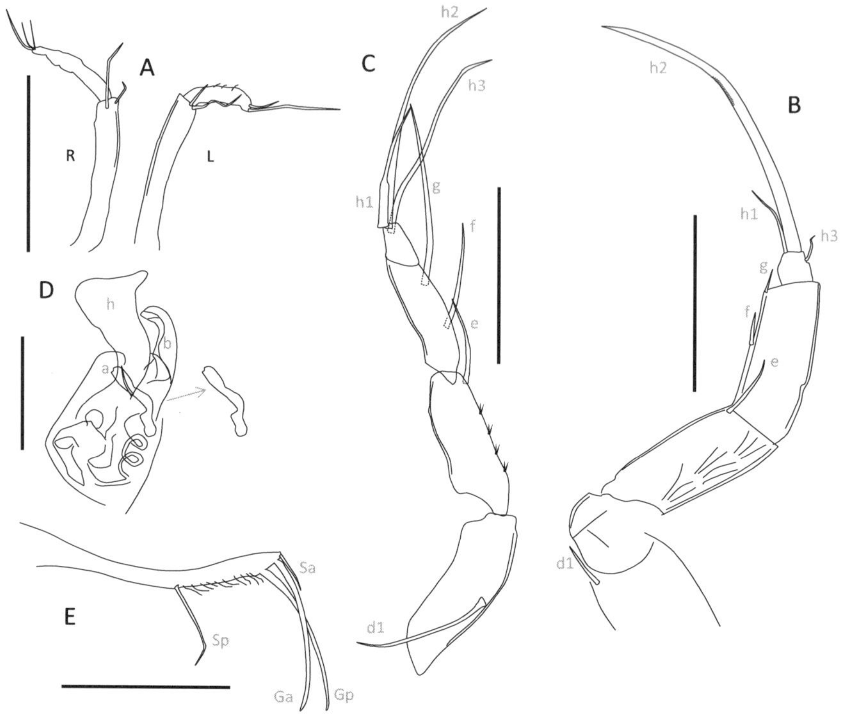

Fig. 10. *Ilyocypris angulata* Sars, 1903, male. A, L5 palps (dyzoc389); B, L6 (dyzoc389); C, L7 (dyzoc389); D, Hp (dyzoc389); E, Ur (dyzoc383). Scale bars = 100 μm. This figure is published in colour in the online edition of this journal, which can be accessed via http://booksandjournals.brillonline.com/content/journals/15685403.

end. Internal end of lobe h pointed. A fourth bar-like structure positioned across the Hp with a small mouth-like opening at the distal extremity.

ZO (fig. 11) with ca. 22-23 rows of spines. Distal end slightly inflated.

Comparisons with the type specimens.— Victor & Fernando (1981a) figured female *I. angulata* (as *I. identifera* form *angulata*) from the type series. Karanovic & Lee (2013) found males from the type material and presented SEM images of

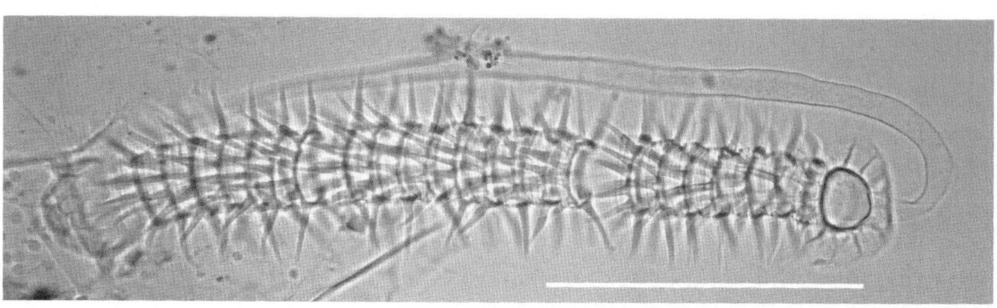

Fig. 11. *Ilyocypris angulata* Sars, 1903, male ZO (dyzoc383). Scale bar = 100 μm.

valves of both sexes, as well as line drawings of some of the limbs of female and Hp of male. It is unfortunate that much of the chaetotaxy of the A1, A2, and male L5 palps, and the morphology of male ZO of the type specimens have not been described, which prevents a complete comparison here. Nonetheless, some differences can be found between the Beijing specimens and the types. Females measured by Victor & Fernando (1981a) have a mean H/L ratio of 0.58 ($n = 20$; probably measured on whole carapaces). The LV of female paralectotype examined by Karanovic & Lee (2013) has a H/L ratio of 0.56 (measured from fig. 11A of that paper). The females of *I. angulata* from Beijing (this paper) have a mean H/L ratio of 0.53 ($n = 5$), being more elongated than the type specimens. The males from Beijing have a mean H/L ratio of 0.52 ($n = 5$), which is slightly more elongated than the females. The only male carapace illustrated by Karanovic & Lee (2013) (fig. 11I therein) was probably decalcified and deformed (Ivana Karanovic, personal communication) and thus the H/L ratio of the male type specimens is not available at the present. Female L6 terminal claw of the type specimen (fig. 35 of Victor & Fernando (1981a)) is significantly shorter than those of both sexes of the Beijing specimens (fig. 10). Besides, both sexes of the Beijing specimens lack the small apical spinule on the L7 penultimate segment, which was illustrated for the female in fig. 36 of Victor & Fernando (1981a).

Distribution.— *Ilyocypris angulata* was originally described from Fujian Province of southern China (Sars, 1903), and was also found in the nearby Hunan (Chen, 1990) and Jiangsu (Yu et al., 2005) Provinces. This is the first report in the Palaearctic part of the country (i.e., north of Qinling Mountain). Okubo (1974) reported both sexes of *I. angulata* from Japan. However, Karanovic & Lee (2013) pointed out that the Hp of the so-called *I. angulata* by Okubo (1974) is different from that of the males in the type series, and suggested that the Japanese specimens belong to a different species.

Ilyocypris mongolica Martens, 1991

1991 *Ilyocypris mongolica* n. sp. Martens: 55; fig. 1

Material examined.— Three dissected males (dyzoc319-321) and two dissected females (dyzoc322, dyzoc323) from a puddle (X36, fig. 1, table I) in Inner Mongolia. Two dissected females (dyzoc341, dyzoc342) from a small lake formed by a dead arm of a river (X4, fig. 1, table I) in Inner Mongolia.

Dimensions.— Male length (LV, $n = 3$) 800-815 μm, H/L ratio 0.52-0.53. Female length (LV, $n = 4$) 760-890 μm, H/L ratio 0.52-0.54.

Diagnosis (modified after Martens (1991)).— A medium-sized *Ilyocypris* (fig. 12). Lacking nodes (at least for specimens in this study and those found in Martens (1991)). Shell surface covered with well-developed, rounded pits. Very small, short spines present along the posterior margin (sometimes absent). In int., under the SEM, LV with no obvious marginal ripplets (fig. 12). A2 natatory se-

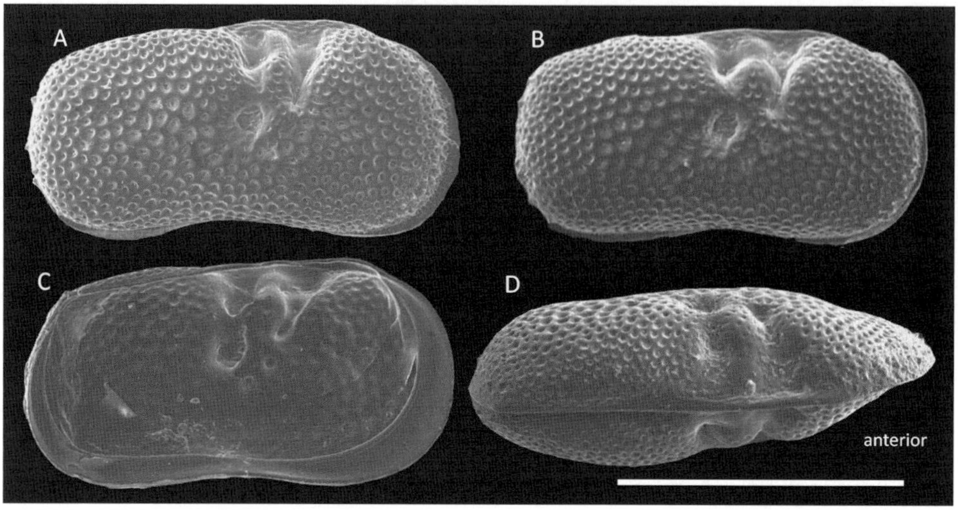

Fig. 12. *Ilyocypris mongolica* Martens, 1991, male and female. A, male RV, ext. (dyzoc320); B, female RV, ext. (dyzoc322); C, female LV, int. (dyzoc322); D, male carapace, oblique-dor. (dyzoc319). Scale bar = 500 μm.

tae extending beyond terminal claws with ca. 50% of their lengths (fig. 13). Male L5 palps sub-equal, slender, with sub-apical part of the fingers slightly inflated (fig. 14). L6 penultimate segment divided. L7 penultimate segment with two setae (f and g). This segment apically bearing 3-6 hook-like structures. Hp with short, broad lobe a, slender lobe b, triangular lobe h (i.e., medial lobe) and one slender, pointed stylus.

Re-description.— Carapace sub-reniform in lat. (fig. 12). Dorsal margin straight, sloping posteriorly. Highest point situated at first 1/6 of valve. Anterior margin broadly rounded. Posterior margin nearly straight. Ventral margin concave. Anterior and postero-ventral margins surrounded by narrow compressed zones. Two transverse sulci starting from the anterior part of dorsal margin, directed slightly posteriorly, terminated at middle of the shell. A sub-ovate depression situated below the posterior sulcus. Carapace sub-elliptical in dor., both ends bluntly pointed. Shell surface densely covered with well-developed, rounded pits. In int., under the SEM, LV with no obvious ripplets, sometimes with dense, fine striations in the medial area of the valve margin.

A1 (fig. 13) with eight segments, first two of which fused forming a large base, supporting one dorsal seta and two long ventral setae. Third segment quadrate with one dorso-apical seta extending beyond the next segment. Fourth segment elongated with one dorso-apical seta and one shorter ventro-apical seta. Fifth segment quadrate with two long dorso-apical setae and two unequally long ventro-apical setae. Sixth segment quadrate with two dorso-apical setae and two ventro-apical setae, one of which intermediately long, the other one shorter and claw-like.

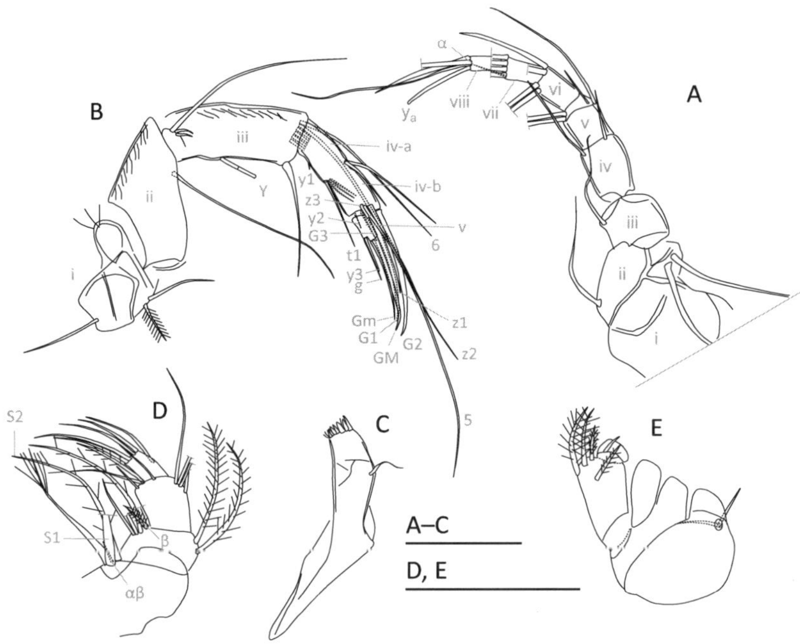

Fig. 13. *Ilyocypris mongolica* Martens, 1991, male (dyzoc320). A, A1; B, A2; C, coxa of Md; D, palp of Md; E, Mx endites and palp. Scale bars = 100 μm. This figure is published in colour in the online edition of this journal, which can be accessed via http://booksandjournals.brillonline.com/content/journals/15685403.

Seventh segment elongated with five apical setae, the shortest one (α seta) situated at the external side of the segment. Eighth segment elongated with four apical setae, the aesthetasc y_a situated at the dorsal corner, a claw-like seta at the ventral corner, and two unequally long ones situated between the above two.

A2 (fig. 13) penultimate segment undivided in both sexes, producing a total of five segments. Swimming setae reaching beyond the terminal claws with ca. 50% of their length, while the sixth seta only to the middle part of the claws. Seta t1 extending somewhat beyond the terminal segment. Setae t2 to t4 sub-equal, ca. 3/5 the length of the iv-b segment. Seta z1 of male claw-like, slightly shorter than the G-claws. Setae z2 and z3 not transformed, extending beyond the claws. Female z-setae untransformed. Male G3 very small, seta-like. Female G3 claw untransformed. Terminal segment with the GM claw slightly shorter than Gm. The aesthetasc y3 ca. half the length of Gm, slightly shorter than g seta.

Md (fig. 13) coxa with short teeth, typical of the genus. Palp 4-segmented. Unnamed seta on the first segment with a swollen, rounded base and strongly setulose (or branchial) distal part. S1 and S2 sub-equally long, both sparsely plumose, basally not swollen. α seta smooth, ca. 1/5 the length of accompanying setae. Second segment with $3 + 1 + \beta$ setae. The grouped setae reaching beyond

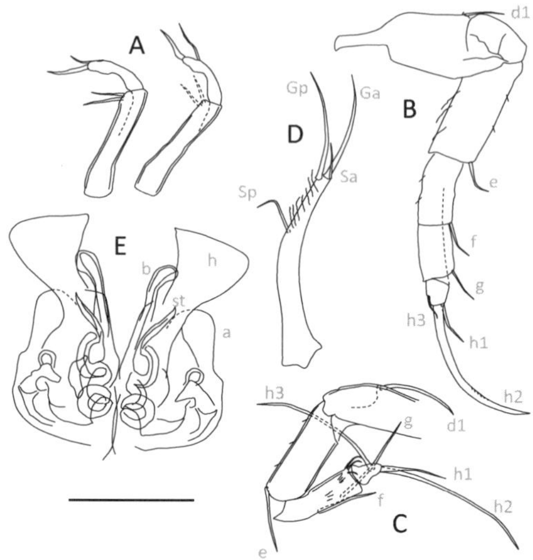

Fig. 14. *Ilyocypris mongolica* Martens, 1991, male (dyzoc320). A, L5 palps; B, L6; C, L7; D, Ur; E, penis. Scale bar = 100 μm. This figure is published in colour in the online edition of this journal, which can be accessed via http://booksandjournals.brillonline.com/content/journals/15685403.

mid-way of terminal claws, sparsely setulose. The fourth seta much shorter. Seta β short, slender, plumose. One of the apical setae on the external side of third segment plumose. Terminal segment with one medio-dorsal claw and four unequal terminal claws.

Mx (fig. 13) palp 2-segmented, the second segment stout, relatively short. Endite lobes progressively shorter.

Male L5 palps (fig. 14) sub-equal. Basal segment (i.e., trunk) elongated and slender with two apical setae. Terminal segment (i.e., finger) narrower at the middle, carrying one strong apical seta and one thin sub-apical seta.

L6 (fig. 14) penultimate segment divided. d1 short, d2 absent, e, f and g sub-equally long. h1 longer than h3. h2 claw smooth.

L7 (fig. 14) penultimate segment undivided, carrying two long setae (f and g). d1 somewhat claw-like, d2 and dp absent, e and f sub-equally long and shorter than g. h1 shorter than h2 and h3. h2 longer and slightly stronger than h3. Penultimate segment apically with a group of hook-like structures. Number of hook-like structures: 3 in male, 4 to 8 in female. Penultimate segment sometimes with ca. 2 rows of fine pseudochaetae at the sub-apical part.

Note: For a more detailed illustration of the "hook-like structures" see Martens (1991). Martens (1991) observed 2 stout, 1 intermediate and 1 slender structures in the holotype male, and 3 + 1 + 2 structures in the allotype female. In the Chinese specimens, the males usually have 2 stout and 1 slender structures. Some females have 7-8 progressively shorter structures (similar to the pattern observed in female allotype), but some carry 3 stout and 1 slender structures (similar to male holotype). It seems that these hook-like structures are variable among individuals of both sexes.

Ur (fig. 14) with slender ramus evenly bent posteriorly. Claws Ga and Gp subequally long. Seta Sp longer than Sa and extending to the base of the claws. Distal half of the ramus bearing fine pseudochaetae.

Hp (fig. 14) consisting of 4 distal lobes, namely outer lobe a, inner lobe b, medial lobe h and the stylus. Lobe a wide, distally broadly rounded. Lobe b slender, longer than a, distally swollen. Lobe h being the longest, distally enlarged forming a nearly perfect symmetrical triangle. Stylus slender, reaching terminal of lobe a, distally pointed.

ZO (fig. 15) with ca. 13-14 rows of spines, both ends spherically enlarged but distal end much larger.

Distribution.— This species was originally reported from a saline ditch in the flood area of the Herlen River in eastern Mongolia. Aside from the type locality, it was also reported from a lake in western Mongolia (Van der Meeren et al., 2010). In the present study, a dense bisexual population was found in site X36 of Inner Mongolia, and several females were also found at site X4 (table I).

Remarks.— Wrozyna et al. (2009) reported subfossil valves and a mummified male of *I.* cf. *mongolica* from Nam Co Lake in southern Tibet, with the identification done mainly based on shell morphology. Carapaces of those specimens are more elongated and the posterior margin of the valve is more arched compared with *I. mongolica*. The compressed zones surrounding the anterior and the postero-ventral margins in *I. mongolica* (fig. 12) are missing in *I.* cf. *mongolica*. And marginal ripplets are clearly seen along the distal part of the postero-ventral margin of LV in *I.* cf. *mongolica*. Moreover the pits on the shell surface are small

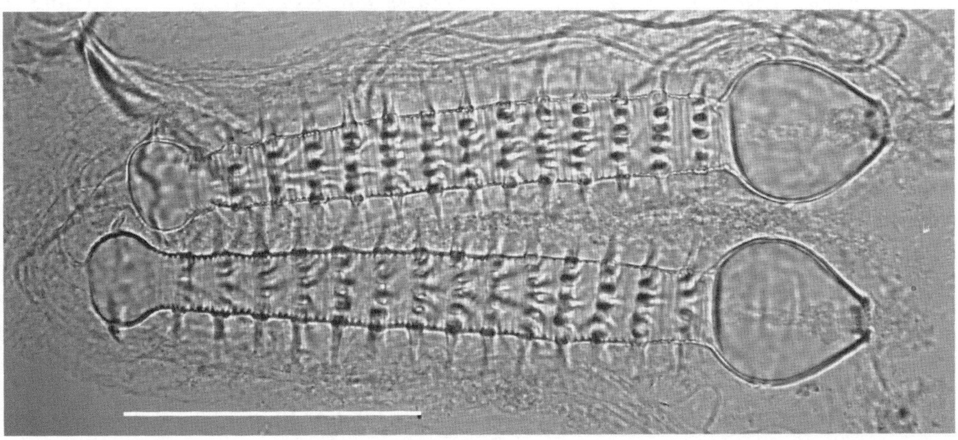

Fig. 15. *Ilyocypris mongolica* Martens, 1991, male ZOs (dyzoc320). Scale bar = 100 μm.

and shallow in *I.* cf. *mongolica*. Based on the above observations, it is here suggested that the *I.* cf. *mongolica* reported by Wrozyna et al. (2009) was probably not "true" *I. mongolica*, and that detailed examinations on the soft parts of those specimens are needed to confirm their taxonomic status.

Family CYPRIDIDAE Baird, 1845
Subfamily CYPRINOTINAE Bronshtein, 1947
Genus *Heterocypris* Claus, 1892
Heterocypris incongruens (Ramdohr, 1808)

1808 *Cypris incongruens* n. sp. Ramdohr: 86; pl. 3
1892 *Heterocypris incongruens* nov. comb. — Claus: 161
1900 *Heterocypris incongruens* var. *elongata* — Kaufmann: 269; pls. 15, 13
1920 *Cyprinotus hertwigi* — Lindner: 37
1938 *Cyprinotus incongruens* s.sp. *attenuata* — Gauthier: 55; figs. 6, 7

Material examined.— Thirty-five dissected parthenogenetic females (dyzoc252, dyzoc256-262, dyzoc264-266, dyzoc268, dyzoc271-275, dyzoc279-282, dyzoc285-290, dyzoc294-300) from a puddle (X35, fig. 1, table I) in Inner Mongolia. Twenty-four dissected parthenogentic females (dyzoc278, dyzoc303-318, dyzoc324-330) from a small pond (X48, fig. 1, table I) in Inner Mongolia.

Dimensions.— Length (LV) 1200-1500 μm ($n = 56$), H/L ratio 0.56-0.62 ($n = 11$).

Brief description.— Carapace shape variable, but generally slightly stout in lat. (fig. 16). Anterior and posterior margin sub-equally rounded. RV with tiny pustules on both anterior and posterior margins; presence of pustules variable among individuals (fig. 16), sometimes missing. In dor. anterior end pointed, or (rarely) slightly beak-shaped with LV overlapping RV. A1 swimming setae long. A2 swimming setae extending well beyond terminal claws (fig. 17). G1, G3 and GM sub-equally long, G2 shorter. All these claws serrated. Md palp with slender α and β setae and strong, setulose γ seta. Third endite of Mx with two serrated tooth bristles. L6 without d2 seta. Gp of Ur slightly shorter than Ga. Sp only slightly shorter than Gp.

Comparisons with *H. incongruens* specimens reported elsewhere.— The present specimens are slightly smaller than the average level of the European specimens (usually 1.4-1.6 mm) examined by Meisch (2000), and are also smaller than those from Dianchi Lake in SW China (1.49-1.61 mm) (Yu et al., 2010). Additionally, the tiny spinules on the anterior part of the shell (Meisch, 2000; Yu et al., 2010) are only present in some of the specimens examined by the present authors.

Comparisons with other species.— (1) The posterior margin of *H. rotundata* (Bronshtein, 1928) is distinctly more broadly rounded than the anterior margin (Meisch, 2000). In *H. incongruens*, both margins are sub-equally broad. In fron., carapace of *H. rotundata* is oblique, with the greatest width of LV situated below

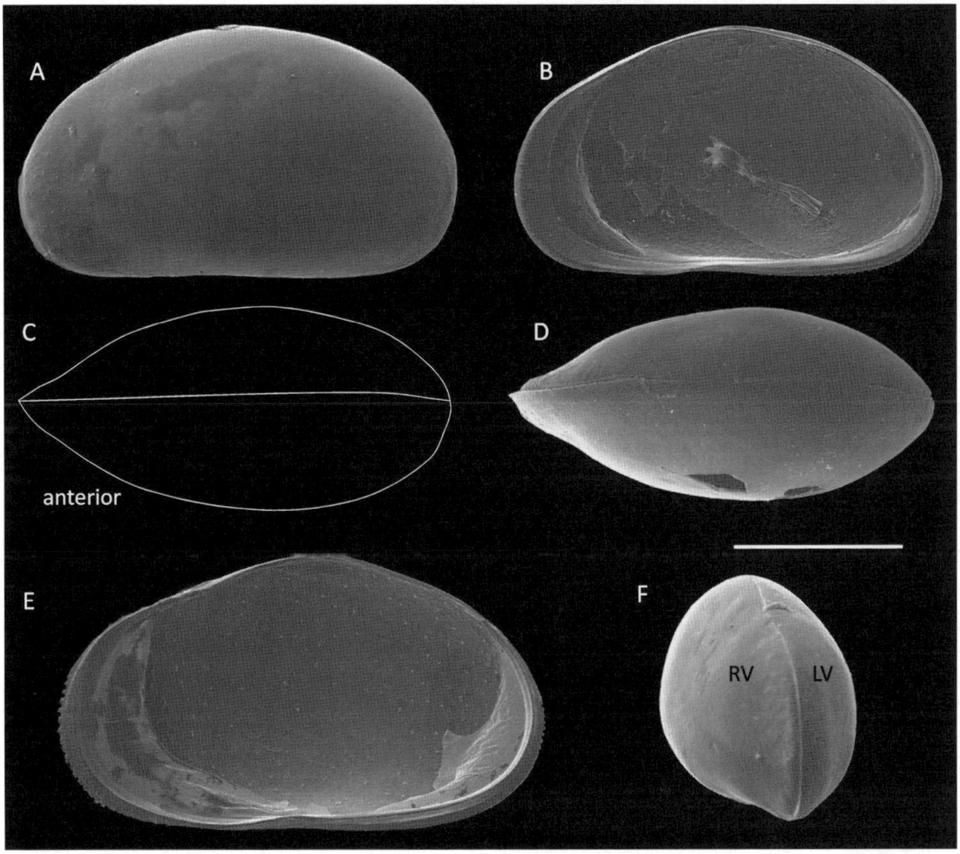

Fig. 16. *Heterocypris incongruens* (Ramdohr, 1808), female. A, LV, ext. (dyzoc252); B, RV, int. (dyzoc252); C, carapace, dor. (dyzoc252); D, carapace, dor. (dyzoc260); E, RV, int. (dyzoc311); F, carapace, oblique-fron. (dyzoc259). Scale bar = 500 μm.

mid-height. *H. incongruens* lacks this feature (fig. 16F). Meisch (2000) suggested that the beak-shaped anterior end in dor. of *H. rotundata* can also help discriminate between the two. However, while most of the present specimens have pointed anterior in dor. (fig. 16C), a few of them have slightly beak-shaped anterior ends (fig. 16D). It is suggested that the shape of the anterior end of carapace in dor. is variable within the species. (2) Males are usually present in *H. barbara* (Gauthier & Brehm, 1928) and the anterior end of this species is always distinctly beak-shaped (Meisch, 2000). (3) *H. salina* (Brady, 1868) has a distinctly arched (angled) dorsal margin, with a dorsal hump on the LV. Moreover the carapace of *H. salina* usually has a dark-brown colour pattern.

Distribution.— This is a cosmopolitan species occurring in various geographical areas, including Europe, North Africa, North America and Japan (Meisch, 2000; Smith et al., 2011). In China, subfossil and living populations were reported from

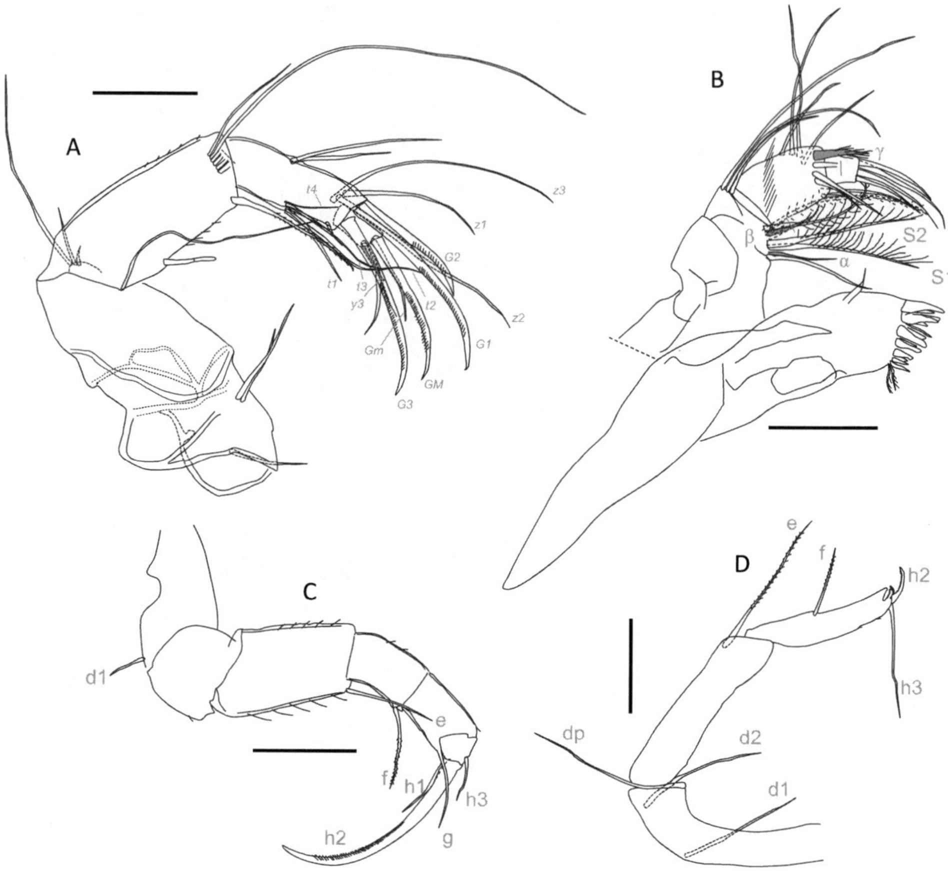

Fig. 17. *Heterocypris incongruens* (Ramdohr, 1808), female (dyzoc252). A, A2; B, Md palp and coxa; C, L6; D, L7. Scale bars = 100 μm. This figure is published in colour in the online edition of this journal, which can be accessed via http://booksandjournals.brillonline.com/content/journals/15685403.

Qilian Mountain areas (Mischke et al., 2003), Qinghai Province (Li et al., 2010), Shanghai City and Yunnan Province (Yu et al., 2010).

Remarks.— In the present study, at least two forms of *H. incongruens* were found. The specimens in site X35 are smaller than those from site X48, and the ratios between lengths of setae and carapace are also smaller than those of X48 population (unpublished data).

Heterocypris auricularis sp. nov.

2004 *Heterocypris sanukiensis* nomen nudum — Okubo: 25; fig. 10

Type locality.— A small lake formed by the dead arm of a river in Inner Mongolia (X4, fig. 1, table I).

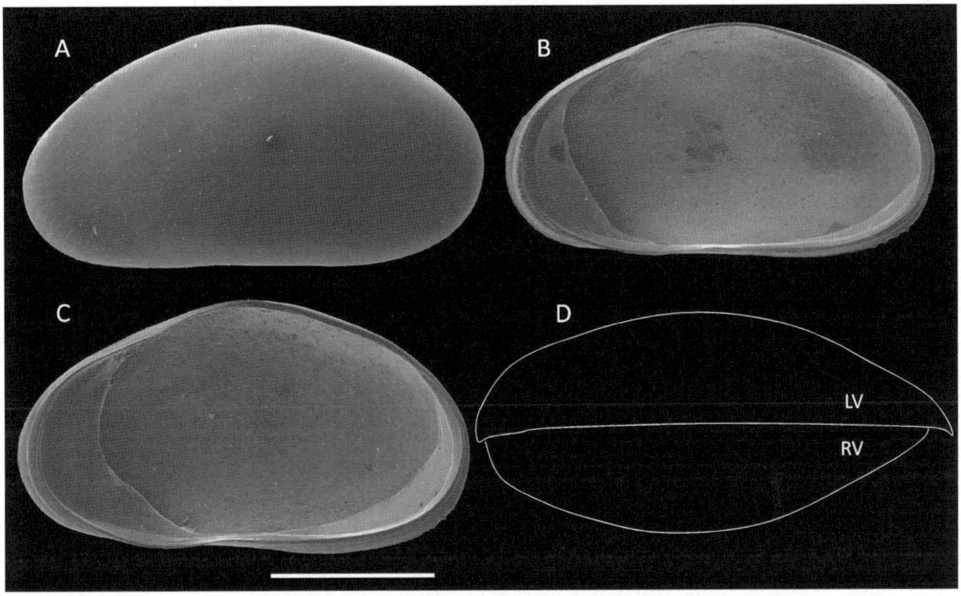

Fig. 18. *Heterocypris auricularis* sp. nov., male. A, LV, ext. (dyzoc331); B, RV, int. (dyzoc331); C, RV, int. (dyzoc350); D, carapace, dor. (dyzoc331). Scale bar = 500 μm.

Type material.— Holotype: one dissected male (dyzoc350) from the type locality. Paratype: one dissected male (dyzoc331) from the type locality.

Dimensions.— Male (LV, $n = 2$) length 1420-1450 μm, H/L ratio 0.54-0.57.

Derivation of name.— From the Latin *auricula*, referring to the small auricular-like expansion on the lobe a of Hp.

Diagnosis (based on male).— Carapace sub-ovate in lat., dorsal margin arched with greatest height situated at about mid-length (fig. 18). Sub-elliptical in dor., anterior end slightly compressed, beak-shaped with LV overlapping RV. In LV, selvage, fused zone and calcified inner lamella inclined toward the outer lamella. In RV, selvage and fused zone protruding to fit LV in closure. Postero-ventral margin of RV with 3-4 (may be more variable) small pustules. Anterior margin of RV without pustules. A2 swimming setae extending slightly beyond terminal claws (fig. 19). L5 palps unequal (fig. 20). Left trunk becoming narrower distally. Right trunk conical. Left finger slender, hook-like. Right finger inflated, triangle-shaped. Lobe a of Hp hammer-like, with a small auricular-like expansion at the outer-distal corner. Lobe b very small. Lobe h large, narrowly rounded.

Description of male.— Shell sub-ovate in lat., slightly elongated (fig. 18). Dorsal margin arched, highest point situated at mid-length or slightly behind. Both ends broadly rounded but anterior margin slightly narrower. Ventral margin slightly concave. In int., selvage significantly displaced inwardly and valve margin defined by the flange at the anterior, ventral and posterior. Fused zone narrow with

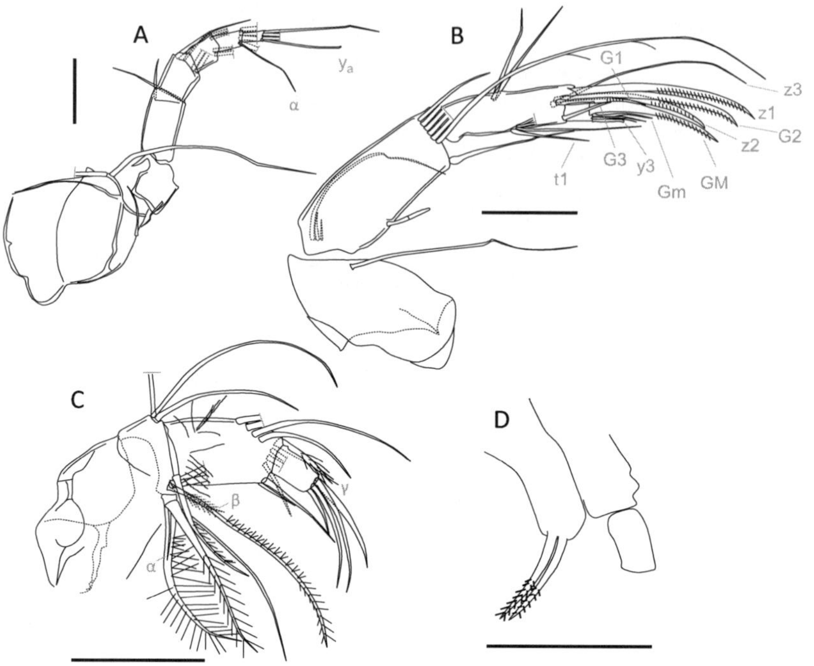

Fig. 19. *Heterocypris auricularis* sp. nov., male (dyzoc350). A, A1; B, A2; C, palp of Md; D, Mx endites and palp. Scale bars = 100 μm. This figure is published in colour in the online edition of this journal, which can be accessed via http://booksandjournals.brillonline.com/content/journals/15685403.

inconspicuous radial pore canals. In LV, selvage, fused zone and calcified inner lamella inclined toward the outer lamella. In RV, selvage and fused zone protruding to fit LV upon closure. Postero-ventral margin of RV with 3-4 small pustules. Anterior margin of RV without such pustules. In dor. carapace sub-elliptical, anterior end slightly compressed, beak-shaped with LV conspicuously longer. Posterior end blunt. Greatest width situated at about mid-length. LV overlapping RV at the anterior, ventral and posterior margins. Shell surface smooth, sparsely covered with fine, soft, short setae.

A1 (fig. 19) 8-segmented. First two segments fused forming a large, quadrate base supporting 1 dorsal seta and 2 unequal, long ventral setae. Third segment quadrate, with 1 short dorsal seta. Fourth segment elongated, with 1 dorso-apical seta and 1 shorter ventro-apical seta. Fifth segment quadrate, with 2 long, strong dorso-apical setae and 2 short, unequal ventro-apical setae. Sixth segment quadrate, with 2 long, strong dorso-apical setae and 2 short, unequal ventro-apical setae. Seventh segment elongated, with 4 strong, long apical setae and a short α seta ca. 3 times the length of terminal segment. Terminal segment slender, ventro-apical seta not claw-like.

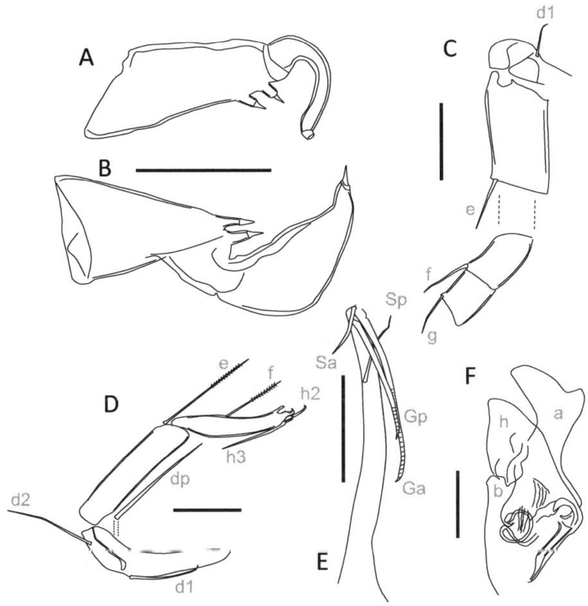

Fig. 20. *Heterocypris auricularis* sp. nov., male (dyzoc350). A, left L5 palp; B, right L5 palp; C, L6; D, L7; E, Ur; F, Hp. Scale bars = 100 μm. This figure is published in colour in the online edition of this journal, which can be accessed via http://booksandjournals.brillonline.com/content/journals/15685403.

A2 (fig. 19) longest exopodal seta extending beyond first endopodal segment. Swimming setae extending slightly beyond the claws. Sixth seta extending to middle of penultimate segment. t1 and t4 extending to end of terminal segment, t2 and t3 to mid-way of claws. G1 short, ca. 1/2 the length of the longest claws, slightly S-shaped. G3 short, ca. 3/5 of G1. G2 slightly longer than GM. z1 transformed, being the longest claw. z2 transformed, claw-like, slender, slightly shorter than GM. z3 untransformed, extending slightly beyond the claws. Gm situated to the internal of GM. z1, G2, GM strongly serrated, comb-like. z2 and G1 slightly serrated.

Md (fig. 19) coxa with narrow proximal end. Vibratory plate with ca. 6 setae. Unnamed seta on first palp segment slender, nearly smooth. S1 and S2 thick, strongly plumose. α seta slim, smooth. Setal group with 3 plumose setae. Fourth seta shorter, plumose. β short, plumose. Second segment with 3 unequal, smooth antero-apical setae. Third segment with long lateral pseudochaetae resembling those of *H. incongruens*. Third segment with 4 sub-apical setae, strong, serrated γ seta and another 3 latero-apical setae and 2 unequal postero-apical setae. Terminal segment with 3 claws and 1 short postero-apical seta.

Mx (fig. 19) palp and endite lobes elongated. Tooth bristles on third endite serrated, with the proximal one more strongly serrated.

L5 palps (fig. 20) asymmetric, with finger of right organ slender and that of left organ inflated. Left trunk elongated, with two short apical bristles tips of which poorly sclerotized. Finger slender, curved, distal end poorly chitinized. Right trunk conical, becoming narrower toward the distal part, with 2 short apical bristles tips of which poorly chitinized. Finger inflated, sub-triangular, ventral margin straight, dorsal margin curved, distal end poorly sclerotized.

L6 as in fig. 20. Setae e, f and g sub-equal. h2 (not shown) serrated along the distal half.

L7 (fig. 20) chaetotaxy typical of the genus. dp and e long. e and f setulose along the distal half.

Ur (fig. 20) with arched ramus. Ga longer than Gp, both serrated. Sa short. Sp extending to about base of claws. Posterior margin of ramus with very short, inconspicuous pseudochaetae (not shown).

ZO (not shown) considerably long, with ca. 39 rows of spines. Neither end enlarged.

Hp (fig. 20) lobe a hammer-like, outer corner of the distal part with an auricular-like expansion. Lobe b short. Lobe h narrowly rounded, shorter than a. b and h partly fused.

Remarks.— Judging by the valve outline, the number of pustules along the postero-ventral margin of RV, as well as the morphology of Hp and Ur, the present specimens are most likely conspecific with the Japanese species *Heterocypris sanukiensis* Okubo, 2004 nomen nudum. Okubo (2004) presented the photos of valves (both sexes), penis and Ur, but did not designate any type specimen for *H. sanukiensis*. Moreover, repository of those specimens (coded as 1131, 1134 and 1139 in that publication) is unknown. Therefore the species name *Heterocypris sanukiensis* Okubo, 2004 would be invalid according to the International Code on Zoological Nomenclature (article 16; International Commission on Zoological Nomenclature, 2000). In the present paper, *Heterocypris auricularis* is described as a new species, with *Heterocypris sanukiensis* Okubo, 2004 nomen nudum listed as a synonym.

Differential diagnosis.— The new species has some unique characters, e.g., the number of marginal pustules on RV, the shape of trunks of male L5 palps, and the small auricular-shaped expansion on lobe a of Hp. These features allow easy distinction with most of the other species in this genus. Hp of *H. bosniaca* Petkovski et al., 2000 also has a similar expansion, which is less prominent. Distal part of the lobe h of *H. bosniaca* is quite broad and bears characteristic striations (Petkovski et al., 2000). Moreover, finger of male right L5 palp of *H. bosniaca* has a rounded ventral margin and is not triangle-shaped as that of *H. auricularis*. Although *H. auricularis* resembles *Heterocypris barbara* (Gauthier & Brehm, 1928) in the shapes of the carapace and the male L5 palps, it can be easily distinguished from the latter by the auricular-shaped expansion on lobe a of Hp, as well as the much narrower lobe h of Hp. Besides, the number of pustles on RV is 3-4 in the new species, while that of *H. barbara* is much larger (var. *barbara*) or absent (var. *inermis*) (Meisch, 2000).

Distribution.— At the present, the species is only known from Japan and China.

Subfamily CYPRICERCINAE McKenzie, 1971

Genus *Bradleycypris* McKenzie, 1982

Bradleycypris vittata (Sars, 1903)

1903 *Cypris vittata* n. sp. Sars: 22; pl. II, fig. 3
1916 *Cypris* (*Strandesia*) *vittata* — Gurney: 339
1937 *Strandesia vittata* nov. comb. — Tressler: 197
2009 *Bradleycypris vittata* nov. comb. — Savatenalinton & Martens: 3

Material examined.— One dissected female (dyzoc360) captured alive from an artificial pond in the Olympic Park of Beijing (BNS, fig. 1, table I). One dissected female (dyzoc381) from the same locality, dead, shell whitish, blackish bands faded, LV broken, soft parts partly deteriorated. One dissected female (dyzoc404) from a small lake in Tsinghua University (Tsh, fig. 1, table I), dead, LV broken, soft parts partly deteriorated.

Dimensions.— Female (LV) length 810-840 μm (n = 3), H/L ratio 0.60-0.62 (n = 2).

Re-description of female.— Carapace (fig. 21) sub-ovate in lat. Dorsal margin arched, slightly angled at the middle. Highest point situated at or slightly in front of mid-length. Postero-dorsal margin long and straight. Anterior end broadly rounded. Posterior end narrow. Ventral margin nearly straight. In dor. sub-elliptical with greatest width at or slightly behind mid-length. In int., anterior duplicature wide, posterior duplicature narrow. Carapace asymmetric both in fron. and in caudal view, with LV lower than RV (i.e., the greatest width of LV situated near the ventral margin, while the greatest width of RV situated near the mid-height). Shell surface densely covered with small rounded pits, bearing fine, short and soft setae. Carapace greyish when alive, with 4 transverse blackish bands in lat.

A1 (fig. 22) with 8 segments. First two segments fused forming a large base. Fourth through eighth segments slender. Setae considerably thin. A1 base with one short dorsal seta and two long ventral setae. Third segment quadrate with one dorso-apical seta extending to middle of fourth segment. Fourth segment long, elongated, with one dorso-apical seta and one ventro-apical seta. Fifth segment with two long dorso-apical setae and two short ventro-apical setae. Segmentation lines between the fifth through the eighth segments oblique, directed dorso-proximally. Sixth segment with three dorso-apical setae and one short ventro-apical seta. Seventh segment with four long apical setae. α seta not seen. Eighth segment with relatively long y_a, two long medio-apical setae and one ventral seta slightly shorter than y_a, not claw-like.

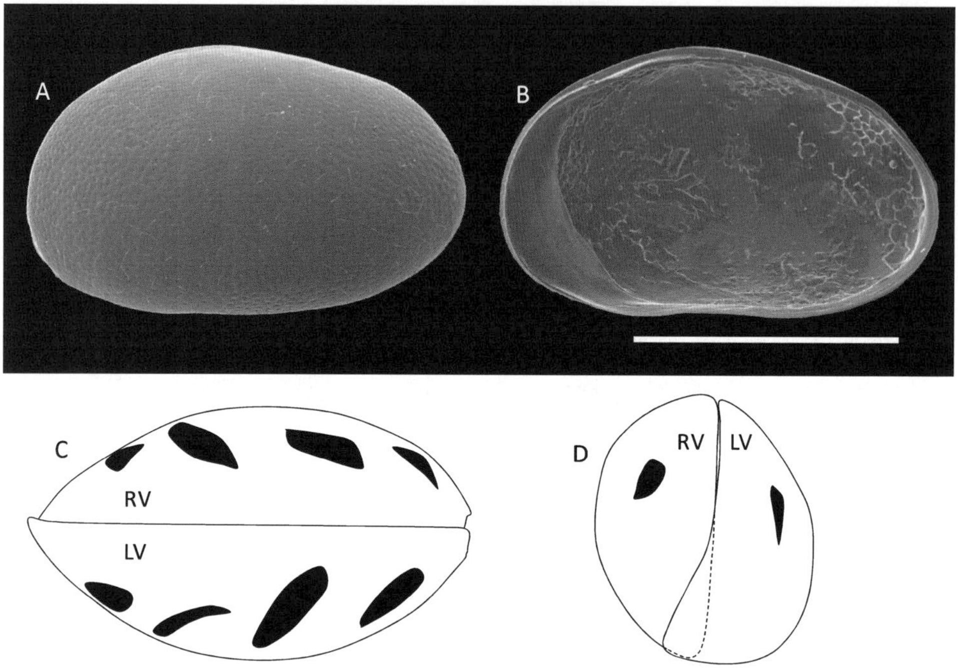

Fig. 21. *Bradleycypris vittata* (Sars, 1903), female (dyzoc360). A, LV, ext.; B, RV, int.; C, carapace, dor.; D, carapace, fron. Scale bar = 500 μm.

A2 (fig. 22) with 5 segments, penultimate segment undivided. Segments and claws slender. First segment with two sub-equal ventro-apical setae. Second segment with one ventro-subapical seta extending to middle of penultimate segment. Longest exopodal seta reaching beyond the first endopodal segment with ca. 30% of its length. The rest two exopodal setae very short. Aesthetasc Y situated slightly proximal to the middle of the third segment, relatively long. First five swimming setae sub-equally long, extending to about tips of the claws. In right A2 each of these setae bearing one pseudochaeta at the mid-length. Sixth seta not extending beyond mid-way of penultimate segment. Ventro-apical seta of third segment extending to middle of terminal segment, with a somewhat swollen base. The iv-a segment with 2 unequally long dorso-apical setae. t-setae unequally long, the longest ones (t2 and t3) extending to the tip of the claws, the shortest one (t1) slightly beyond the terminal segment. z-setae extending to about tip of the claws. G1, G3 and GM claws sub-equally long. G2 slightly shorter. Gm shorter than G2.

Md coxa with a swollen middle part (not shown). Teeth branched. Vibratory plate with 5 setae. Palp (fig. 23) with 4 segments. The unnamed seta on first segment smooth, slender. S1 and S2 sub-equally long, strong, plumose. α seta ca. 1/3 the length of S1, smooth. Setal group with three sub-equal setae extending to tips of terminal claws. The fourth seta much shorter. β seta short, strong, pointed,

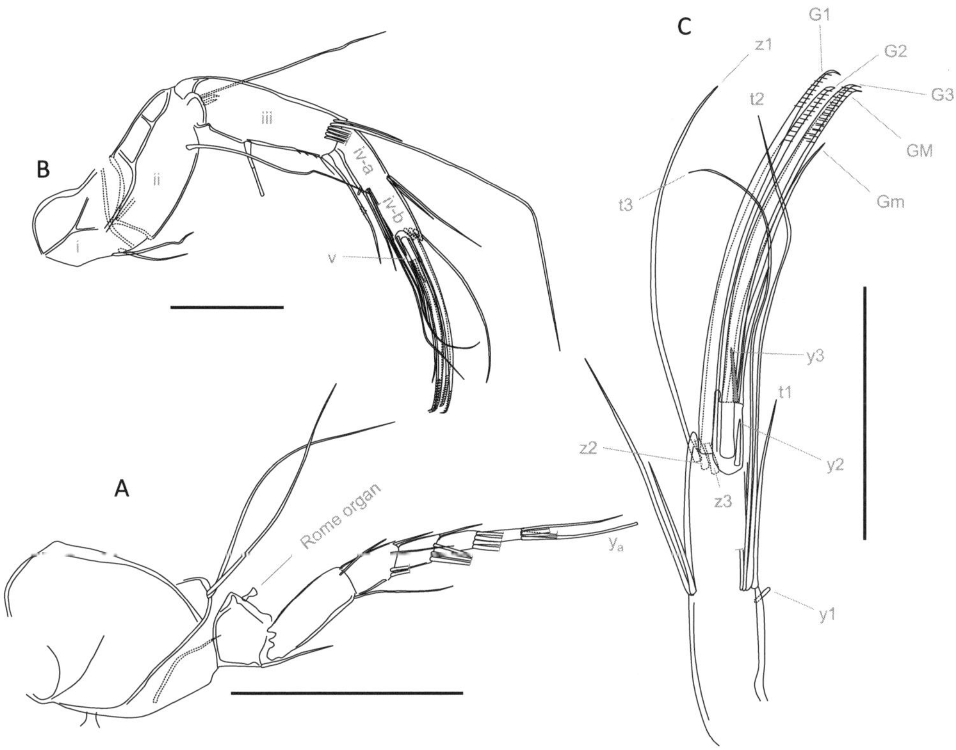

Fig. 22. *Bradleycypris vittata* (Sars, 1903), female (dyzoc360). A, A1; B, A2; C, details of distal part of A2. Scale bars = 100 μm. This figure is published in colour in the online edition of this journal, which can be accessed via http://booksandjournals.brillonline.com/content/journals/15685403.

strongly plumose. Second segment anteriorly with three unequally long smooth setae. Third segment with 4 sub-equally long sub-apical setae, as well as 5 sub-equally long apical setae, including a strong, smooth (or nearly smooth) γ seta.

Mx (fig. 23) palp 2-segmented. The two large teeth on third endite smooth (specimen dyzoc360) or slightly serrated (specimen dyzoc381). Vibratory plate with ca. 16 setae.

L5 (not shown) vibratory plate with 6 setae. Palp bearing three long terminal setae, with h2 being the longest.

L6 (fig. 23) with 6 segments. Segmentation line between second and third endopodal segments oblique. d1 significantly longer than d2. f longer than e. e longer than g. g sub-equal to h1. h1 longer than h3. All setae smooth. h2 serrated distally.

L7 (fig. 23) d1 and d2 sub-equally long. dp longer than d1 and d2, situated at sub-apical part of protopod, extending slightly beyond first endopodal segment. e extending to base of f seta. Penultimate segment not divided. f relatively short,

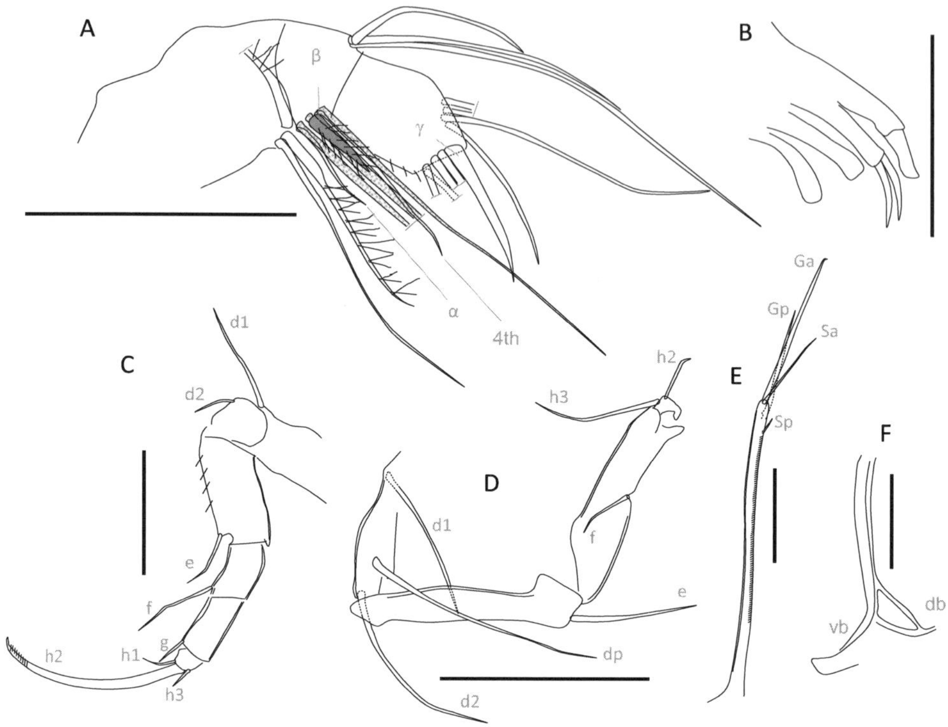

Fig. 23. *Bradleycypris vittata* (Sars, 1903), female (dyzoc360). A, palp of Md (terminal segment not shown); B, Mx endites and palp; C, L6; D, L7; E, Ur; F, Ur attachment. Scale bars = 100 μm. This figure is published in colour in the online edition of this journal, which can be accessed via http://booksandjournals.brillonline.com/content/journals/15685403.

usually inversely directed. h3 intermediately long. h2 short and straight. Pincer organ typical of the family.

Ur ramus very slender, nearly straight (fig. 23, where an approximately anterior view is given). Sp very short, never extending beyond tip of ramus. Ga strong, length ca. 1/2 of the ramus. Gp nearly as thin as Sa, slightly longer than 1/2 of Ga. Sa long, only slightly shorter than Gp. Posterior margin of ramus finely serrated. Triebel loop of attachment large, sub-triangular, situated in front of the branch, typical of the genus. Median shaft strong.

Comparisons with *B. vittata* specimens reported elsewhere.— (1) Type specimens re-described by Victor & Fernando (1981b). The present specimens match the type material in many details including the shape of carapace in lat. and in dor., the position and length of the aesthetasc Y, the morphology of A2 claws, the tooth bristles on Mx, as well as the chaetotaxy of L7 and Ur. Nonetheless, the type specimens are larger (averaging 0.93 mm in length, $n = 20$), and the boundary lines between the A1 segments are generally vertical to the dorsal and the ventral margins (Victor & Fernando, 1981b). (2) *B. vittata* from Thailand (Savatenalinton & Martens, 2010), which belong to bisexual populations, differ from the type specimens and the present ones in many aspects, although most of those illustrated limbs are based on females. For the most impressing feature, the longest A2 exopodal seta of the Thailand *B. vittata* only reaches mid-way of the first endopodal segment and has a sinuated distal part (Savatenalinton & Martens, 2010). It is not known whether this is also the case in the type material, but in the present specimens, the longest A2 exopodal seta reaches well beyond the first endopodal segment (fig. 22). It is also noted that the L6 segments of the Thailand specimens are stouter, and the setae f and h3 of this limb are longer, while the h2 claw is shorter than both the type specimen illustrated by Victor & Fernando (1981b, fig. 228) and the Beijing specimens. Ur Sp of Thailand *B. vittata* extend beyond the ramus, while those of the type and the present specimens never extend beyond the ramus. Additionally, the Thailand females also differ from the Beijing females at least in the length of the ventro-apical seta of A1 terminal segments, the ventro-apical seta of the second A2 segments, the A2 claws, d1, e, f and h1 setae of L6, f seta of L7, the Ur claws, as well as the shape and length ratios of segments on A1, A2 and L6.

Distribution.— This species is currently known from both the Palaearctic and the Oriental Regions (Martens & Savatenalinton, 2011). Previous records in China were restricted to the southern provinces (Yu et al., 2009), and this is the first record in the Palaearctic part of the country.

Subfamily CYPRIDOPSINAE Kaufmann, 1900

Genus *Cypridopsis* Brady, 1867

Cypridopsis vidua (O. F. Müller, 1776)

1776 *Cypris vidua* n. sp. O. F. Müller: 199
1821 *Cypris pincta* — Straus: 59; figs. 17-19
1869 *Cypridopsis obesa* — Brady & Robertson: 364; pl. 18
1900 *Cypridopsella helvatica* — Kaufmann: 310; pls. 19, 22
1900 *Cypridopsella tumida* — Kaufmann: 313; pls. 19, 22
1900 *Cypridopsis parva* — G. W. Müller: 81; pl. 18
1943 *Cypridopsis mariae* — Rome: 6
1977 *Cypridopsis parvoides* — J. M. Martens: 87; figs. 1-8; pls. 1, 2

Material examined.— Two dissected females (dyzoc105-106) from a medium-sized lake (KM, fig. 1, table I) in Beijing. Three dissected females (dyzoc361-363) from an artificial pond (BNS, fig. 1, table I) in Beijing.

Dimensions.— Female (LV, $n = 4$) length 610-690 μm, H/L ratio 0.58-0.66.

Brief description.— When alive, this species can be identified based on the carapace alone (Meisch, 2000). Carapace stout (fig. 24) carrying four transverse blackish bands on each valve that meet at the dorsal margin. Posterior duplicature of both valves with double folded inner list (fig. 24B). Tiny pustules present along the anterior margin of RV (fig. 24D). A2 swimming setae extending somewhat

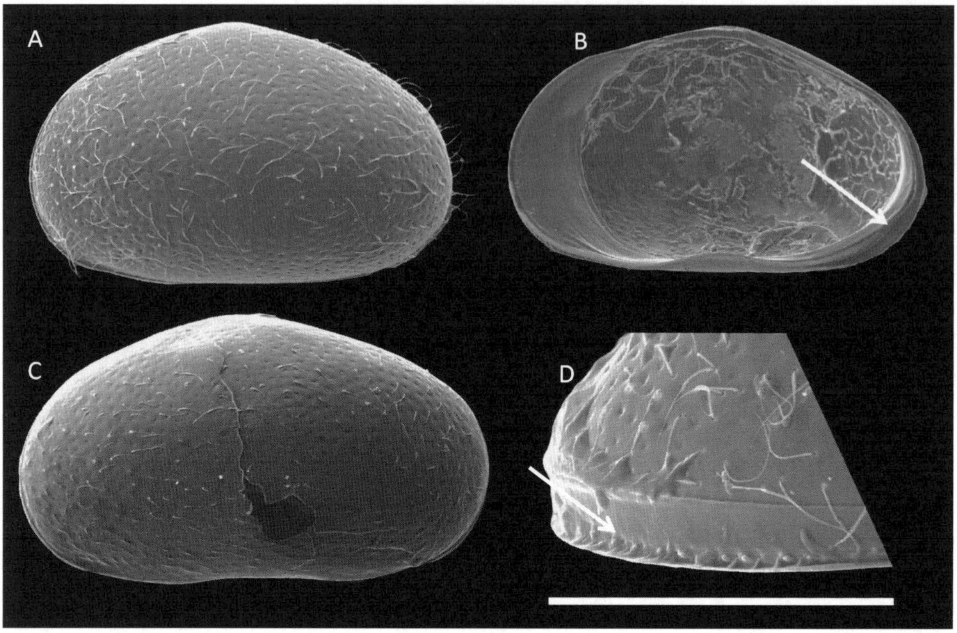

Fig. 24. *Cypridopsis vidua* (O. F. Müller, 1776), female. A, LV, ext. (dyzoc362); B, RV, int. (dyzoc362), note the inner list; C, LV, ext. (dyzoc105); D, RV, dor. (dyzoc363), showing small pustles on the extro-anterior margin; Scale bar = 500 μm for A, B and C, 141 μm for D.

beyond the terminal claws. Md palp with plumose γ seta. L6 d1 absent. h3 absent or very short (fig. 25).

Distribution.— This is a cosmopolitan species (Meisch, 2000). In China, *C. vidua* was reported from the western (e.g., Qilian Mountains, Mischke et al., 2003), north-central (Beijing, Pang, 1996) and southeastern (Fujian, Sars, 1903) areas, but most of the records were sub-recent.

Genus *Plesiocypridopsis* Rome, 1965

Plesiocypridopsis newtoni (Brady & Robertson, 1870)

1870 *Cypridopsis newtoni* n. sp. Brady & Robertson: 14; pl. 7
1898 *Cypridopsis aldabrae* — G. W. Müller: 281; pl. 18
1965 *Cypridopsis* (*Plesiocypridopsis*) *newtoni* nov. comb. — Rome: 51
1971 *Plesiocypridopsis newtoni* nov. comb. — McKenzie: 164

Material examined.— Four dissected females (dyzoc339, dyzoc340, dyzoc345, dyzoc346) from a small lake formed by the dead arm of a river (X4, fig. 1, table I).

Dimensions.— Female (LV) length 920-950 μm ($n = 4$), H/L ratio 0.59-0.61 ($n = 2$).

Brief description.— Carapace sub-triangular in lat. (fig. 26). Dorsal margin strongly arched with the highest point situated anterior to middle of the shell.

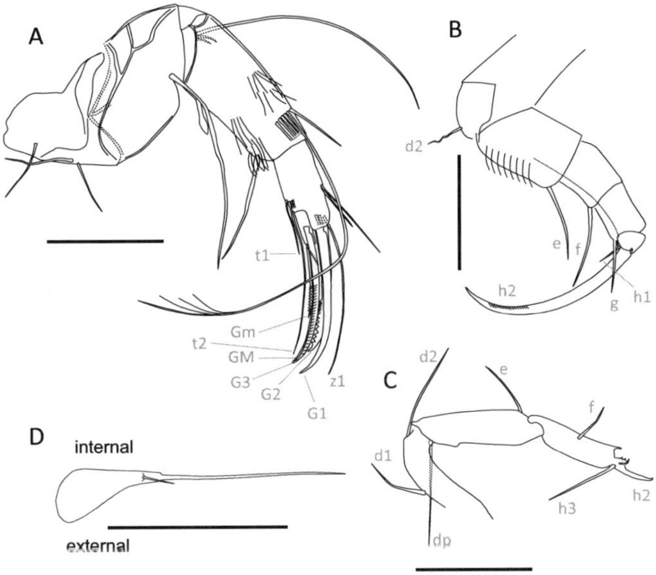

Fig. 25. *Cypridopsis vidua* (O. F. Müller, 1776), female. A, A2 (dyzoc105); B, L6 (dyzoc105); C, L7 (dyzoc105); D, Ur (dyzoc361). Scale bars = 100 μm. This figure is published in colour in the online edition of this journal, which can be accessed via http://booksandjournals.brillonline.com/content/journals/15685403.

Anterior margin broad. Posterior margin narrow, sometimes slightly pointed. Ventral margin concave, with an inward lobe-like expansion at the middle. Shell surface densely pitted, covered with many setae. Extero-anterior margin of RV with small pustules. Anterior duplicature broad. Posterior duplicature narrow. A1 with very large basal segment and long swimming setae. y_a very long. A2 with plumose swimming setae reaching beyond the terminal claws (fig. 27). G1 significantly longer and stronger than the rest claws. G3 usually slightly dorsally directed. Setal group of Md palp with 3 smooth setae. α seta short and smooth. β short, strong and setulose. γ very strong. Third endite of Mx bearing two serrated teeth. L6 d1 absent, d2 short, f setulose, h2 claw serrated, h3very short. L7 with d1 shorter than d2 and dp. e and f relatively short. Pincer organ typical of the family. Ur consiting of a cylindrical base with a short sub-apical seta and distally a long flagelliform structure. An unnamed strong hook structure situated adjacent to Ur.

Distribution.— *Plesiocypridopsis newtoni* is widespread across Eurasia and northern Africa, but it has not been reported from China before. This is also the first report of the genus *Plesiocypridopsis* in this country.

Remarks.— Zhai (2010) and Zhai et al. (2011) reported *Cypridopsis* sp. from the Holocene sediment of Hulun Lake of Inner Mongolia. The identification was based on shell morphology. Based on careful re-examination of the shells, some of these specimens probably belong to *Plesiocypridopsis newtoni*, or at least belong to

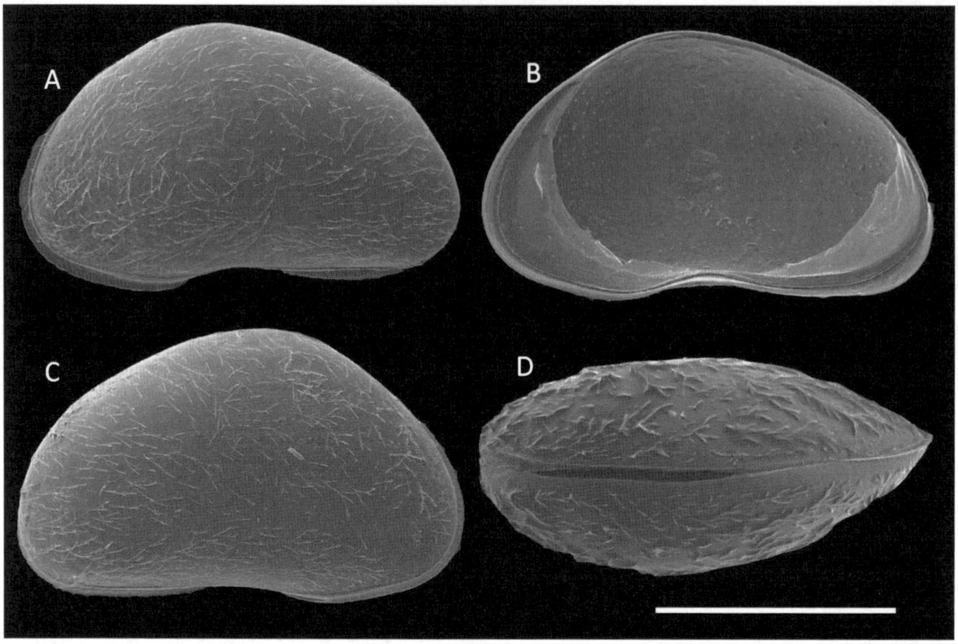

Fig. 26. *Plesiocypridopsis newtoni* (Brady & Robertson, 1870), female. A, LV, ext. (dyzoc339); B, RV, int. (dyzoc339); C, RV, ext. (dyzoc340); D, carapace, oblique-dor. (dyzoc346). Scale bar = 500 μm.

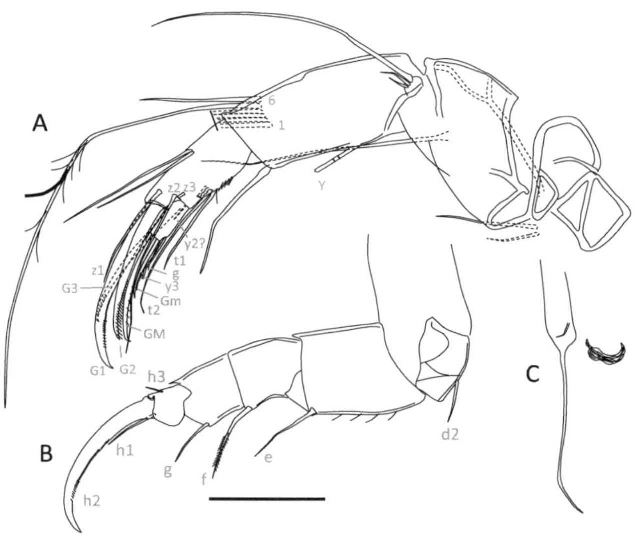

Fig. 27. *Plesiocypridopsis newtoni* (Brady & Robertson, 1870), female (dyzoc339). A, A2; B, L6; C, Ur and a hook structure adjacent to it. Scale bar = 100 μm. This figure is published in colour in the online edition of this journal, which can be accessed via http://booksandjournals.brillonline.com/content/journals/15685403.

the genus *Plesiocypridopsis*. Further study of the soft part morphology are needed to confirm the taxonomy of the Cypridopsinae specimens from Hulun Lake.

Superfamily CYTHEROIDEA Baird, 1850
Family LIMNOCYTHERIDAE Klie, 1938
Subfamily LIMNOCYTHERINAE Klie, 1938
Genus *Limnocythere* Brady, 1868
Limnocythere inopinata (Baird, 1843)

1843 *Cythere inopinata* n. sp. Baird: 195
1867 *Limnocythere inopinata* nov. comb. — Brady: 121
1888 *Limnocythere incisa* — Dahl: 617; pl. 17
1963 *Limnocythere sappaensis* — Staplin: 1197; pl. 160
1964 *Limnocythere binoda* — Huang: 252; pl. 5
1972 *Limnocythere inopinata* s.sp. *orientalis* — Ghetti: 140; pl. 11
1987 *Limnocythere inopinata pleistocenica* — Krstić: 216; pls. 1-2

Material examined.— One dissected female (dyzoc380) from a small lake in Tsinghua University (Tsh, fig. 1, table I) in Beijing. One dissected female (dyzoc406) from a small lake formed by a dead river (X4, fig. 1, table I) in Inner Mongolia.

Dimensions.— Female (LV, $n = 2$) length 500-540 μm, H/L ratio 0.54-0.55.

Brief description.— Female carapace sub-rectangular in lat. (fig. 28). Dorsal margin slightly arched or straight. Ventral margin concave. Anterior end more broadly rounded than posterior end. The fused zone highly laterally compressed, bearing long radial-pore canals, distributed along the whole anterior margin, the antero-ventral, postero-ventral and part of posterior margins. Shell surface more or less reticulated. Adductor muscle scars typical of the family, situated in an elongated depression below the transverse sulcus. Male carapace longer and more inflated than female in the posterior half, especially at the postero-dorsal part (see Yin et al., 1999; Zhai et al., 2010). Both male and female with variable noding

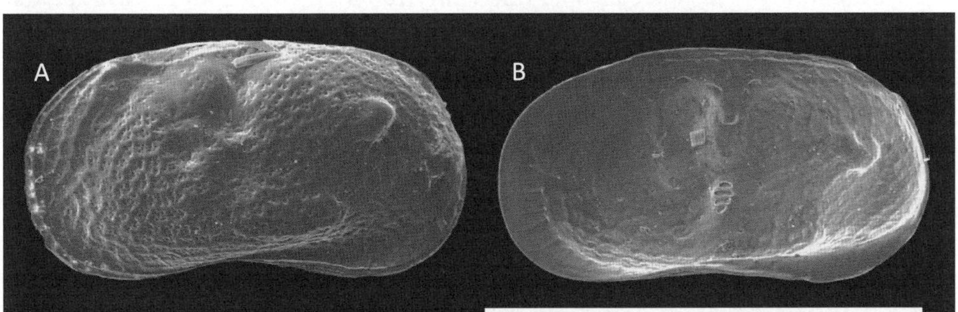

Fig. 28. *Limnocythere inopinata* (Baird, 1843), female (dyzoc380). A, LV, ext.; B, RV, int. Scale bar = 500 μm.

patterns (Yin et al., 1999; Zhai et al., 2010). Female limbs with less diagnostic features, typical of the genus.

Distribution.— This species is widespread across Eurasia (Meisch, 2000) and was found in North America (Forester et al., 2005). It was also found in Africa south of Sahara (Martens, 1994). In China, *L. inopinata* has been reported from various localities, including Qilian Mountains (Mischke et al., 2003) and Lake Qinghai (Yin et al., 1999; Li et al., 2010) in western provinces, Inner Mongolia (Wang & Ji, 1995; Zhai et al., 2010) in northeastern areas, as well as Taihu Lake (Zhao & Wang, 1998) in southern China. In general, however, *L. inopinata* is more abundant in arid and semi-arid regions (western and northern China) than in humid regions (southern and eastern China).

Limnocythere stationis Vávra, 1891

1891 *Limnocythere stationis* n. sp. Vávra: 108; fig. 38

Material examined.— One dissected female (dyzoc189) from a medium-sized lake in Beijing (KM, fig. 1, table I).

Dimensions.— The specimen (dyzoc189) found from Beijing has a length (RV) of 430 μm and a height of 250 μm, which is close to the average size of European specimens (Meisch, 2000). Those from Taihu Lake (Yu et al., 2005) (ranging from 0.33 to 0.35 mm), and the syntype described by Vávra (1891) (0.36 mm) were considerably smaller.

Brief description (slightly modified from Meisch (2000)).— Carapace very small for the genus, stout in lat. Anterior higher than posterior. Dorsal margin straight, sloping posteriorly. Postero-dorsal part of RV usually with 1-3 denticles directed posteriorly or upwards. Each valve usually with two dorsal tubercles, one postero-ventral ala, and one central tubercle near the central muscle scars. Valve surface reticulated. L7 strikingly small, shorter than L5 and L6.

Distribution.— This species was frequently reported from Europe, Russia and Africa (Martens, 1984; Meisch, 2000). In China, it was only reported from Taihu Lake of Jiangsu Province (Yu et al., 2005). This species is very rare in the site KM (fig. 1, table I) investigated in the present study. It is unclear whether it has colonized this lake successfully. Alternatively, it might be an occasional visitor to the lake.

DISCUSSION

The 12 species recorded from northern China in this study provide new information on the zoogeography of living non-marine ostracods. Despite that half of the species (*Cypridopsis vidua* (O. F. Müller, 1776), *Heterocypris incongruens*

(Ramdohr, 1808), *Limnocythere inopinata* (Baird, 1843), *L. stationis* Vávra, 1891, *Physocypria kraepelini* G. W. Müller, 1903 and *Plesiocypridopsis newtoni* (Brady & Robertson, 1870)) have been recorded from various other parts of the world and display wide geographical distribution (Martens & Savatenalinton, 2011), some of them have been rarely, or not at all found in China before. For example, previous data indicated that *P. kraepelini* is widespread across Eurasia and North America (Meisch, 2000; Yu et al., 2005, 2010; Martens & Savatenalinton, 2011), but it was not known from the Chinese territory north of Qinling Mountain (i.e., the Palaearctic part of China) until this study. The present study also presents the first reports of *B. vittata*, *I. angulata*, and *L. stationis* in Palaearctic China. *P. newtoni*, another widespread species in the Palaearctic and Afrotropical Regions, is reported from China for the first time.

Four species known from other East Asian countries are found in China. *Candona quasiakaina* Karanovic & Lee, 2012 is reported for the first time outside South Korea. *Fabaeformiscandona myllaina* Smith & Kamiya, 2007 and *Heterocypris sanukiensis* Okubo, 2004 nomen nudum (as *Heterocypris auricularis* sp. nov. in this paper) are new for areas outside of Japan. *Ilyocypris mongolica* Martens, 1991 is reported for the first time outside of Mongolia. It is possible that these ostracods are endemic to East Asia, but more data are needed to define the geographical occurrence of these relatively less known species.

It should be noted that the occurrence of the ostracods recorded in this study may not faithfully represent their natural distributions. The Beijing specimens were collected from artificial lakes in the city. The water in these lakes comes from man-made dams to the north and northwest of the city, which were in turn fed by brooks and rivers arising from nearby mountain areas. The aquatic plants and animals in these urban lakes may thus have come from the above water-bodies, or even have been introduced from more distant places. The sampling area in Inner Mongolia is very sparsely populated, but the possibility of man-conducted dispersal cannot be ruled out, since the sampling sites were close to the paved road.

China has a vast continental territory and has diverse geographical conditions. Various types of water-bodies, ranging from small puddles to large lakes, from underground flows to great rivers, from fresh to hyper-saline waters are present. The ostracod fauna residing in this land is thus expected to be diverse. Many of these species, however, are poorly known to science and a more intense investigation of China's extant ostracod fauna is urgently required.

ACKNOWLEDGEMENTS

We dedicate this paper to Prof. Eugen Karl Kempf, in honour of his great efforts during the past decades in compiling and updating the Kempf Ostracoda

Database. Dr. Renate Matzke-Karasz, Dr. Robin J. Smith, Dr. Steffen Mischke and one anonymous reviewer contributed important comments and suggestions. Dr. Finn A. Viehberg and Dr. Renate Matzke-Karasz arranged the reviews. The authors are thankful to Dr. J. C. von Vaupel Klein and Dr. M. Jarmuszewski for their patience in preparing the proof of this paper. Deep thanks are due to Dr. Ivana Karanovic and Dr. Robin J. Smith for sharing important references and taxonomic data. Dr. Na Yu and her student Qiang Kong from the East China Normal University communicated with us on some taxonomic issues, and provided important references. This research is supported by Grants NSFC41102110 and NSFC41130101.

REFERENCES

BAIRD, W., 1843. Notes on the British Entomostraca. Zoologist, **1**: 193-197.

BRADY, G. S., 1867. A synopsis of the recent British Ostracoda. The Intellectual Observer, **12**: 110-130.

BRADY, G. S. & D. ROBERTSON, 1869. Notes of a week's dredging in the west of Ireland. Annals and Magazine of Natural History, series 4, **3**: 353-374.

— — & — —, 1870. The Ostracoda and Foraminifera of tidal rivers, part 1. Annals and Magazine of Natural History, series 4, **6**: 1-33.

CHEN, S., 1990. Description of a male *Ilyocypris angulata* Sars (Crustacea: Ostracoda). Acta Micropalaeontologica Sinica, **7**: 253-256. [In Chinese.]

CLAUS, C., 1892. Beiträge zur Kenntnis der Süßwasser-Ostracoden: I. Über den Körper- und Gliedmaßenbau der Cypriden nebst Bemerkungen über einzelne innere Organe derselben. Arbeiten aus dem Zoologischen Institut der Universität Wien und der Zoologischen Station in Triest, **10**: 147-216.

DAHL, F., 1888. Die Cytheriden der westlichen Ostsee. Zoologische Jahrbücher. Abteilung für Systematik, Geographie und Biologie der Tiere, **3**: 597-638.

DUBOWSKY, A., 1927. Über zwei neue Ostracoden im Bassin des Sewerny-(Nord)-Donetz (Nebenfluss des Don). Zoologischer Anzeiger, **73**: 81-88.

FORESTER, R. M., A. J. SMITH, D. F. PALMER & B. B. CURRY, 2005. North American Non-Marine Ostracode Database "NANODe" Version 1, available online at http://www.kent.edu/NANODe. (Kent State University, Kent, OH).

GAUTHIER, H., 1938. Ostracodes continentaux récoltés par M Monod au Sahara occidental et en Mauritanie. Bulletin de la Société des Sciences Naturelles du Maroc, **18**: 39-61.

GAUTHIER, H. & V. BREHM, 1928. Ostracodes et Cladocères de l'Algérie et de la Tunisie. (3 note). Bulletin de la Société d'Histoire Naturelle de l'Afrique du Nord, **19**: 114-121.

GHETTI, P. F., 1972. L'ostracodofauna di alcune risaie turche e persiane. L Atenec Parmense, **8**: 117-147.

GURNEY, R., 1916. On some fresh-water Entomostraca from Ceylon. Proceedings of the Zoological Society of London, **86**: 333-343.

HORNE, D. J., B. B. CURRY, L. D. DELORME, K. MARTENS, A. J. SMITH & R. J. SMITH, 2011. OMEGA: the ostracod metadatabase of environmental and geographical attributes. Joannea Geologie und Paläontologie, **11**: 80-84.

HOU, Y. & Y. GOU, 2007. Fossil Ostracoda of China, **2**: 1-798. (Science Press, Beijing). [In Chinese.]

HOU, Y., Y. GOU & D. CHEN, 2002. Fossil Ostracoda of China, **1**: 1-1090. (Science Press, Beijing). [In Chinese.]

HSIAO, Y. & S. YOUNG, 2001. Freshwater ostracods from Taiwan. Abstract of the Fisheries Society of Taiwan 2001 Annual Conference: **54**. [In Chinese.]

HUANG, B. R., 1964. Ostracods from Gansen District, Chaidam Basin. Acta Palaeontologica Sinica, **12**: 241-270. [In Chinese.]

INTERNATIONAL COMMISSION ON ZOOLOGICAL NOMENCLATURE, 2000. International Code on Zoological Nomenclature, Fourth Edition, available online at http://www.nhm.ac.uk/hosted-sites/iczn/code/.

KARANOVIC, I., 2006. Recent Candoninae (Crustacea, Ostracoda, Candonidae) of North America. Records of the Western Australian Museum, **71**(Supplement): 1-75.

— —, 2012. Recent freshwater ostracods of the World: 1-608. (Springer-Verlag, Berlin).

KARANOVIC, I. & W. LEE, 2012. A review of candonid ostracods (Crustacea: Ostracoda: Podocopida) from East Asia, with descriptions of five new species from South Korea. Zootaxa, **3368**: 7-49.

— — & — —, 2013. On the ostracod genus *Ilyocypris*, with description of one new species from Korea and the first report of males of *I. bradyi* (Crustacea: Ostracoda: Podocopida). Proceedings of the Biological Society of Washington, **126**: 39-71.

KAUFMANN, A., 1900. Cypriden und Darwinuliden der Schweiz. Revue Suisse de Zoologie, **8**: 209-423.

KLIE, W., 1941. Süßwasserostracoden aus Südosteuropa. Zoologischer Anzeiger, **133**: 233-244.

KRSTIĆ, N., 1987. Vrsta *Limnocythere inopinata* 1 Nene Podvrste (*Limnocythere inopinata* species and its subspecies). Radovi Geoinstituta, **20**: 213-222.

LI, X., W. LIU, L. ZHANG & Z. SUN, 2010. Distribution of Recent ostracod species in the Lake Qinghai area in northwestern China and its ecological significance. Ecological Indicators, **10**: 880-890.

LINDNER, E., 1920. Neue mazedonische Ostracoden. Zoologischer Anzeiger, **52**: 37-41.

MARTENS, J. M., 1977. *Cypridopsis parvoides* n. sp., ein neuer Ostracode aus Norddeutschland (Crust.: Ostracoda: Podocopida). Abhandlungen und Verhandlungen des Naturwissenschaftlichen Vereins in Hamburg, **20**: 87-98.

MARTENS, K., 1984. On the freshwater ostracods (Crustacea, Ostracoda) of the Sudan, with special reference to the Red Sea Hills, including a description of a new species. Hydrobiologia, **110**: 137-161.

— —, 1987. Homology and functional morphology of the sexual dimorphism in the antenna of *Sclerocypris* Sars, 1924 (Crustacea, Ostracoda, Megalocypridinae). Bijdragen tot de Dierkunde, **57**: 183-190.

— —, 1991. On a small collection of non-marine ostracods from Mongolia, with the description of a new species (Crustacea, Ostracoda). Miscellanea Zoologica Hungarica, **6**: 53-60.

— —, 1994. Summary of the morphology, taxonomy and distribution of *Limnocythere inopinata* (Baird, 1843) (Ostracoda, Limnocytherinae). In: D. J. HORNE & K. MARTENS (eds.), The evolutionary ecology of reproductive modes in non-marine Ostracoda: 17-22. (Greenwich University Press, Greenwich).

MARTENS, K. & H. J. DUMONT, 1984. The ostracod fauna (Crustacea, Ostracoda) of Lake Donk (Flanders): a comparison between two surveys 20 years apart. Biological Journal of Dodonaea, **52**: 95-111.

MARTENS, K. & S. SAVATENALINTON, 2011. A subjective checklist of the Recent, free-living, non-marine Ostracoda (Crustacea). Zootaxa, **2855**: 1-79.

MCKENZIE, K. G., 1971. Species list of South African freshwater Ostracoda with an appendix listing museum collections and some further determinations. Annals of the South African Museum, **57**: 157-213.

MEISCH, C., 2000. Freshwater Ostracoda of western and central Europe. In: J. SCHWOERBEL & P. ZWICK (eds.), Süßwasserfauna von Mitteleuropa, **8**: 1-522. (Spektrum Akademischer Verlag, Heidelberg).

MISCHKE, S., U. BÖSSNECK, B. DIEKMANN, U. HERZSCHUH, H. JIN, A. KRAMER, B. WÜNNEMANN & C. ZHANG, 2010. Quantitative relationship between water-depth and sub-fossil ostracod assemblages in Lake Donggi Cona, Qinghai Province, China. Journal of Paleolimnology, **43**: 589-608.

MISCHKE, S., U. HERZSCHUH, H. KÜRSCHNER, D. FUCHS, J. ZHANG, F. MENG & Z. SUN, 2003. Sub-recent Ostracoda from Qilian Mountains (NW China) and their ecological significance. Limnologica, **33**: 280-292.

MÜLLER, G. W., 1898. Ergebnisse einer zoologischen Forschungsreise in Madagaskar und Ost-Afrika 1889-1895 von Dr A Voeltzkow: Die Ostracoden. Abhandlungen der Senckenbergischen Naturforschenden Gesellschaft, **21**: 255-296.

— —, 1900. Deutschlands Süßwasser-Ostracoden. Zoologica, Original-abhandlungen aus dem Gesamtgebiete der Zoologie, **12**: 1-112.

— —, 1903. Hamburgische Elbuntersuchungen — 3: Ostracoden. Mitteilungen aus dem Naturhistorischen Museum in Hamburg, **19**: 163-167.

MÜLLER, O. F., 1776. Zoologiae danicae prodromus, seu animalium daniae et norvegiae indigenarum characters, nomina, et synonyma imprimis popularium: I-XXXII, 1-282. (Typis Hallageriis).

NEALE, J. W. & Y. ZHAO, 1991. A new species of the freshwater genus *Notodromas* (Ostracoda) from Heilongjiang, Northeast China. Acta Micropalaeontologica Sinica, **8**: 153-170. [In Chinese.]

OKUBO, I., 1974. Freshwater Ostracoda from Japan, XI. *Ilyocypris angulata* Sars, 1903. The Research Bulletin of Okayama Shujitsu Junior College, **4**: 46-50.

— —, 2004. Nihon tansui san kaimijinko rui ni tsuite: 1-72. (Kabushikigaisha Sanmon Insatsusho, Okayama). [In Japanese.]

PANG, Q., 1996. Ostracoda. In: C. HUANG, Z. KONG, Q. PU, L. MIN, C. JI, G. SHEN, Q. PANG, C. LIU, Y. LIU & X. ZHAI (eds.), Kunming Lake Sediments (over 3500 yrs), Summer Palace, Beijing: 77-90. (China Ocean Press, Beijing). [In Chinese.]

PETKOVSKI, T., B. SCHARF & D. KEYSER, 2000. New and little known ostracods of the genus *Heterocypris* (Crustacea, Ostracoda) from the Balkan Peninsula. Limnologica, **30**: 45-57.

RAMDOHR, F. A., 1808. Über die Gattung *Cypris* Müll. und drei zu derselben gehörige neue Arten. Magazin der Gesellschaft Naturforschender Freunde zu Berlin für die Neuesten Entdeckungen in der Gesammten Naturkunde, **2**: 85-93.

ROME, D. R., 1943. *Physocypria perlata*, *Cythereis rivulorum*, *Cypridopsis mariae*, trois espèces nouvelles d'Ostracodes d'eau douce. Bulletin du Musée Royal d'Histoire Naturelle de Belgique, **19**: 1-8.

— —, 1965. Crustacea: Ostracoda. In: B. HANSTROEM, P. BRINCK & G. RUDEBECK (eds.), South African animal life. Results of the Lund University Expedition in 1950-1951, **11**(1): 9-58. (Swedish Natural Science Research Council, Stockholm).

SARS, G. O., 1903. Freshwater Entomostraca from China and Sumatra. Archiv for Mathematik og Naturvidenskab, **25**: 1-44.

SAVATENALINTON, S. & K. MARTENS, 2009. Generic revision of Cypricercinae McKenzie, 1971 (Crustacea, Ostracoda), with the description of three new genera and one new species and a phylogenetic analysis of the subfamily. Hydrobiologia, **632**: 1-48.

— — & — —, 2010. On the subfamily Cypricercinae McKenzie, 1971 (Crustacea, Ostracoda) from Thailand, with the description of six new species. Zootaxa, **2379**: 1-77.

SCHÄFER, H.-W., 1934. Beiträge zur Ostracodenkunde. Bulletin de la Société des Sciences de Cluj (Roumanie), **7**: 559-565.

SMITH, R. J. & H. JANZ, 2008. Recent species of the Family Candonidae (Ostracoda, Crustacea) from the ancient Lake Biwa, Central Japan. Journal of Natural History, **42**: 2865-2922.

SMITH, R. J., H. JANZ & I. OKUBO, 2011. Recent Cyprididae and Ilyocyprididae (Crustacea: Ostracoda) from Lake Biwa, Japan, including a summary of the lake's ostracod fauna. Zootaxa, **2874**: 1-37.

SMITH, R. J. & T. KAMIYA, 2007. Copulatory behaviour and sexual morphology of three *Fabaeformiscandona* Krstić, 1972 (Candoninae, Ostracoda, Crustacea) species from Japan, including descriptions of two new species. Hydrobiologia, **585**: 225-248.

STAPLIN, F. L., 1963. Pleistocene Ostracoda of Illinois. Part 2: subfamilies Cyclocypridinae, Cypridopinae, Ilyocyprinae; families Darwinulidae and Cytheridae. Journal of Paleontology, **37**: 1164-1203.

STRAUS, H. E., 1821. Mémoire sur les *Cypris*, de la classe des crustacés. Memoires du Museum d'Histoire Naturelle, **7**: 33-61.

TRESSLER, W. L., 1937. Ostracoda. Internationale Revue der gesamten Hydrobiologie und Hydrographie, **34**: 188-207.

VAN DER MEEREN, T., J. E. ALMENDINGER, E. ITO & K. MARTENS, 2010. The ecology of ostracodes (Ostracoda, Crustacea) in western Mongolia. Hydrobiologia, **641**: 253-273.

VÁVRA, W., 1891. Monographie der Ostracoden Böhmens. Archiv der Naturwissensch. Landesdurchforschung von Böhmen, **8**: 1-116.

VICTOR, R. & C. H. FERNANDO, 1981a. *Ilyocypris dentifera* Sars, 1903 versus *Ilyocypris angulata* Sars, 1903. A taxonomic study of two species of *Ilyocypris* Brady and Norman, 1889 (Crustacea, Ostracoda) described from China. Canadian Journal of Zoology, **59**: 1103-1110.

— — & — —, 1981b. Freshwater ostracods (Crustacea: Ostracoda) of the genus *Strandesia* Vávra, 1895 from Malaysia, Indonesia and the Philippines. Archiv für Hydrobiologie, **58**(Supplement): 469-522.

WANG, S. & L. JI, 1995. Paleolimnology of Hulun Lake: 1-125. (University of Science and Technology of China Press, Hefei). [In Chinese.]

WROZYNA, C., P. FRENZEL, P. STEEB, L. ZHU & A. SCHWALB, 2009. Recent lacustrine Ostracoda and a first transfer function for palaeo-water depth estimation in Nam Co, southern Tibetan Plateau. Revista Española de Micropaleontología, **41**: 1-20.

YIN, Y., W. GEIGER & K. MARTENS, 1999. Effects of genotype and environment on phenotypic variability in *Limnocythere inopinata* (Crustacea: Ostracoda). Hydrobiologia, **400**: 85-114.

YIN, Y. & K. MARTENS, 1997. On a new species of *Fabaeformiscandona* Krstic, 1972 (Crustacea, Ostracoda) from China, with a preliminary checklist of recent Chinese nonmarine ostracods. Hydrobiologia, **357**: 117-128.

YU, N., Q. ZHAO & X. CHENG, 2010. New records of freshwater Ostracoda from the Dianchi Lake, Yunnan Province, SW China. Acta Micropalaeontologica Sinica, **27**: 344-350. [In Chinese, with English abstract.]

YU, N., Q. ZHAO, E. LI, S. CHEN & L. CHEN, 2009. An undated and annotated checklist of recent nonmarine ostracods from China. Zootaxa, **2067**: 29-50.

YU, N., Q. ZHAO, E. L. SCHORNIKOV & L. CHEN, 2005. Recent ostracods from the Taihu Lake. Acta Micropalaeontologica Sinica, **22**: 143-151. [In Chinese.]

ZHAI, D., 2010. Holocene hydrological and climatic changes reflected by the ostracode record of Hulun Lake, northeastern Inner Mongolia. (Unpublished Ph.D. Thesis, Institute of Geology and Geophysics, Chinese Academy of Sciences). [In Chinese.]

ZHAI, D., J. XIAO, L. ZHOU, R. WEN, Z. CHANG & Q. PANG, 2010. Similar distribution pattern of different phenotypes of *Limnocythere inopinata* (Baird) in a brackish-water lake in Inner Mongolia. Hydrobiologia, **651**: 185-197.

ZHAI, D., J. XIAO, L. ZHOU, R. WEN, Z. CHANG, X. WANG, X. JIN, Q. PANG & S. ITOH, 2011. Holocene East Asian monsoon variation inferred from species assemblage and shell chemistry of the ostracodes from Hulun Lake, Inner Mongolia. Quaternary Research, **75**: 512-522.

ZHAO, Q. & P. WANG, 1998. Biotas and thanatocoenosis of micropalaeontology. In: C. LI & P. WANG (eds.), Researches on stratigraphy of the Late Quaternary period in Yangtze River Mouth: 29-68. (Science Press, Beijing). [In Chinese.]

First received 14 December 2012.
Final version accepted 4 September 2013.

[When citing this chapter, refer to Crustaceana 87 (2014) 1027–1035]

This paper is dedicated to Prof. Dr. Eugen Kempf on the occasion of his 80th birthday (16 April 2012)

BATHYCONCHOECIA LIUI N. SP., A NEW SPECIES OF OSTRACOD (MYODOCOPA, HALOCYPRIDIDAE) FROM THE SOUTH CHINA SEA

BY

JIAN-QIANG YIN[1]), QING-CHAO CHEN and KAI-ZHI LI

Key Laboratory of Marine Bio-resources Sustainable Utilization, South China Sea Institute of Oceanology, Chinese Academy of Sciences, Guangzhou 510301, P.R. China

ABSTRACT

A new species of halocyprid ostracod, *Bathyconchoecia liui*, from the bathypelagic zone of the Nansha Trough (North-West Borneo Trough, southern South China Sea) is described and illustrated. It is compared with the closely similar species *B. angeli* George, 1977. The present species is easily distinguished from *B. angeli* by the structures of the toothed edges of the mandibular basis and coxa, and by the segmentation of the fifth limb.

Key words. — Pacific Ocean, zooplankton; deep sea, systematics, Bathyconchoeciinae

ZUSAMMENFASSUNG

Eine neue Art der Halocypriden (Ostrakoden), *Bathyconchoecia liui*, wird aus der bathypelagischen Zone des Nansha Trough (Nord-West-Borneo Trough) im südlichen Südchinesischen Meer beschrieben und dargestellt. Sie wird mit der ähnlichen Art *B. angeli* George, 1977 verglichen. Die vorliegende Spezies ist von *B. angeli* durch die Strukturen der gezähnten Ränder der Mandibelbasis und -coxa und durch die Segmentierung des fünften Extremität leicht zu unterscheiden.

Schlüsselworte. — Pazifischer Ozean, Zooplankton, Tiefsee, Systematik, Bathyconchoeciinae

INTRODUCTION

The genus *Bathyconchoecia* was established by Deevey (1968) to include six new species from the Gulf of Mexico and two species originally attributed to the genus *Euconchoecia*. Since then 15 more species have been described (Kornicker, 1969, 1981; Poulsen, 1969a, b, 1972; Angel, 1970, 1973, 1976; George, 1971, 1977; Deevey, 1975; Kornicker & Angel, 1975; Ellis, 1987, 1989; Kornicker &

[1]) Corresponding author; e-mail: jqyin@scsio.ac.cn

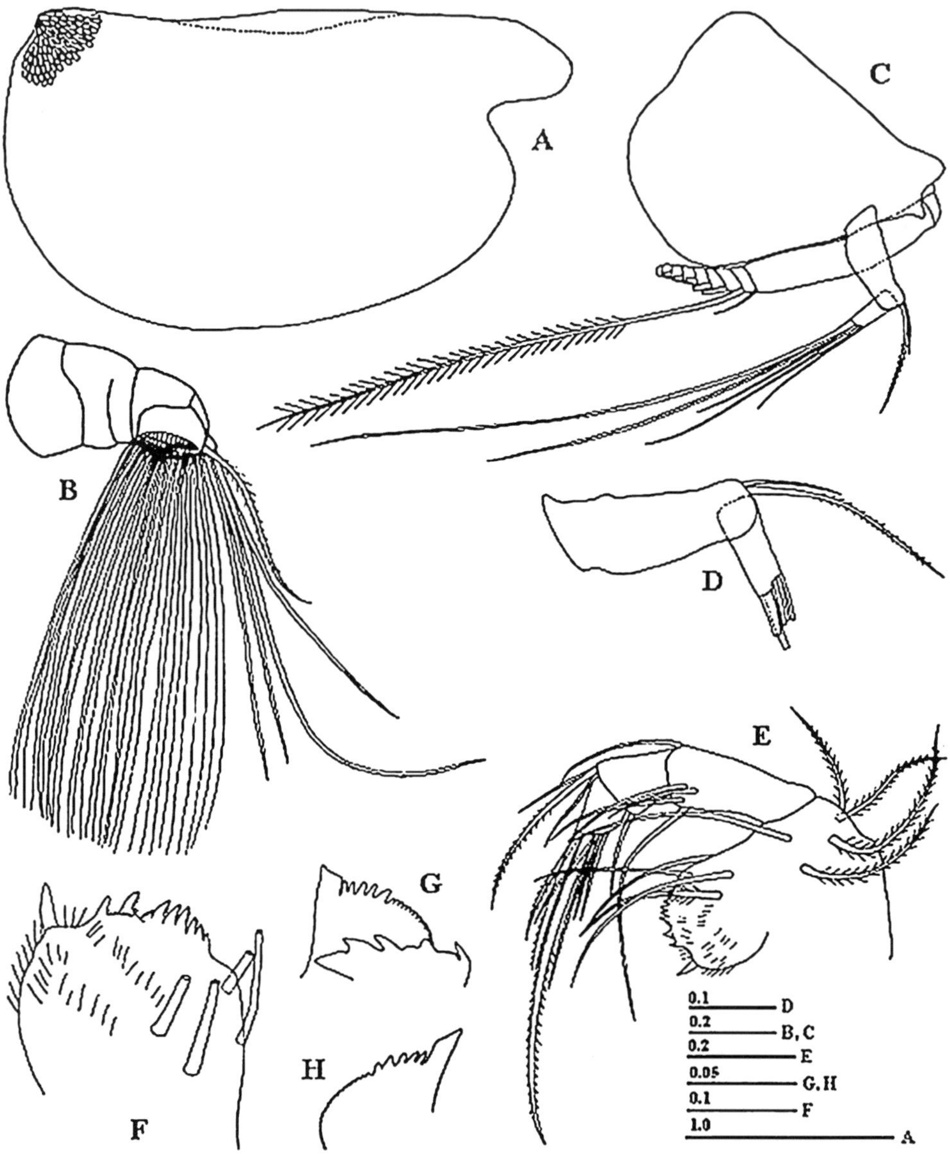

Fig. 1. *Bathyconchoecia liui* n. sp., female, holotype. A, carapace viewed laterally; B, first antenna; C, second antenna; D, endopodite of second antenna; E, basis and endopodite of mandible; F, toothed edge of basis of mandible; G, toothed edge and tooth list of coxa of left mandible; H, toothed edge of coxa of right mandible. Scale bars in mm.

Rudjakov, 2004). Recently, Angel (2012) has established a new genus *Scottoecia* to include a new species, *S. arabica* (= *B. lacunosa* sensu James, 1973), and other four species formerly assigned to the genus *Bathyconchoecia*. Angel & Graves (2013) have established a new subfamily, Bathyconchoeciinae, to include the genera *Bathyconchoecia* and *Scottoecia*. Most species of *Bathyconchoecia* have been recorded from the Atlantic, and few species have been found in the Pacific (Kornicker, 1969, 1991; Poulsen, 1969a). However, this is unlikely to representative of the true zoogeographical distribution of the genus (Angel, 1973). Yin & Chen (1991) described three new recorded species of *Bathyconchoecia* from the South China Sea, and Angel (personal communication) believes that the specimen identified as *B. paulula* is a new species. In the 1980s, the expedition of the Multidisciplinary Oceanographic Expedition Team of the Academia Sinica to the Nansha Islands collected a number of bathypelagic, vertical trawl samples from the southern South China Sea. In one of these samples, an adult female specimen of a previously undescribed *Bathyconchoecia* species was found; it is described as a new species of *Bathyconchoecia* below.

MATERIAL AND METHODS

The carapace length (CL) and height of the specimen were measured under a stereomicroscope at 25× magnification using a calibrated eyepiece micrometer. The specimen was dissected, and the limbs were placed on a series of slides under a stereomicroscope. The limbs and setae were measured using a microscope and an eyepiece micrometer. These measurements were standardised by expressing them as percentages of the carapace length (% CL). The complete animal and the limbs were drawn using camera lucidas. The figures for publication were inked from the pencilled camera lucida drawings.

DESCRIPTION

Systematics

Class OSTRACODA Latreille, 1802
Subclass MYODOCOPA Sars, 1866
Order HALOCYPRIDA Dana, 1853
Suborder HALOCYPRIDINA Dana, 1853
Family HALOCYPRIDIDAE Dana, 1853
Subfamily BATHYCONCHOECIINAE Angel & Graves, 2013
Genus *Bathyconchoecia* Deevey, 1968

Bathyconchoecia liui n. sp. (figs. 1 and 2)

Holotype.— Unique specimen, adult female, has been deposited in the South China Sea Institute of Oceanology, Chinese Academy of Sciences, Guangzhou, China. Reg. number SCSIOCAS-B001.

Type-locality.— Southern South China Sea, 6°04.55′N 113°35.83′W; June 5, 1985; sounding 2626 m; vertical haul from 2600 to 1500 m, 0.505 mm mesh net.

Etymology.— The species is named in honour of Rui-Yu Liu, Institute of Oceanology, Chinese Academy of Sciences, Qingdao, P.R. China, who was a famous marine biologist and carcinologist.

Female. Carapace (fig. 1A).— The carapace length (CL) is 2.46 mm with a height 1.41 mm. The anteroventral and posteroventral corners are well rounded. The posterodorsal corner is bluntly rounded. The anterior, ventral and posterior margins are rounded. The dorsal margin is slightly curved, and there is a shallow sulcus on the hinge line at about mid-length. The rostrum is large, straight, rounded, and accounts for 14.2% of the total length beyond the anterior margin. The incisure is shallow and open (14.3% CL from the tip of the rostrum to the inner margin). The asymmetrical glands open symmetrically at posterior margin close to the posterodorsal corner. The sculpturing of carapace is distinct. The entire surface of carapace is entirely covered with tiny pits, but otherwise there are no striations or reticulations.

Frontal organ.— Frontal organ is absent.

First antenna (fig. 1B).— The limb is short and bare (approx. 21.1% CL), and its segmentation is indistinct. The fourth segment (but see Poulsen, 1969a) ventrally carries a cluster of 200-240 sensory filaments, which are arranged in 10-12 rows, each row consisting of about 20 filaments. The fifth segment carries a plumose distodorsal seta that is about half the lengths of the sensory filaments. The terminal segment carries 4 setae. The principal seta is slightly longer than the sensory filaments, and remaining 3 setae are almost subequal and shorter than the sensory filaments. The sensory filaments are about 38.2% CL, the distodorsal seta is 19.1% CL, the principal seta is 46.3% CL and other three terminal setae are 31.7, 30.5 and 31.7% CL respectively.

Second antenna (fig. 1C, D).— The protopodite is pear-shaped (30.5% CL), and is only slightly more long than broad. The exopodite is 9-segmented. The first exopodite segment (19.1% CL) is straight and is twice as long as segments 2-9. It carries a small terminal seta just reaching to the fourth segment. Segments 2-8 each carry a long swimming seta. The terminal segment carries 3 spinose setae of different lengths (not shown). The endopodite is 2-segmented. The first endopodite segment is almost rectangular and one-third of the length of the protopodite (10.1% CL). It carries 2 distodorsal setae. The a-seta (4.9% CL) is bare and slightly less than half of the length of the b-seta (11.0% CL). The length of the b-seta is approximately equal to length of the segment, and distally has marginal spinules. The second endopodite segment carries 5 setae. The f-, g-, h-, i- and j-setae are 38.6, 54.5, 17.9, 25.2 and 12.6% CL, respectively. They all are longer than the b-seta of the first endopodite segment, the longest one of which is a little longer than the swimming setae. The swimming seta of the second exopodite segment is 48.8% CL.

Mandible (fig. 1E-H).— The toothed edge of the coxal endite has 5 roughly even blunt teeth on the ventral edge, and the anterior edge is serrated. The proximal and distal tooth lists are fused into a single list of about 5 uneven teeth. The basal endite is broad and robust. The toothed edge has 2 spine teeth, 6 triangular cutting teeth armed with secondary serrations, and an "outer tooth" (outermost one, see Angel & Graves, 2013). The first spine tooth is stout, pointed, and well offset from all other teeth. The second spine tooth is short and blunt. The basis has 4 plumose and 1 bare setae near the insertion of the endopodite, and 4 bare setae on the anterior edge and lateral surface of the endite. The first endopodite segment has 1 distodorsal seta and 4 ventral setae. The second endopodite segment has 3 distodorsal setae and 1 distoventral seta. The end-segment has 7 terminal setae, and one of them is extremely long (24.8% CL) and exceeds the length of the endopodite.

Maxilla (fig. 2A, B).— The precoxal endite has 6 plumose setae. The coxal endite has a total of 22 setae of these 11 are slender and bare, and the others are stout and spinose. The basis has 2 plumose seta. The anterior margin of the first endopodite segment has a group of 4 plumose setae proximally and one distally; the posterior margin has 4 plumose medial setae. There are short hairs on the anterior face. The second endopodite segment has 2 stout claws and 4 slender setae.

Fifth limb (fig. 2C).— The epipodial setae are in three groups of 4, 4, 5 (dorsally). The precoxa and coxa are covered with short fine spinules on the ventral face. The precoxa has 2 plumose ventral setae. The coxa has total of 10 setae (4 plumose, 6 bare) and 2 claws on the ventral face. The basis has 5 ventral setae (3 plumose, 2 bare), 1 pair of plumose lateral setae, and 1 very long distodorsal seta (a remnant of exopodite, see Angel, 2012) that extends past the end of the limb. The endopodite appears to consist of 3 segments. The first endopodite segment is short, with 1 pair of ventral setae and 1 lateral seta. The second endopodite segment has 1 dorsal seta, 1 ventral seta and 1 pair of lateral setae. The third endopodite segment has 3 terminal setae, and two of which are claw-like and stout. The dorsal seta is a little longer than the central and ventral setae (10.2% CL).

Sixth limb (fig. 2D).— The epipodial setae arranged in three groups of 5, 5, 6 (dorsally). The coxa has 6 plumose and 1 bare setae ventrally. The basis has 4 plumose ventral setae, 1 plumose lateral seta, and 1 extremely long distodorsal seta (a residual exopodite, see Angel, 2012) that extends beyond the end of the limb. The endopodite is 3-segmented. The first endopodite segment has 5 ventral setae. The second endopodite segment has 4 ventral setae and 1 dorsal seta. The third endopodite segment has terminally 2 claw-like and 1 slender setae of which the dorsal one is the longest (12.2% CL).

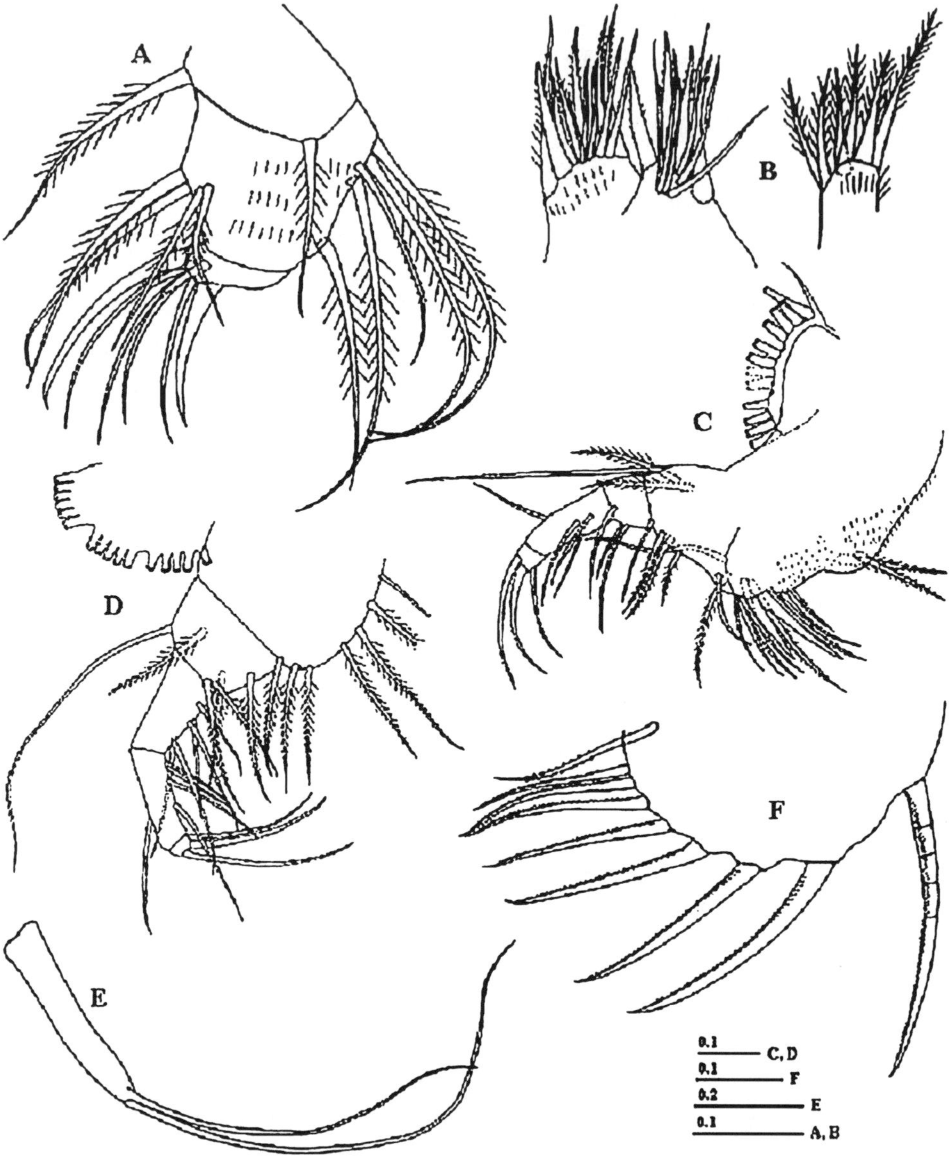

Fig. 2. *Bathyconchoecia liui* n. sp., female, holotype. A, basis and endopodite of maxilla; B, precoxal endite (right) and coxal endite (left) of maxilla; C, fifth limb; D, sixth limb; E, seventh limb; F, caudal furca. Scale bars in mm.

Seventh limb (fig. 2E).— The limb is slender and the suture between the two segments is very indistinct. The terminal segment bears 2 unequal and bare setae terminally.

TABLE I

Comparison of the morphologic characteristics between females *B. liui* n. sp. and *B. angeli* George, 1977 and the depths where these were found

Character	*B. liui*	*B. angeli*
Carapace length (mm)	2.46	2.5
Height/length (%)	57.3	54.0
Rostrum length/ carapace length (%)	14.2	16.0*
Carapace, posterior margin	Rounded	Straight
Rostrum tip	Well rounded	Slightly rounded
A2, first endopodite segment	B-seta equal in length to segment	B-seta longer than segment
Mandible basal toothed edge	1st spine tooth > 2nd one	2 spine teeth equal
Mandible coxal toothed edge	Anterior edge serrated	Anterior edge smooth
Fifth limb	3 endopodite segments	2 endopodite segments
Sixth limb, 1st and 2nd endopodite segment setae	5 + 5	4 + 3
Seventh limb, 2 terminal setae	Unequal	Equal
Depth of net/sounding (m)	2600-1500/2626	200-0/230

* Indicates data derived from measuring drawings in the original description.

Caudal furca (fig. 2F).— Each lamella has 8 long, slim and weak claws all with 2 rows of spines on the posterior edge. The first claw is 5-jointed. There is a long unpaired seta behind claws.

DISCUSSION

The present species belongs to the genus *Bathyconchoecia* rather than to *Scottoecia*, based on its features of the locations of the asymmetrical glands opening and the structures of the toothed edges of the mandible. Angel (2012) has made a comprehensive summary of the characters of carapace of all species described as *Bathyconchoecia*. *B. liui* can be easily distinguished from all other species, except *B. angeli*, *B. arctica* and *B. latirostris*, by the lack of spines or serrations on the carapace and its open incisure. However, its rostrum is very different in form from *B. arctica* and *B. latirostris*. *B. arctica* at the moment is a unique *Bathyconchoecia* species in having a frontal organ. The novel species is most similar to *B. angeli* in the size and shape of the carapace, and also in the locations at which the asymmetrical glands open. The morphological characteristics and the depths where these two species were found are compared in table I. Nevertheless, there are differences between them notably the structures of the toothed edges of the mandibles and the segmentations of the fifth limbs are particularly significant.

The depths at which *B. angeli* and the present species have been taken are also observably different. George's specimen was collected in the epipelagic zone of Malacca Strait, Indian Ocean (George, 1977), while the present specimen was collected in the bathypelagic zone of Nansha Trough (North-West Borneo Trough, southern South China Sea). Most species of *Bathyconchoecia* had been collected in the bathypelagic, abyssopelagic and benthopelagic zones (>1000 m) (Deevey, 1968, 1975; Poulsen, 1969a, 1972; Angel, 1970, 1973; Kornicker & Angel, 1975; Ellis, 1987, 1989; Kornicker & Rudjakov, 2004).

ACKNOWLEDGEMENTS

We would like to thank Gu-Xian Zhang for helping to collect samples. We especially give our thanks to Dr. M. V. Angel, National Oceanography Centre, Southampton, U.K. for reviewing a draft of the paper and for his valuable suggestions, and the reviewers for their comments that helped to improve the manuscript. This study was supported by the Natural Science Foundation of China (Nos. 31101619 and 41130855).

REFERENCES

ANGEL, M. V., 1970. *Bathyconchoecia subrufa* n. sp. and *B. septemspinosa* n. sp., two new halocyprids (Ostracoda, Myodocopida) from the tropical North Atlantic and the description of the larval development of *B. subrufa*. Crustaceana, **19**: 181-199.

— —, 1973. The description of the female of the ostracod *Bathyconchoecia sagittarius* Deevey, 1968 (Myodocopida, Halocyprididae). Crustaceana, **25**: 211-219.

— —, 1976. *Bathyconchoecia arctica* n. sp., a new species of ostracod (Halocyprididae, Myodocopidae) from the Arctic. Crustaceana, **31**: 59-65.

— —, 2012. *Scottoecia* — a new genus of halocyprid ostracod, with the description of *Scottoecia arabica* nov. sp. and the redescription of *Bathyconchoecia darcythompsoni* (Scott, 1909). Zootaxa, **3254**: 32-54.

ANGEL, M. V. & C. GRAVES, 2013. Bathyconchoeciinae, a new subfamily of deep oceanic planktonic halocyprid Ostracod (Myodocopa, Ostracoda). Zootaxa, **3630**: 243-269.

DEEVEY, G. B., 1968. *Bathyconchoecia*, a new genus of pelagic ostracods (Myodocopa Halocyprididae) with six new species from the deeper waters of the Gulf of Mexico. Proceedings of the Biological Society of Washington, **81**: 539-570.

— —, 1975. Two new species of *Bathyconchoecia* (Myodocopa Halocyprididae) from the Caribbean Sea. Proceedings of the Biological Society of Washington, **87**: 351-380.

ELLIS, C. J., 1987. *Bathyconchoecia longispinata* n. sp., a new species of halocyprid ostracod with seven carapace spines. Crustaceana, **53**: 83-93.

— —, 1989. The vertical distribution of the halocyprid ostracod genus *Bathyconchoecia* at 31°N 25°W, including a description of a new species. Biological Journal of the Linnean Society, **36**: 31-48.

GEORGE, J., 1971. On the occurrence of *Bathyconchoecia deeveyae* Kornicker (Ostracoda, Halocypridae) in the Indian Ocean. Crustaceana, **21**: 141-144.

— —, 1977. *Bathyconchoecia angeli* sp. nov., a new halocyprid ostracod from the Malacca Strait, Indian Ocean. Crustaceana, **33**: 70-74.

JAMES, C. M., 1973. On *Bathyconchoecia lacunosa* (Müller), a rare halocyprid ostracod from the Arabian Sea. Journal of the Marine Biological Association of India, **15**: 433-447.

KORNICKER, L. S., 1969. *Bathyconchoecia deeveyae*, a highly ornamented new species of Ostracoda (Halocyprididae) from the Peru-Chile Trench system. Proceedings of the Biological Society of Washington, **82**: 403-408.

— —, 1981. Range extension and supplementary description of *Bathyconchoecia deeveyae* (Ostracoda: Halocyprididae). Proceedings of the Biological Society of Washington, **94**: 1237-1243.

— —, 1991. Myodocopid Ostracoda of hydrothermal vents in the eastern Pacific Ocean. Smithsonian Contributions to Zoology, **516**: 1-46.

KORNICKER, L. S. & M. V. ANGEL, 1975. Morphology and ontogeny of *Bathyconchoecia septemspinosa* Angel, 1970 (Ostracoda: Halocyprididae). Smithsonian Contributions to Zoology, **195**: 1-21.

KORNICKER, L. S. & J. A. RUDJAKOV, 2004. Two new species of seven-spined *Bathyconchoecia* from the North Atlantic and Indian Oceans (Crustacea: Ostracoda: Halocypridae). Proceedings of the Biological Society of Washington, **117**: 412-421.

POULSEN, E. M., 1969a. Ostracoda-Myodocopa. Part IIIA Halocypriformes-Thaumatocypridae and Halocypridae. Dana Report, **75**: 1-100.

— —, 1969b. Ostracoda-Myodocopa from the eastern tropical Atlantic. Vidensk. Meddr. Dansk natuh. Foren., **132**: 129-197.

— —, 1972. On the *Bathyconchoecia* (Ostracoda Myodocopa) from the Azores collected by the Bathyscaph "Archimedein" 1969. Tethys, **4**: 445-455.

YIN, J. Q. & Q. C. CHEN, 1991. Species, fauna and zoogeography of the pelagic ostracods in the waters around the Nansha Islands. In: THE MULTIDISCIPLINARY OCEANOGRAPHIC EXPEDITION TEAM OF ACADEMIA SINICA TO NANSHA ISLANDS (Ed.), Marine fauna and zoogeography of the Nansha Islands and neighboring waters: 64-139. (Ocean Press, Beijing). [In Chinese with English abstract.]

First received 28 December 2012.
Final version accepted 29 July 2013.

[When citing this chapter, refer to Crustaceana 87 (2014) 1036–1042]

This paper is dedicated to Prof. Dr. Eugen Kempf on the occasion of his 80th birthday (16 April 2012)

ADENCYTHERE KODALNAMERENSIS, A NEW GENUS AND A NEW SPECIES (OSTRACODA) FROM KOD AL-NAMER BEACH, ADEN CITY, YEMEN

BY

MUNEF MOHAMMED[1]) and DIETMAR KEYSER[2,3,4])

[1]) Department of Earth and Environmental Sciences, Faculty of Science, Sana'a University, Sana'a, Yemen

[2]) Zoological Institute and Museum, Hamburg University, Hamburg, Germany

[3]) Biozentrum Grindel, Martin-Luther-King Platz 3, D-20146 Hamburg, Germany

ABSTRACT

A new marine genus of Ostracoda from the tidal flats of Aden City is erected and one new species is described. The systematic position of the present taxon is discussed.

Key words. — *Adencythere kodalnamerensis*, marine, Ostracoda, Aden City

ZUSAMMENFASSUNG

Eine neue marine Ostracodengattung von den Wattflächen in Aden City und eine neue Art werden beschrieben. Die systematische Stellung dieses Taxons wird diskutiert.

Key words. — *Adencythere kodalnamerensis*, marine, Ostracoda, Aden City

INTRODUCTION

Aden City is located in the south-western part of Yemen (fig. 1). Knowledge of the ostracoda from this region is scarce (Schornikov, 1980; Bonaduce, 1985; Malz, 1986, 1987; Bonaduce & Russo, 1990); therefore, it is not unusual to find new taxa in this fauna. A detailed study of the morphologic features of the collected species led the authors to separate it from its closest relatives and to erect a new genus. Based on the elongated and subrectangular shape of the carapace (fig. 2) and the muscle scar pattern (fig. 3) as well as the missing sieve structure in the lateral

[4]) Corresponding author; e-mail: keyser@uni-hamburg.de

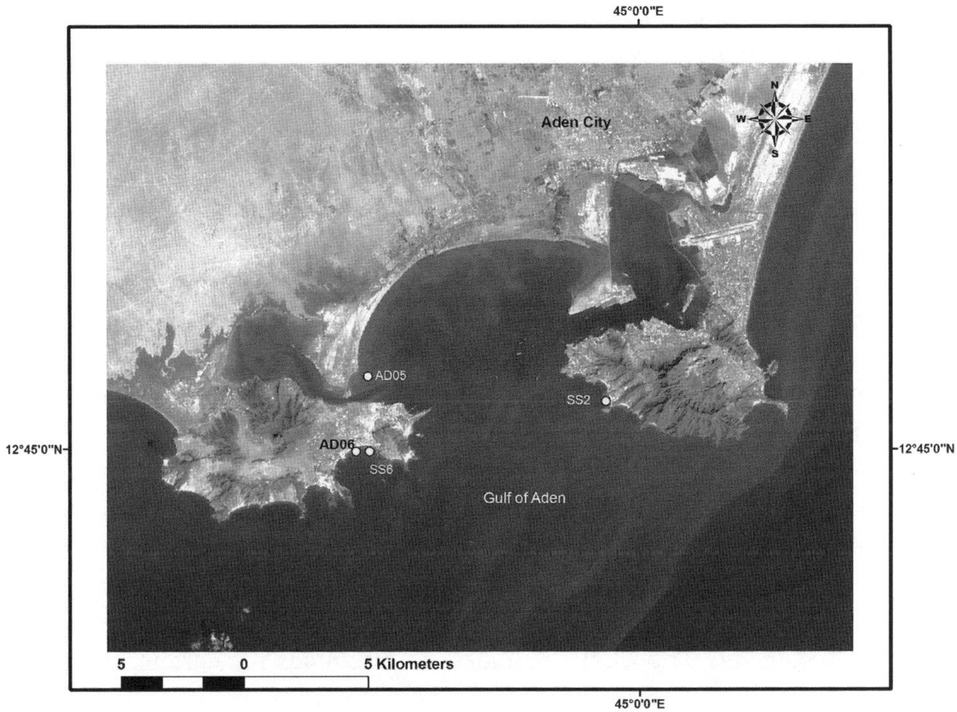

Fig. 1. Location map with sampling station. This figure is published in colour in the online edition of this journal, which can be accessed via http://booksandjournals.brillonline.com/content/journals/15685403.

pores, it has been placed under the family Trachyleberididae. The material used in the current paper was collected in 2001 and 2007 from the tidal flat area of the city of Aden; it has been studied previously by Mohammed (2004) and Mohammed & Keyser (2012). The abbreviations used are: SS, Sana'a sample (sample number); AD, Aden (sample number); ZMH, Zoologisches Museum Hamburg (sample number).

SYSTEMATIC DESCRIPTION

Class OSTRACODA Latreille, 1802
Order PODOCOPIDA Sars, 1866
Family TRACHYLEBERIDIDAE Sylvester-Bradley, 1948
Subfamily TRACHYLEBERIDINAE Sylvester-Bradley, 1948
Genus **Adencythere** n. gen.

Type-species.— *Adencythere kodalnamerensis* n. sp.

Etymology.— A combination of the place of the collection (Aden City) and the Greek word for "*Cytherea*", synonymous to "Aphrodite", the Goddess of love and beauty.

Diagnosis.— Medium-sized, elongated carapace, ornamented with frontal thick rib and three lateral ones, transverse secondary short ribs fuse with the principal ribs, wide shallow foveolae and small pits giving a spongy appearance to the carapace. Sub-central tubercle and two knobs are developed, the two knobs occur on the connecting areas between the terminals of median and ventral ribs with a small transverse curved rib, the lower knob forming a small alar prolongation. Hinge holamphidont. A distinct protuberance developed under the front element of the hinge. Muscle scars are four adductor non-split scars with frontal one V-shaped (fig. 3).

Affinities.— The present genus is close to *Hermanites* Puri, 1955 in general shape, both of them having nearly the same arrow-shape from ventral view, but the latter lacking the median rib and the internal protuberance. It also resembles *Costa* Neviani, 1928 in outline and in the presence of three lateral ridges, but the latter lacking the alar prolongation and internal protuberance. The general shape of *Isocythereis* Triebel, 1940 somewhat resembles the current genus, but the former has divided muscle scars and wider duplicature, it differs also in hingement. *Occultocythereis* Howe, 1951 could be compared to the present genus but differs in size and ornamentation. The presence of the internal protuberance and the shape of muscle scars in this genus are the same as in the genus *Kingmania* Keij, 1957 but the other features of the carapace are different. It also resembles the genera *Patagonacythere* Hartmann, 1962, *Coquimba* Ohmert, 1962, *Cornucoquimba* Ohmert, 1968 and *Neohornibrookella* Jellinek, 1993 of the family Hemicytheridae in general shape and some characters of ornamentation, but they have double frontal scars, lateral sieve pores and the absence of a caudal process which serves here to separate the current genus from Hemicytheridae.

Adencythere kodalnamerensis n. sp. (fig. 2)

2004 *Hermanites* aff. sp. Mohammed, p. 46, pl. 3, figs. 9-11.
2012 ? *Neohornibrookella* sp. A. Mohammed & Keyser, p. 269, fig. 10: 145-149.

Etymology.— The species was named after its type locality, the beach of Kod Al-Namer on Little Aden peninsula. The name thus is an adjective agreeing in gender with the generic name.

Holotype.— An adult female carapace, SS6.1.
Paratypes.— One complete carapace and 4 valves, SS6.2 and ZMH K-42566.
Type locality.— Sample SS6. Kod Al-Namer Beach 12°44′57″N 44°53′39″E (fig. 1, table I).
Size.— Holotype female L: 0.60 mm, H: 0.30 mm; Paratype male L: 0.63 mm, H: 0.31 mm.

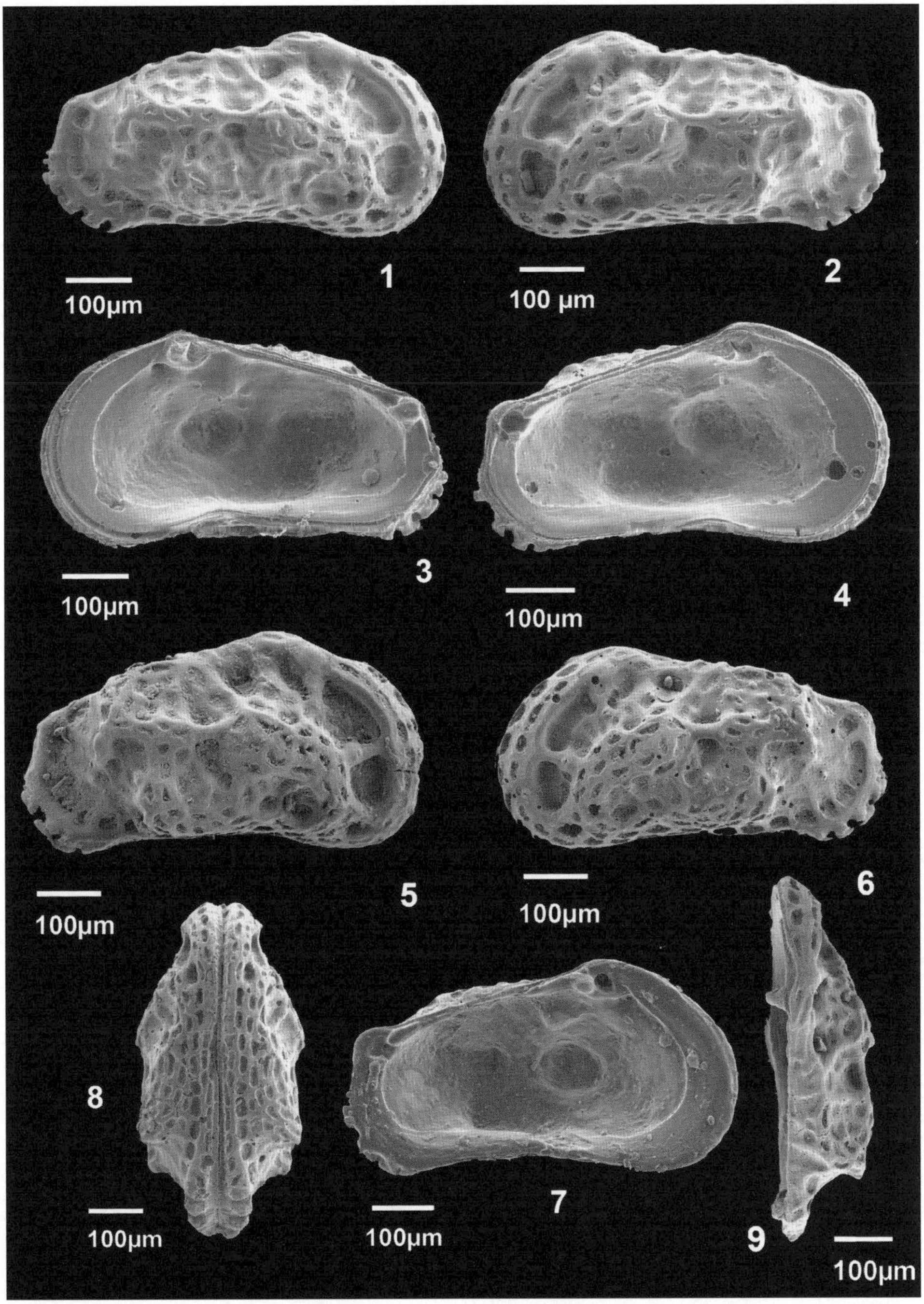

Fig. 2. *Adencythere kodalnamerensis* n. gen., n. sp. 1, ♂, RV, external view; 2, ♂, LV, external view; 3, ♂, RV, internal view; 4, ?♂, LV, internal view; 5, ♀, RV, external view; 6, ♀, LV, external view; 7, ♀, LV, internal view; 8, complete carapace, ventral view; 9, RV, dorsal view.

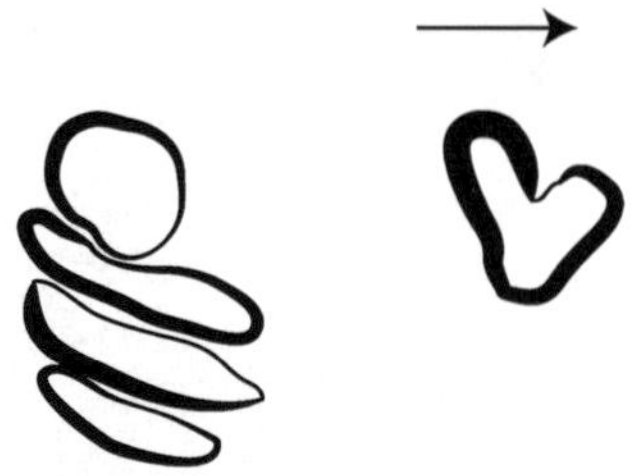

Fig. 3. *Adencythere kodalnamerensis* n. gen., n. sp., muscle scars of RV.

Material.— Two complete carapaces, 6 right valves and 5 left valves.

Occurrence.— SS2, SS6, AD05, AD06.

Description.— Medium-sized, elongate, sub-rectangular carapace, arrow-shaped in ventral view; highest in front of the median, anterior margin broadly rounded, posterior margin narrower than the anterior one, truncated at the postero-dorsal margin and nearly well rounded at the mid-posterior and the postero-ventral margin, possessing four broad smooth spines and a smaller pointed one; no caudal process. Dorsal and ventral margins nearly straight, converging towards the posterior margin; left valve larger than the right valve. Hinge holamphidont, the hinge of the right valve consists of a high and conical anterior tooth, a large adjacent socket which merges into a crenulated median groove that terminally deepens and widens and a large posterior weakly lobed tooth. Muscle scars consist of four slightly oblique unequal adductor scars, with a V-shaped frontal one located in the sub-central tubercle (fig. 3); a well-developed protuberance could be seen in the internal view of both valves, located in area under the front element of the hinge and the sub-central tubercle which probably represents a point of muscle attachment. The surface is ornamented with three ribs running along the lateral surface of the carapace with a relatively thicker frontal marginal one, the frontal rib starts running from the eye tubercle, parallel to the anterior margin, connected in the antero-ventral part with the ventro-lateral rib which is running parallel to the ventral margin and ends in a pronounced knob, forming a wing-like small alar prolongation; the median rib starts running obliquely from the antero-ventral margin, thickens around the subcentral tubercle and terminates in a small knob without

TABLE I
Sampling stations and geographical position

Station	Latitude (N)	Longitude (E)
SS6	12°44′57″	44°53′39″
SS2	12°46′03″	44°59′14″
AD05	12°46′36″	44°53′55″
AD06	12°44′57″	44°53′39″

reaching the posterior margin; there are nearly eight small minor ribs interconnecting the median rib with the frontal, ventral and dorsal ribs. The surface of both valves is also covered by a wide shallow open foveolae and secondary small pits which gives a spongy appearance to the carapace. Inner lamella moderately wide, selvage well developed; vestibule lacking, marginal pore canals not visible, normal pore canals few and scattered. Eye tubercle developed. Males slightly longer than females.

Environment.— The present species was recorded from the tidal flats area with a salinity of normal sea water (34‰) and water temperature of 31°C, substrate of medium to fine sand consisting of quartz, feldspar and high amount of shell fragments giving a pale yellow colour to the sand.

Remarks.— The general shape of this species resembles many taxa, but a close examination of the characteristics of the fauna led us finally to separate it from the other similar taxa, although frequently no soft parts could be gathered.

REFERENCES

BONADUCE, G., 1985. *Bosasella*, a new marine ostracode genus from Somalia. Bull. Soc. Natur. Napoli, **93**: 3-11, 2 pls.

BONADUCE, G. & A. RUSSO, 1990. *Patrizia*, a new marine ostracode genus (Ostracoda: Podocopa) from the Gulf of Aden (Somalia) and Patriziainae new subfamily. Bull. Soc. Paleontol. Ital., **29**: 273-277, 1 pl.

HARTMANN, G., 1962. Ostracoden des Eulitorals. Mit. hamb. zool. Mus. Inst., Ergänzungsband, **60**: 169-270.

HOWE, H. V., 1951. New Tertiary ostracode fauna from Levy County, Florida. Bull. Geol. Surv. Florida, **34**: 1-43.

JELLINEK, T., 1993. Zur Oekologie und Systematic rezenter Ostracoden aus dem Bereich des Kenianisschen Barriereriffs. Senckenbergiana lethaea, **73**: 83-225.

KEIJ, A. J., 1957. Eocene and Oligocene Ostracoda of Belgium. Institut Royal des Sciènces Naturelles de Belgique, Memoires, **136**: 1-210.

MALZ, H., 1986. Eine rezente Ostracoden-fauna von Hurghada am RotenMeer (Ägypten). Cour. Forsch. Inst. Senckenberg, **85**: 283-287, 1 pl.

— —, 1987. Tiefsee-Ostracoden aus dem Golf von Aden. Nat. Mus., **117**: 397-399.

MOHAMMED, M., 2004. The study of Ostracoda in recent deposits — South West Yemen. (Ph.D. Thesis, Baghdad University, Baghdad).

MOHAMMED, M. & D. KEYSER, 2012. Recent ostracods from the tidal flats of the coast of Aden City, Yemen. Mar. Biodiv., **42**: 247-280.

MOORE, R. C., 1961. Treatise on invertebrate Paleontology, Part Q, Arthropoda, (Ostracoda). Geol. Soc. Am. Univ. Kansas Press **422**. (University of Kansas Press, Lawrence, KS).

NEVIANI, A., 1928. Ostracodi fossili d'Italia I. Vallebiaja (Calabriano). Mem. Pontif. Accad. Nuovi Lincei, **11**: 1-120.

OHMERT, V. W., 1968. Die Coquimbinae, eine neue Unterfamilie der Hemicytheridae (Ostracoda) aus dem Pliozän von Chile. Mitt. Bayer. Staatsslg. Paläont. hist. Geol., **8**: 127-165.

PURI, H. S., 1955. *Hermanites*, new name for *Hermania* PURI 1954. J. Paleontol., **29**: 558.

SCHORNIKOV, E. I., 1980. [New ostracod species of coral reefs of the Red Sea and Gulf of Aden.]. In: B. V. PREOBRAZHENSKIJ & E. V. KRASNOV (eds.), Biology of the coral reefs (Morphology, Systematics, Ecology): 131-158. (Akad. Nauk USSR, Moscow).

TRIEBEL, E., 1940. Die Ostracoden der Deutschen Kreide. 3. Cytherideinae und Cytherinae aus der Unteren Kreide. Senckenbergiana, **22**: 160-227.

First received 15 December 2012.
Final version accepted 5 January 2014.

[When citing this chapter, refer to Crustaceana 87 (2014) 1043–1071]

This paper is dedicated to Prof. Dr. Eugen Kempf on the occasion of his 80th birthday (16 April 2012)

MORPHOTYPES OF *CYTHERIDELLA ILOSVAYI* (OSTRACODA) DETECTED BY SOFT AND HARD PART ANALYSES

BY

CLAUDIA WROZYNA[1,3]), WERNER E. PILLER[1]) and MARTIN GROSS[2])

[1]) University of Graz, Institute of Earth Sciences, NAWI Graz, Heinrichstrasse 26, A-8010 Graz, Austria

[2]) Universalmuseum Joanneum, Department for Geology & Palaeontology, Weinzöttlstrasse 16, A-8045 Graz, Austria

ABSTRACT

Quantitative approaches dealing with soft- and hard-part morphological variability of ostracods are very rare for most species. Here we present intraspecific limb and shell variability of the Neotropical freshwater ostracod species assumed as *Cytheridella ilosvayi*. For the first time such an analysis considered adult females and males, as well as juveniles. Variability of the appendages included the antennule, second antenna as well as the first and second thoracopod and was analysed in terms of ratios of podomeres, setae and limbs. Limb variability of adult and juvenile individuals (down to A-3) is generally low. The highest variation is shown in podomere proportions of the antennas, while thoracopods and setae provide minor and/or non-significant influence on the variability. Based on discrimination analyses shell parameters (i.e., shell length, position of the transversal sulcus) emerge to be more important for differentiation of groups than limb ratios. Adult females exhibit a large size range in which two clearly separated morphotypes exist. Although differences in limbs are small their existence can be proven also on the basis of proportions between podomeres of the antennas. Males and juveniles, however, do not display a separation into corresponding subgroups or morphotypes. The presence of two morphologically similar females and only one type of males indicates the coexistence of female morphotypes which may represent either two (cryptic) species or a mixed reproduction population in which parthenogenetic and sexual reproduction coexists.

Key words. — Ostracods, Neotropics, *Cytheridella*, morphological variability, morphotypes

ZUSAMMENFASSUNG

Quantitative Ansätze die darauf zielen die morphologische Variabilität der Weich- und Hartteile von Ostrakoden zu bestimmen sind für die meisten Arten sehr selten. Wir haben die intraspezifische Variabilität von Gliedmaßen und Klappen der neotropischen Ostrakodenart *Cytheridella*

[3]) Corresponding author; e-mail: claudia.wrozyna@uni-graz.at

ilosvayi untersucht. Erstmals werden bei einer derartigen Studie, adulte Weibchen und Männchen sowie Juvenile berücksichtigt. Die Variabilität der Gliedmaßen wurde anhand von Verhältnissen von Podomeren, Setae und Gliedmaßen der Antennula und der zweiten Antenne sowie dem ersten und zweiten Thoracopoden ermittelt. Die Variabilität von Adulten und Juvenilen (bis A-3) ist generell niedrig. Dabei tritt die größte Variabilität in den Proportionen der Segmente der Antenna auf, während Thoracopoden und Setae geringste und/oder nicht signifikante Anteile an der Variabilität aufweisen. Basierend auf Diskriminanzanalysen, konnten Klappenparameter (Länge, Position des transversalen Sulcus) als wichtigste Faktoren für die Abtrennung von Gruppen identifiziert werden. Adulte Weibchen weisen einen weiten Größenbereich auf, in dem zwei deutlich voneinander getrennte Morphotypen auftreten. Obwohl die Unterschiede in den Gliedmaßen gering sind, konnte die Existenz der zwei Weibchen auch anhand von Proportionen zwischen Segmenten der Antennen nachgewiesen werden. Männchen und Juvenile zeigen keine Aufspaltung in entsprechende Morphotypen. Das Vorkommen von morphologisch sehr ähnlichen Weibchen bei nur einem Typ Männchen könnte entweder auf zwei (kryptische) Arten oder eine gemischte Reproduktionspopulation deuten, in der parthenogenetische und sexuelle Reproduktion koexistieren.

Key words. — Ostracods, Neotropics, *Cytheridella*, morphological variability, morphotypes

INTRODUCTION

Ostracods represent one of the most diverse groups, with an estimated >20 000 living species, of which only approx. 10 000 have been described (Rodriguez-Lazaro & Ruiz-Muñoz, 2012). At present, ostracods are popular model organisms for research on the evolution of reproductive modes (e.g., Martens, 1998; Schön & Martens, 2012) and used as proxies for climate and ecosystem changes (e.g., De Deckker & Forester, 1988; Horne, 2007; Wrozyna et al., 2010). They are one of the few invertebrate groups with an extensive fossil record, but are also rather unique as a fossil group with such a diverse modern representation (Schön et al., 1998).

Systematic, ecologic and evolutionary studies of ostracods rely heavily on morphological information. Traditionally, morphological analyses of ostracods are focused either on shell characteristics or on the appendages, but with limited linkage between them. In ostracod taxonomy, dimensions of the carapace (usually length and height) are commonly considered (e.g., Meisch, 2000) but not of the limbs. In contrast to shell variability, soft parts of ostracods are disregarded and lack sound documentation. Commonly, variability of the appendages was studied either due to ontogenetic concerns (e.g., Smith & Kamiya, 2003, 2005) or focused on specific features such as the male copulatory organs (e.g., Martens, 2000). Moreover, most studies refer to qualitative data only from adults and predominantly of females.

Thus, investigations related to intraspecific variation of ostracod taxa are based on shells and do not include information on the appendages (Neil, 2000). First approaches focused on individual limbs such as the first (Maddocks, 2000) or second antenna (Danielopol, 1978). Although efforts aligned for comparative description of ostracod soft parts have increased, these data mostly deal with qualitative differences (Martens et al., 1998a).

In this study we present data on morphological variability of both soft and hard parts of the ostracod species *Cytheridella ilosvayi* Daday, 1905, common in Central and tropical South America. Generally, little is known about the distribution of ostracod species, their biology and ecology in the Neotropics. The genus *Cytheridella* (Limnocytheridae, Timiriaseviinae) has a global distribution during the Cenozoic ranging from central Europe, India and Africa to North and South America (Purper, 1974; Colin et al., 1997). Modern species occur in Africa (Karanovic, 2012) and the Neotropics (Purper, 1974), but information on species biogeography and ecology is restricted to the two species *C. ilosvayi* and *C. boldii* Purper, 1974 from North, Central and South America. Modern (American) *Cytheridella* has a wide geographical range covering the subtropics and tropics from approximately 30°N to 30°S (Sarasota County, Alvarez Zarikian et al., 2005; Tampa Bay, Willard et al., 2007; Florida; Quintana Roo, Mexico, Gabriel et al., 2009; Yucatan peninsula including Lago Petén Itzá, Guatemala, Peréz et al., 2010a, b; Hispaniola, Perez-Gelabert, 2008; Brazil, Upper Paraná floodplain, Higuti et al., 2007; Mormul et al., 2010; Rio Grande do Sul, Purper, 1974). *Cytheridella* occurs in various aquatic ecosystems, but shows preferences for rivers or at least slightly flowing conditions (Peréz et al., 2010a).

METHODS

Sampling localities and study organism

Two surface samples were taken in September 2009 from two lake habitats in western Brazil, State of Amazonas near the city of Eirunepé (fig. 1). The localities Barro Branco (6°50′18.3″S 69°45′37.0″W) and Lago Comprido (6°43′52.7″S 69°44′33.9″W) are abandoned channels of the Tarauacá River. Samples were taken with a hand net from the margins of the water bodies. At Barro Branco samples were taken between roots of floating plants (*Eichhornia*) (BB01-09) and from the surface sediments of the lake shore (BB02-09). Sample BB02-09 consisted of whitish mud with abundant leaf litter. Lago Comprido samples comprise a surface sample (LC01-09) of whitish mud and abundant leaf litter and a surface sample (LC02-09) covering the uppermost 10 cm. This sample consisted of whitish mud. Barro Branco samples contained a high number of ostracods with >80% living (appendages visible) specimens. Less ostracod material was available in Lago Comprido samples. The samples were clearly dominated by *Cytheridella ilosvayi*. Additional taxa were *Darwinula stevensoni* (Brady & Robertson, 1870) and *Stenocypris* sp. The population of *C. ilosvayi* was dominated by females.

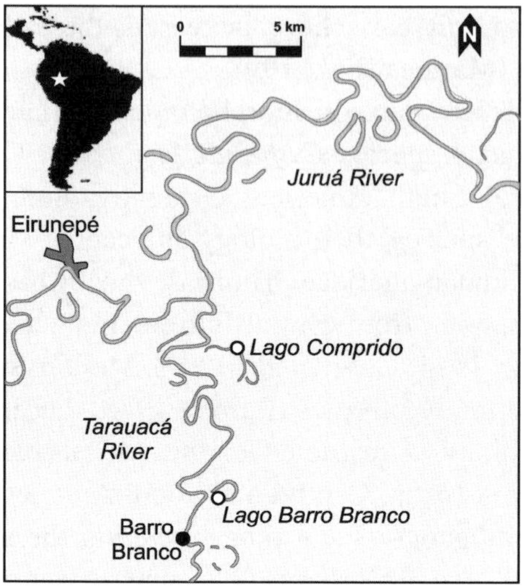

Fig. 1. Map of sampling localities Barro Branco and Lago Comprido (white circles), State of Amazonas, western, Brazil.

All figured specimens are stored at the Museu Paraense Emílio Goeldi, Belém. Supplementary material is stored at the Universalmuseum Joanneum, Graz.

Sample preparation

All samples were washed in the field through 200 μm sieves. The sieve residuals were stored in ethanol (96%) immediately after sieving. Ostracods were completely picked from the samples under a binocular (Zeiss Discovery.V8). Specimens with opened valves were selected randomly for soft part dissection. Soft parts were detached from both valves using entomological needles, dissected, and mounted with Hydro-Matrix® (Micro-Tech-Lab, Graz, Austria) onto cover slides. Photographs for analysis of soft and hard parts were taken using a Zeiss Axioplan 2 and a digital camera Leica DFC 320. Stacked pictures were compiled with the software Helicon Focus Pro (Helicon Soft, Kharkov, Ukraine). Species identification of *Cytheridella ilosvayi* was carried out in accordance with Löffler (1961), Purper (1974) and Karanovic (2012).

Abbreviations applied to limb morphology

A1, first antenna; A1EV-EIII, fifth to third podomere of the first antenna; A1EIVe2, second anterior seta on the fourth podomere of A1; A2, second antenna; A2EIV-EII, fourth to second podomere of A2; A2 EIIIGM, apical claw of second antenna; T1, first thoracopod; T1EIV-EII, fourth to second podomere of T1; T1EIIe, antero-distally e-seta on the second podomere of T1; T2, second

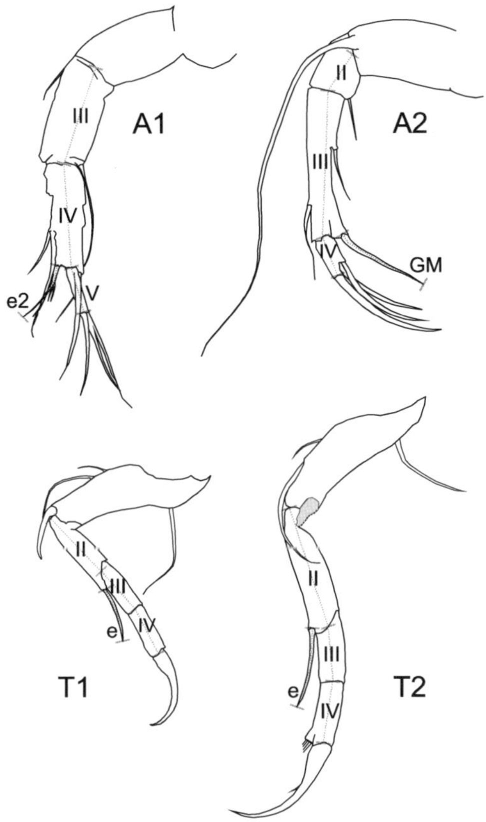

Fig. 2. Simplified appendages of *Cytheridella ilosvayi* Daday, 1905 used for morphological analyses: the antennule (A1), the second antenna (A2), the first and second thoracopod (T1, T2). Indicated are the measured lengths (dashed line) of podomeres and setae, respectively. Roman numerals refer to fifth to third and fourth to second podomeres, respectively, of the respective limb. For abbreviations see also abbreviation list in the text.

thoracopod; T2EIV-EII, fourth to second podomere of T2; T2EIIe, antero-distally e-seta on the second podomere of T2 (based on Meisch, 2000; Karanovic, 2012). See also fig. 2.

Biometric analyses of limbs

The morphology of limbs of 38 specimens were investigated using a transmitted light microscope at 100× magnification. A total of 16 linear distances (lengths of segments and setae mentioned in the abbreviation list above) were taken on each of the 38 specimens using the Image Access 3.0 software (Imagic Bildverarbeitungs AG). In order to analyse symmetry between left and right limbs, each measurement was performed on both sides of 20 specimens. The distance measurements were further converted into 24 ratios: A1EIV/EIVe2, A1EV/EIV,

A1EV/EIII, *A1EIV/EIII*, *A2EIII/EIIIGM*, *A2EIV/EIII*, A2EIV/EII, *A2EIII/EII*, T1EI/EIe, T1EIV/EIII, T1EIV/EII, *T1EIII/EII*, T2EII/EIIe, T2EIV/EIII, T2EIV/EII, *T2EIII/EII*, *A1/A2*, *T1/T2*, A1EIVe2/A2EIIIGM, T1EIIe/T2EIIe, *A1/T1*, *A2/T2*, *A1/T2*, *A2/T1* (ratios in italics were selected for further statistical analyses). In order to include the proportion of the limbs among each other, we added the length of each considered podomere referred to as the length of the limb (e.g., A2/T2 refers to the sum of A2EIV, EIII and EII vs. sum of T2EIV, EIII and EII). Drawings of the limbs were performed with CorelDraw® X4 (Corel), based on photographs of the limbs.

Male copulatory organs were also separated and checked for their morphological variability.

Analysis of shell variability

In order to analyse variability of the shells, we have measured length, height and the length between the anterior margin and the transversal sulcus (referred to hereafter as 'length sulcus'). SEM pictures were checked for ornamentation, width and hinge.

Statistical analyses

First, the differences of the limb ratios between left and right limbs from 20 specimens were tested with two-sample *t*-tests performed with the Microsoft Excel® 2007. Multivariate statistical analyses were performed with SPSS 19 (IBM, 2010). For all 24 limb ratios the mean, range, variance and standard deviation were calculated. Prior to statistical analyses the linear measurements of the shell parameters were log-transformed. Out of all ratios measured, thirteen were selected (see above) for discrimination analyses based on normal distribution, homogeneity of variance (Levene's test) and significance of differences of variable means (Tukey's range test in conjunction with an ANOVA). Discrimination analysis was performed for (a) limb ratios, (b) shell parameters (length, height, L/H, length sulcus) and (c) both limb ratios and shell parameters.

RESULTS

Shell characteristics

The length-height plot shows ontogenetic growth stages (see fig. 3). Accordingly, our dataset comprises instars down to A-3. Through the presence of fully developed copulatory organs the first three size classes were identified as adult females and males, respectively. Adult females reveal a large size range forming

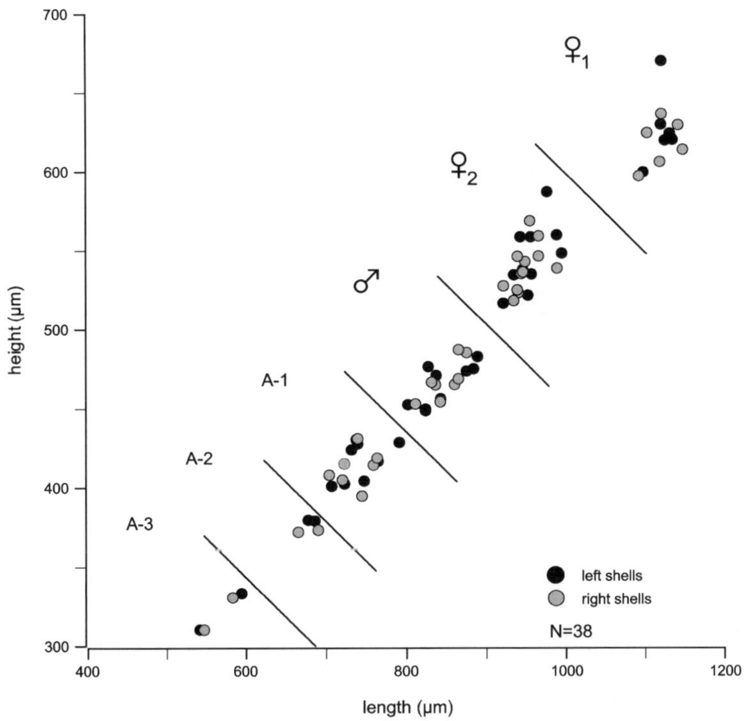

Fig. 3. Carapace size in terms of length and height of *Cytheridella ilosvayi* Daday, 1905 in adult females, males and juvenile instars down to A-3.

two well-separated size clusters. The larger females have lengths ranging from 1110-1140 μm and heights ranging from 613-653 μm; while length of the smaller females ranges between 921-992 μm, and a height between 518-578 μm. Further, mature characteristics such as well-developed inner lamella, pore-canals (fig. 4), and eggs (variable number of 9 to 12) were observed in both female morphotypes. Male shells are smaller than the females (806-881 μm length, 451-485 μm height). Beside size there are no other (qualitative) shell characteristics (e.g., ornamentation) that are different between the two groups of females (for details see figs. 4 and 5).

Variability of the limbs

Analyses of left and right limb ratios revealed that both are significantly symmetrical ($P<0.05$). Thus, the analysis of one limb (left or right) of each pair is sufficient for morphological analysis.

The variability of all morphological traits is displayed in fig. 6. Results of thirteen selected limb ratios that were used for multivariate statistics are summarised

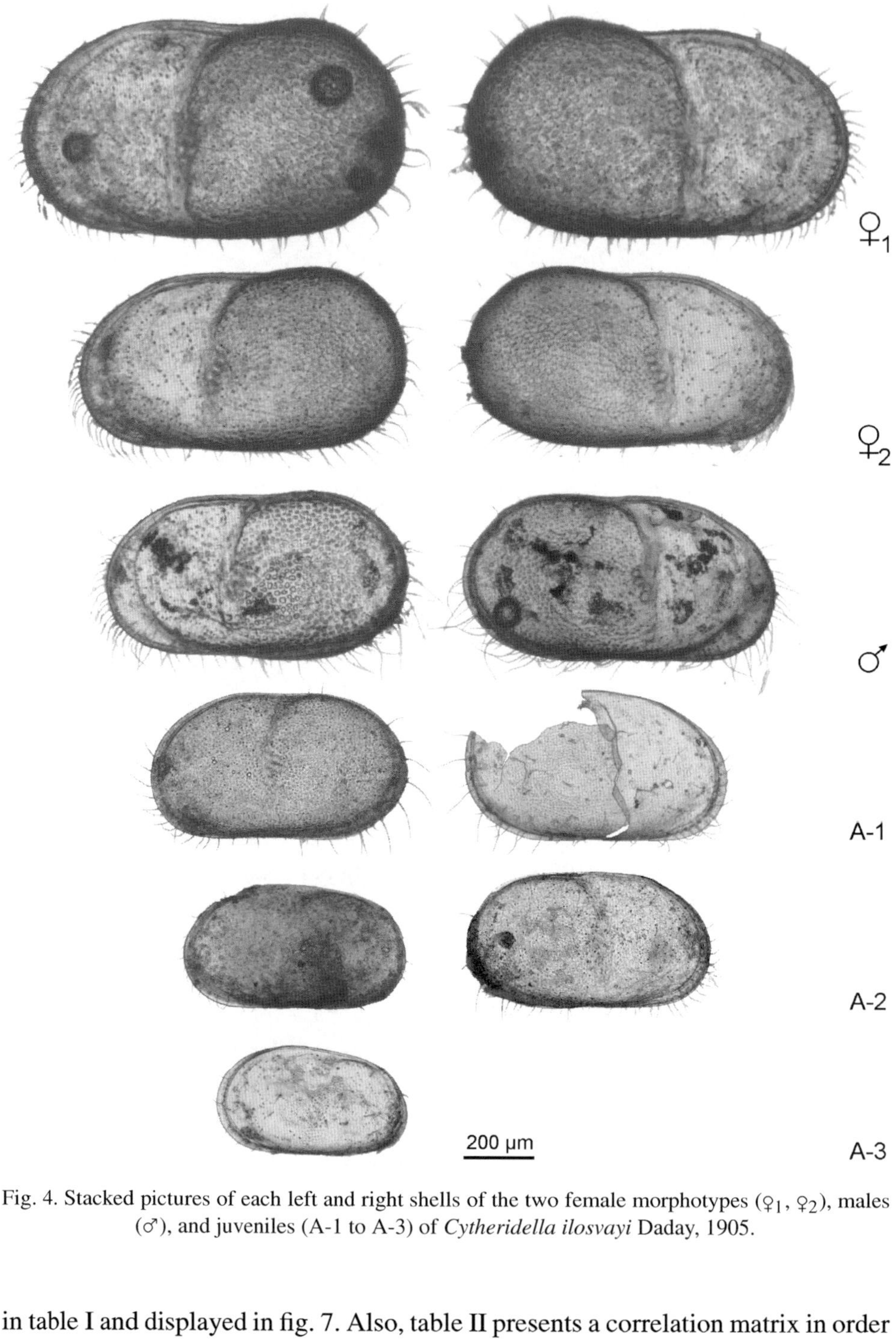

Fig. 4. Stacked pictures of each left and right shells of the two female morphotypes (♀1, ♀2), males (♂), and juveniles (A-1 to A-3) of *Cytheridella ilosvayi* Daday, 1905.

in table I and displayed in fig. 7. Also, table II presents a correlation matrix in order to display correlations between the ratios.

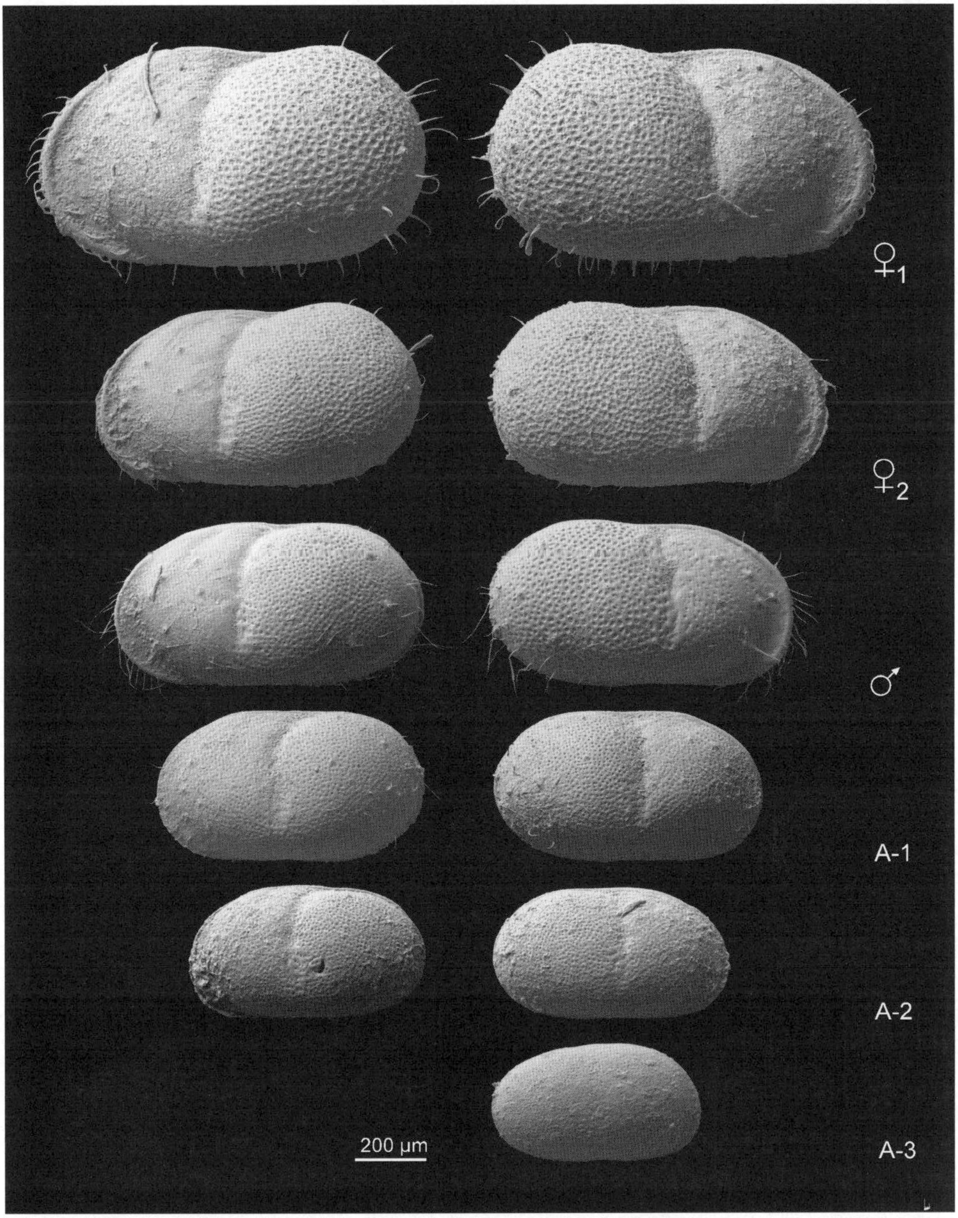

Fig. 5. SEM pictures of left and right shells of females (♀1, ♀2), males (♂), and juveniles (A-1 to A-3) of *Cytheridella ilosvayi* Daday, 1905.

The majority of the limb ratios reveal relatively small ranges of all instars, e.g., A2EIV/EIII. The largest variability occurs at the first antenna (A1) and the second antenna (A2). In particular, the ratios A1EIV/EIVe2, A1EIV/EIII, A2EIII/EII, as well as A1EIVe2/A2EIIIGM reveal large variation. Less variation is displayed by

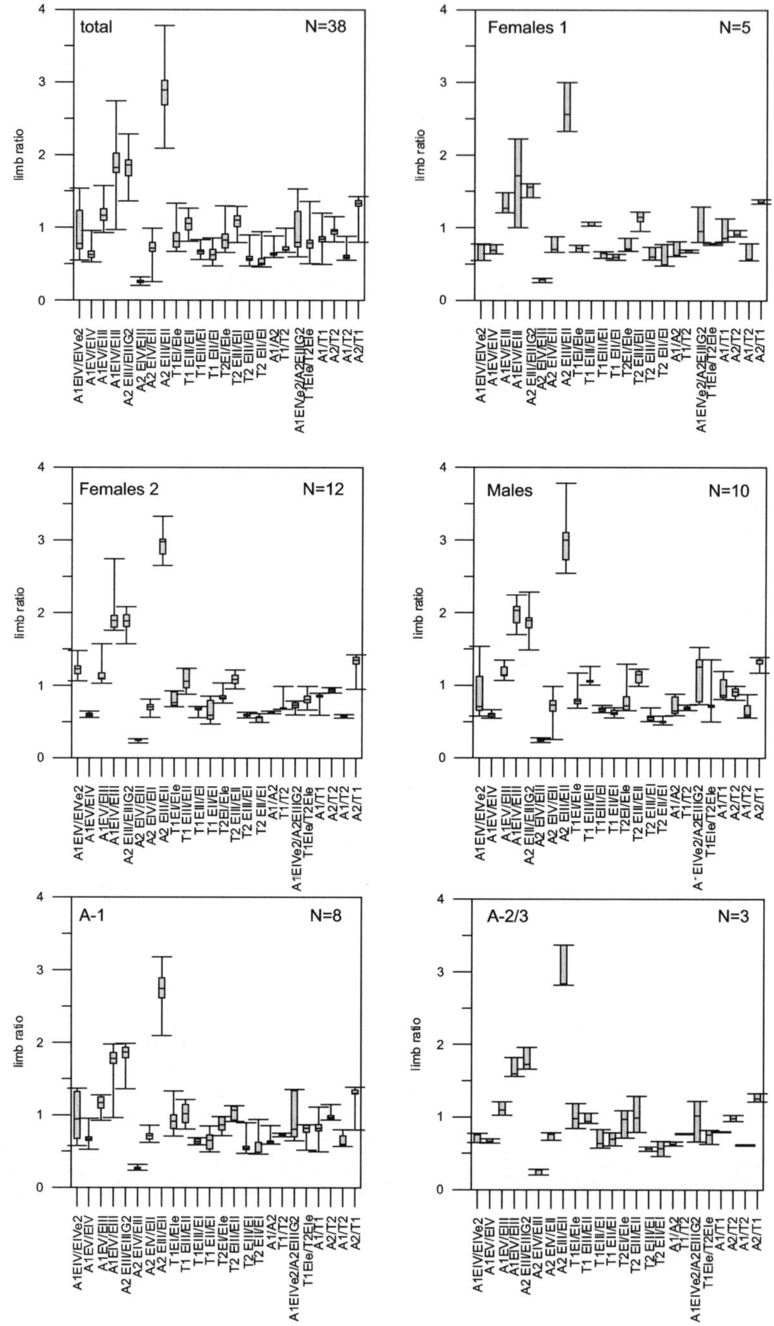

Fig. 6. Variation ranges of the 24 measured limb ratios represented as boxplots for all investigated individuals (total), and separately for instars (A-1 and A-2/3), males and both female morphotypes (females 1 and 2). For codes of the morphological traits see abbreviation list in the text. Black dots represent outliers.

TABLE I

Selected limb ratios and shell parameters of *Cytheridella ilosvayi* Daday used for the discrimination analyses

		A1 EV/ EIII	A1 EIV/ EIII	A2 EIII/ EIIIGM	A2 EIV/ EIII	A2 EIV/ EII	A2 EIII/ EII	T1 EIII/ EII	T2 EIII/ EII	A1/ A2	T1/ T2	A1/ T1	A2/ T2	A1/ T2	Length (μm)	Height (μm)	Length sulcus (μm)	Length/ height
Female 1	Mean	1.28	1.69	1.55	0.28	0.75	2.72	1.04	1.15	0.62	0.68	0.91	0.92	0.57	1125.2	622.3	457.0	1.8
	SD	0.11	0.44	0.08	0.02	0.09	0.35	0.03	0.06	0.02	0.02	0.15	0.04	0.02	18.02	12.12	23.15	0.05
Female 2	Mean	1.17	1.96	1.87	0.24	0.70	2.94	1.08	1.09	0.63	0.67	0.82	0.94	0.58	950.8	542.7	370.8	1.8
	SD	0.17	0.29	0.15	0.02	0.07	0.19	0.13	0.08	0.01	0.05	0.09	0.03	0.02	18.65	13.68	21.99	0.04
Male	Mean	1.17	1.96	1.90	0.25	0.70	3.05	1.08	1.08	0.70	0.70	0.92	0.91	0.64	848.4	510.7	340.8	1.8
	SD	0.11	0.19	0.20	0.02	0.18	0.36	0.07	0.07	0.11	0.02	0.14	0.06	0.10	22.06	131.52	17.77	0.04
A-1	Mean	1.21	1.79	1.85	0.26	0.75	2.86	1.02	1.06	0.66	0.74	0.85	0.96	0.63	736.0	412.6	343.9	1.8
	SD	0.05	0.08	0.07	0.02	0.06	0.19	0.16	0.09	0.09	0.01	0.12	0.02	0.08	23.44	9.24	42.05	0.05
A-2/3	Mean	1.11	1.66	1.78	0.25	0.74	3.01	0.97	1.02	0.63	0.77	0.80	0.99	0.62	646.1	358.9	291.2	1.8
	SD	0.09	0.14	0.16	0.04	0.06	0.31	0.08	0.25	0.03	0.01	0.01	0.04	0.01	55.98	24.23	23.86	0.04

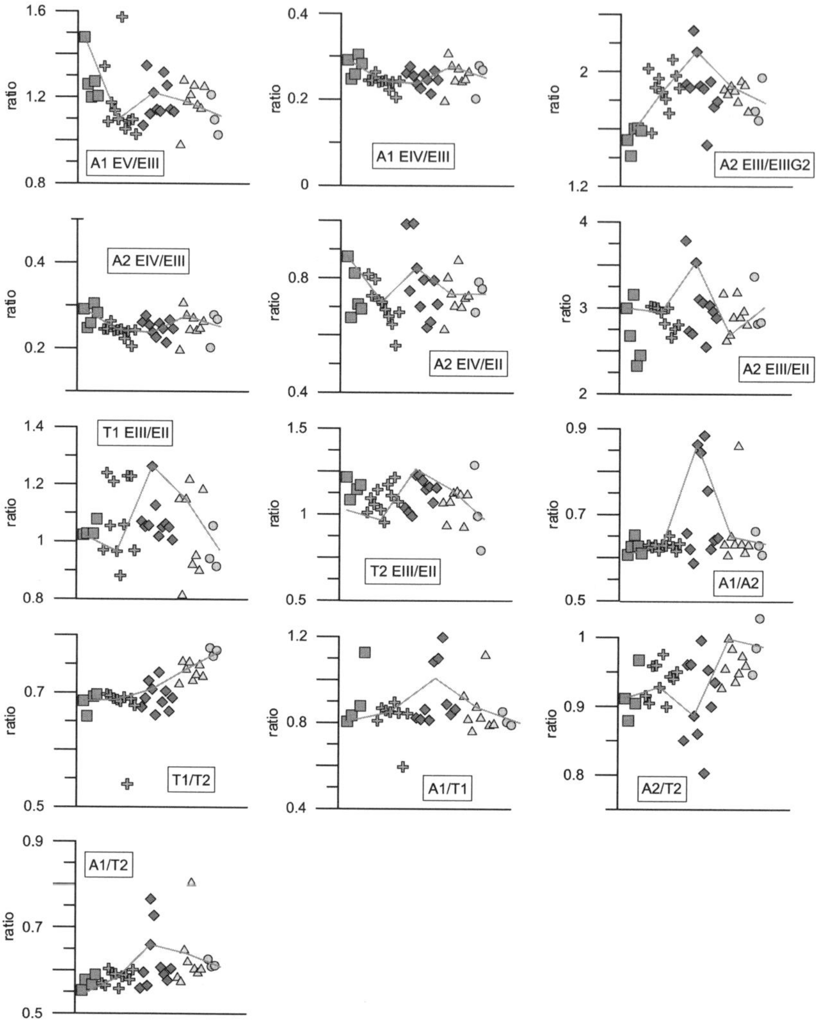

Fig. 7. Selected limb ratios of adult females, males and juveniles of *Cytheridella ilosvayi* Daday, 1905, displaying morphological trajectories (square = females 1, cross = females 2, diamond = males, triangle = A-1, oval = A-2/3). This figure is published in colour in the online edition of this journal, which can be accessed via http://booksandjournals.brillonline.com/content/journals/15685403.

the first and the second thoracopod (T1 and T2) and the terminal podomere of A2. Also, ratios between limbs such as A1/A2, T1/T2, A2/T2 and A1/T2 exhibit only minor variation. Generally, limb ratios are either increasing or decreasing from A-2/3 to adult males and females (fig. 7). In most cases, smaller females and males exhibit similar means of the ratios. Means of limb ratios of the larger females do not continue the previous trend in limb ratios (e.g., A2EIII/EIIIGM, fig. 7). This pattern is most conspicuous in the individual limb ratios (A1EV/EIII to T2EIII/EII). Males show relatively high variability of ratios related to the podomeres EII and EIII of the second antenna and/or second thoracopod. Highest variability is shown for ratios of the second antenna, whereas, the ratios A2EIV/EII and A2EIII/EII are relatively variable. Also, the length of EIII of the second thoracopod shows relatively high variation which becomes noticeable for the 'between limbs' ratios. However, males with, e.g., a short EIII or EII at A2 do not mandatorily have also short EIII at T2.

Ratios considering limbs among each other and those of thoracopods show high correlations (table II). Ratios of the first and second antenna are less frequently correlated. These correlations are in most cases weak (e.g., A2 EIV/EIII to A2 EIII/EII: $r = 0.37$). There is only one negative correlation between A2EIII/EII and A2EIV/EIII ($r = -0.37$).

Discrimination analysis of the limb ratios exhibits a significant separation of the specimens belonging to the different groups (according to the length-height plot): Wilk's Lambda: 0.05, Chi square: 84.39, P: 0.03. Classification results show that 84.2% of the specimens were correctly classified in terms of statistics. In the scatter plot (fig. 8) the large females (Female 1) are assembled in the left lower quadrant associated with negative loadings on axis (function) 1 and 2. Opposite, the smaller females (Female 2) have positive loadings on axis 1 similar to the males, thus, overlapping each other. Juveniles build a separate cluster with higher loadings on axis 2. Loadings of all groups vary around 0 on axis 3 except males which show a trend to more positive values and have a more scattered distribution.

The first canonical function accounted for 53.1% of explained variability and discriminates mostly between the adult females Female 1, Female 2 and males. The second and third functions accounted for 29.9% and 13.3% of explained variability, respectively. Discrimination function 4 explains 3.7% of variability (table IV).

Standardised discrimination function coefficients of function 1 (table IV) do not show a distinct pattern. Except A2EIV/EIII and T1/T2 all variables have relatively high values. Variables with highest coefficients for function 1 are A2/T2 (−2.87), T2EIII/EII (2.09), and A1/A2 (−2.06). Also, A1EV/EIII, A1EV/EIII (1.99), A2EIII/EIIIG2 (1.93) and A1EIV/EIII (−1.89) have high values. The number of high function coefficients of function 2 is lower. The variable with the highest coefficient is T2EIII/EII (3.18). Except A2EIV/EII, variables of antennas have low impact on function 2. Instead, the proportions between limbs are similar to the pattern on function 1, and have very high function coefficients (i.e., A2/T2, A1/T1, and A1/T2). Function 3 is predominantly influenced by variables referring to limb proportions, i.e. A1/T2, A2/T2. Second antenna variables (A2EIV/EII, A2EIII/EII) affect function 3. Limb proportions are important for function 4 with the difference that the variables of A2 contribute more. Canonical loadings, representing correlations between the investigated limb traits and the discrimination functions of

TABLE II

Pearson correlation matrix for selected limb ratios in *Cytheridella ilosvayi* Daday

		A1 EV/ EIII	A1 EIV/ EIII	A2 EIII/ EIIIGM	A2 EIV/ EIII	A2 EIV/ EII	A2 EIII/ EII	T1 EIII/ EII	T2 EIII/ EII	A1/ A2	T1/ T2	A1/ T1	A2/ T2	A1/ T2
A1EV/EIII	Correlation	1	0.704**	−0.060	0.330*	0.205	−0.085	−0.235	0.155	0.116	−0.086	0.003	0.106	0.085
	Significance (two-way)	0.000	0.719	0.043	0.217	0.612	0.156	0.352	0.489	0.608	0.987	0.525	0.614	
	N	38	38	38	38	38	38	38	38	38	38	38	38	38
A1EIV/EIII	Correlation	*0.704***	1	0.135	−0.047	−0.078	0.006	−0.142	0.068	0.160	−0.204	−0.077	−0.014	0.001
	Significance (two-way)	0.000		0.420	0.781	0.642	0.969	0.394	0.685	0.336	0.220	0.646	0.936	0.996
	N	38	38	38	38	38	38	38	38	38	38	38	38	38
A2 EIII/EIIIGM	Correlation	−0.060	0.135	1	−0.064	−0.029	0.262	0.128	0.037	0.707**	−0.016	−0.029	−0.012	0.056
	Significance (two-way)	0.719	0.420		0.704	0.862	0.112	0.445	0.825	0.000	0.923	0.861	0.944	0.736
	N	38	38	38	38	38	38	38	38	38	38	38	38	38
A2 EIV/EIII	Correlation	*0.330**	−0.047	−0.064	1	0.389*	−0.370*	−0.076	−0.139	0.010	0.019	0.016	0.010	−0.030
	Significance (two-way)	0.043	0.781	0.704		0.016	0.022	0.648	0.406	0.952	0.911	0.922	0.951	0.857
	N	38	38	38	38	38	38	38	38	38	38	38	38	38
A2 EIV/EII	Correlation	0.205	−0.078	−0.029	*0.389**	1	0.493**	−0.201	−0.020	0.268	−0.126	−0.015	−0.054	0.081
	Significance (two-way)	0.217	0.642	0.862	0.016		0.002	0.226	0.903	0.103	0.450	0.929	0.749	0.630
	N	38	38	38	38	38	38	38	38	38	38	38	38	38
A2 EIII/EII	Correlation	−0.085	0.006	0.262	*−0.370**	*0.493***	1	−0.186	0.102	0.351*	−0.146	−0.007	−0.027	0.149
	Significance (two-way)	0.612	0.969	0.112	0.022	0.002		0.264	0.541	0.031	0.381	0.965	0.874	0.372
	N	38	38	38	38	38	38	38	38	38	38	38	38	38
T1 EIII/EII	Correlation	−0.235	−0.142	0.128	−0.076	−0.201	−0.186	1	0.379*	0.114	0.771**	0.731**	0.316	0.335*
	Significance (two-way)	0.156	0.394	0.445	0.648	0.226	0.264		0.019	0.496	0.000	0.000	0.054	0.040
	N	38	38	38	38	38	38	38	38	38	38	38	38	38

TABLE II
(Continued)

		A1 EV/ EIII	A1 EIV/ EIII	A2 EIII/ EIIIGM	A2 EIV/ EIII	A2 EIV/ EII	A2 EIII/ EII	T1 EIII/ EII	T2 EIII/ EII	A1/ A2	T1/ T2	A1/ T1	A2/ T2	A1/ T2
T2 EIII/EII	Correlation	0.155	0.068	0.037	−0.139	−0.020	0.102	*0.379**	1	0.198	0.605**	0.642**	0.870**	0.886**
	Significance (two-way)	0.352	0.685	0.825	0.406	0.903	0.541	0.019		0.234	0.000	0.000	0.000	0.000
	N	38	38	38	38	38	38	38	38	38	38	38	38	38
A1/A2	Correlation	0.116	0.160	*0.707***	0.010	0.268	*0.351**	0.114	0.198	1	0.009	0.218	−0.047	0.279
	Significance (two-way)	0.489	0.336	0.000	0.952	0.103	0.031	0.496	0.234		0.957	0.189	0.778	0.090
	N	38	38	38	38	38	38	38	38	38	38	38	38	38
T1/T2	Correlation	−0.086	−0.204	−0.016	0.019	−0.126	−0.146	*0.771***	*0.605***	0.009	1	0.867**	0.679**	0.631**
	Significance (two-way)	0.608	0.220	0.923	0.911	0.450	0.381	0.000	0.000	0.957		0.000	0.000	0.000
	N	38	38	38	38	38	38	38	38	38	38	38	38	38
A1/T1	Correlation	0.003	−0.077	−0.029	0.016	−0.015	−0.007	*0.731***	*0.642***	0.218	*0.867***	1	0.625**	0.737**
	Significance (two-way)	0.987	0.646	0.861	0.922	0.929	0.965	0.000	0.000	0.189	0.000		0.000	0.000
	N	38	38	38	38	38	38	38	38	38	38	38	38	38
A2/T2	Correlation	0.106	−0.014	−0.012	0.010	−0.054	−0.027	0.316	*0.870***	−0.047	*0.679***	*0.625***	1	0.883**
	Significance (two-way)	0.525	0.936	0.944	0.951	0.749	0.874	0.054	0.000	0.778	0.000	0.000		0.000
	N	38	38	38	38	38	38	38	38	38	38	38	38	38
A1/T2	Correlation	0.085	0.001	0.056	−0.030	0.081	0.149	*0.335**	*0.886***	0.279	*0.631***	*0.737***	*0.883***	1
	Significance (two-way)	0.614	0.996	0.736	0.857	0.630	0.372	0.040	0.000	0.090	0.000	0.000	0.000	
	N	38	38	38	38	38	38	38	38	38	38	38	38	38

Italicised values indicate correlation coefficients that represent highest absolute correlations between a variable and a discrimination function.
* Correlation is significant on the 0.05 level (two-way).
** Correlation is significant on the 0.01 level (two-way).

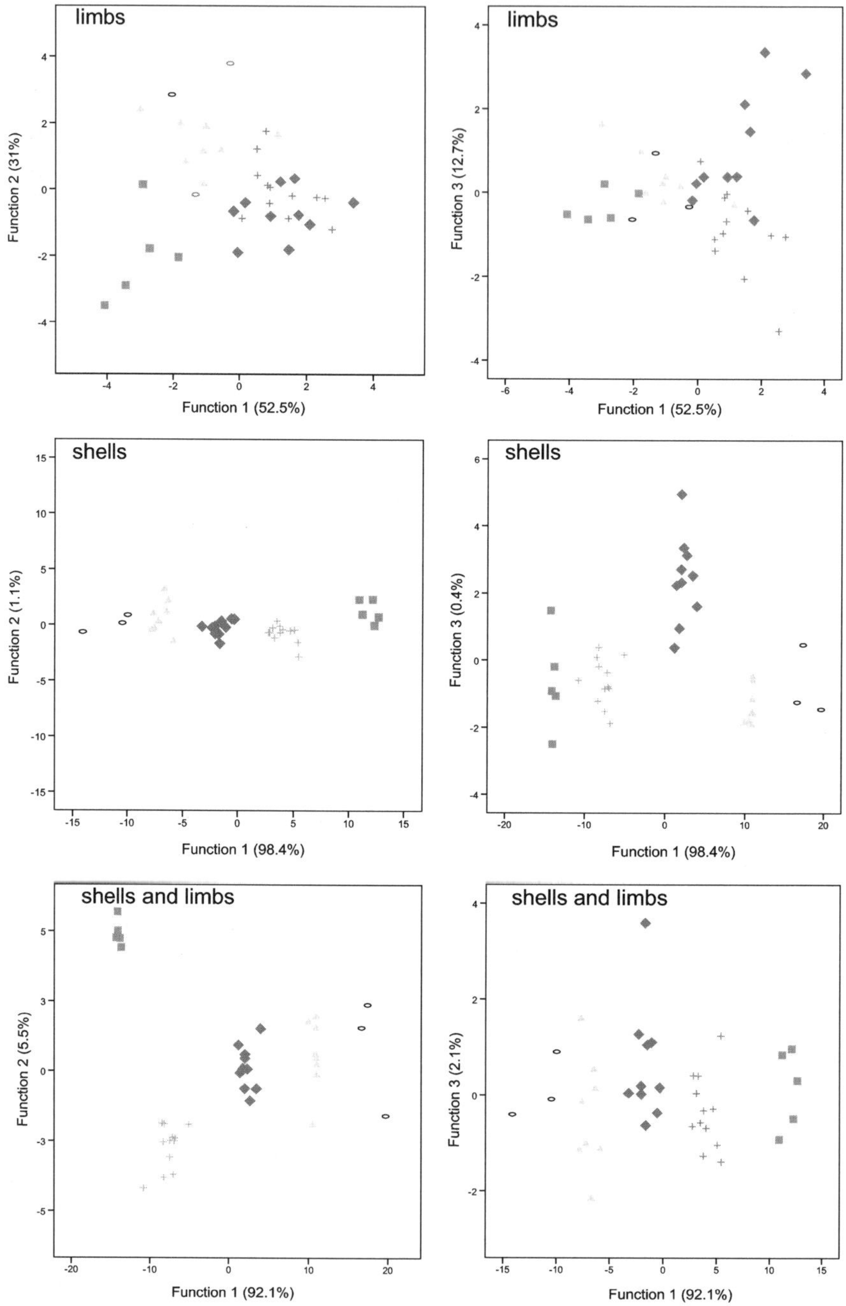

Fig. 8. Scatter plots of the first against the second and third, respectively, discrimination functions for analyses of limb ratios, shell parameters and limb ratios combined with shell parameters (square = females 1, cross = females 2, diamond = males, triangle = A-1, oval = A-2/3). This figure is published in colour in the online edition of this journal, which can be accessed via http://booksandjournals.brillonline.com/content/journals/15685403.

function 1 are highest for A1EIV/EIII (0.32), and A2EIV/EII (−0.09). For the second function highest correlations occur for A2EIV/EIII (0.26) and A2/T2 (−0.25). Highest correlations for function 3 represent the ratios A1/T1 (0.61), A1/T2 (0.47), T1/T2 (0.41) as well as T1EIII/EII (0.3) and T2EIII/EII (0.24). Thus, discrimination of the specimens based on limb traits cannot be described by a single parameter. However, the majority of the explained variability (53.1%) is associated with features of the first and the second antennas. In particular, the ratios A1EIV/EIII and A2 EIV/EII appear to provide a large contribution to the separation of the specimens into different groups.

A highly significant discrimination (Wilk's Lambda = 0.01, Chi square = 150.52, $P = 0.00$) between the groups occurs for the shell parameters. Classification results indicate that 100% of originally grouped specimens were correctly classified. The first canonical function accounts for 98.4% of the explained variability, the second and third for 1.1% and 0.4%, respectively (table II). This distinct separation effect is clearly visible in the scatter plot (fig. 8). The groups are aligned along axis 1, and sorted by instars. Thus, the large females have the most positive loadings and the juveniles A-2/A-3 the most negative loadings on axis 1. The variability range of the loadings of the groups on axis 2 is quite small. Standardised discrimination function coefficients are given in table IV. The highest coefficient on function 1 is length (1.24). The 'length sulcus' (1.02) and L/H (0.85) are highest coefficients on functions 2 and 3, respectively. Accordingly, canonical loading of the shell length was highest for function 1 (0.86), distance to the sulcus had the highest correlation with function 2 (0.87), and L/H ratio showed the highest correlation on function 3. Thus, the length of the shells is the main discriminating parameter. Minor influence is given by the position of the transversal sulcus. The height of the shells has no influence on the separation.

Corresponding to the previous discrimination analyses highly significant separation of the groups is also observable for both limb ratios and shells (Wilk's Lambda = 0.00, Chi square = 216.21, $P = 0.00$; table II). Classification results show that 100% of originally grouped specimens were correctly classified. The first canonical function accounts for 94.1% of explained variability, the second and third functions explain 3.8% and 1.6%, respectively (table II). The scatter plot (fig. 8) shows that the smaller females, males, A-1 and A-2/3 instars are aligned along axis 1 with increasing loadings. The large females are well separated from the other groups associated with highest positive loadings on axis 2. The groups are relatively widely distributed on axis 3. Standardised canonical discriminant function coefficients are summarized in table III. Highest values reach shell length (−1.76), L/H ratio (1.31), ratio between T1/T2 (1.22) as well as A1/T1 (−0.94). All variables of antennas, especially the second (e.g., A2EIV/EII: 6.4), have very high coefficients for function 2. For the first time, shell height (3.58) provides

TABLE III

Eigenvalues and Wilk's Lambda of discrimination analyses for limb ratios, shells, and limb ratios and shells

Function	Eigenvalue	% of variance	Cumulative %	Canonical correlation	Test of functions	Wilk's Lambda	Chi-square	df	Significance
Limb ratios									
1	2.835	53.1	53.1	0.860	1 to 4	0.049	84.385	52	0.003
2	1.593	29.9	83.0	0.784	2 to 4	0.188	46.746	36	0.108
3	0.712	13.3	96.3	0.645	3 to 4	0.488	20.067	22	0.579
4	0.196	3.7	100.0	0.405	4	0.836	5.018	10	0.890
Shells									
1	51.504	98.4	98.4	0.990	1 to 4	0.010	150.515	16	0.000
2	0.586	1.1	99.6	0.608	2 to 4	0.512	21.787	9	0.010
3	0.213	0.4	100.0	0.419	3 to 4	0.811	6.802	4	0.147
4	0.016	0.0	100.0	0.126	4	0.984	0.518	1	0.472
Limb ratios and shells									
1	100.913	94.1	94.1	0.995	1 to 4	0.000	200.621	68	0.000
2	4.030	3.8	97.8	0.895	2 to 4	0.045	80.394	48	0.002
3	1.684	1.6	99.4	0.792	3 to 4	0.228	38.391	30	0.140
4	0.631	0.6	100.0	0.622	4	0.613	12.725	14	0.548

some effect on discrimination. Highest canonical correlation is provided for shell length (−0.61) on function 1. Thus, according to negative correlation coefficients differentiation between the individuals decreases with decreasing size. The highest loading on function 2 is given to length sulcus (−0.36) aside A2EIV/EIII (−0.24) and A1EIV/EIII (0.21). More complex is the situation of the canonical loadings on function 3. The highest loadings on function 3 have L/H (0.31), A1/T1 (0.28) and A1/A2 (0.27).

DISCUSSION

Intraspecific variations of limbs and shells

Intraspecific variability of ostracods mostly covers variations in shape, size and length ratio of claws, setae and segments. These phenotypic variations are thought to reflect either underlying genotypic variability or are environmentally induced (ecophenotypic) or are a combination of interactive factors between both genotype and the environment (Martens et al., 1998a). There are no (qualitative) differences between the detailed descriptions by Purper (1974) and our material, except spines on the claw of T2 of females. In the present material, all females except four specimens of the small female group did not exhibit these spines.

TABLE IV

Standardised canonical discrimination function coefficients for analyses of limb ratios, shells, and limb ratios with shells

	Function			
	1	2	3	4
Limb ratios				
A1EV/EIII	*−1.89*	0.02	0.34	−0.80
A1EIV/EIII	*1.99*	0.12	−0.14	0.38
A2EIII/EIIIGM	*1.93*	0.42	0.34	*−1.65*
A2EIV/EIII	−0.16	*0.91*	0.61	0.59
A2EIV/EII	*0.89*	−0.25	−0.83	−0.87
A2EIII/EII	−0.68	−0.14	0.89	*1.08*
T1EIII/EII	*−0.82*	*−1.01*	0.02	−0.15
T2EIII/EII	*2.09*	*3.18*	−0.41	−0.35
A1/A2	*−2.06*	−0.61	−0.81	*2.76*
T1/T2	−0.36	−0.43	0.17	0.80
A1/T1	*1.05*	*1.73*	0.61	0.59
A2/T2	*−2.87*	*−2.65*	*−1.98*	*3.08*
A1/T2	*1.18*	*−1.28*	*2.25*	*−3.00*
Shells				
Length	*1.24*	−0.24	−0.01	−0.25
Height	−0.13	−0.20	0.39	0.94
Length sulcus	−0.19	*1.02*	−0.23	0.28
Length/height	−0.60	0.53	*0.85*	−0.26
Limb ratios and shells				
A1EV/EIII	0.66	*−1.27*	−0.86	0.85
A1EIV/EIII	0.09	*1.12*	*1.08*	−0.30
A2EIII/EIIIGM	−0.77	*2.15*	0.37	1.78
A2EIV/EIII	0.18	*−3.79*	*1.44*	−6.06
A2EIV/EII	−0.13	*6.40*	−0.49	9.98
A2EIII/EII	0.58	*−4.07*	0.48	−6.81
T1EIII/EII	0.11	−0.50	*−1.37*	0.35
T2EIII/EII	−0.46	*1.55*	*2.34*	1.25
A1/A2	0.06	*−2.04*	−0.65	−2.86
T1/T2	*1.22*	−0.45	0.66	−1.52
A1/T1	*−0.94*	0.46	*1.75*	1.04
A2/T2	0.26	*−2.63*	*−2.89*	−3.41
A1/T2	0.22	*1.89*	−0.50	2.90
Length	*−1.76*	−0.63	0.07	−1.41
Height	0.21	*3.58*	−0.19	5.93
Length sulcus	0.46	−0.30	−0.47	0.97
Length/height	*1.31*	−0.02	0.74	1.22

Ratios between limbs, e.g. A1/A2, include only the last three podomeres. Length sulcus is related to the distance between the margin and the transversal sulcus. Italicised values indicate correlation coefficients that represent highest absolute correlations between a variable and a discrimination function.

No differences in shell or limb morphology between individuals from the different localities appeared. The majority of limb traits of *Cytheridella ilosvayi* have relatively low variability ranges (see fig. 4, table I). However, a large amount of variability is displayed by the antennula and second antenna which is not observed in other ostracod species thus far (e.g., Iepure et al., 2007, 2008). Relative length of setae is considered as an important parameter causing intraspecific variability (e.g., Yin et al., 1999; Iepure et al., 2007, 2008). The performed measurements on *C. ilosvayi* included one seta at each limb. But only one ratio (A2EIII/EIIIGM) has been selected for multivariate analyses according to statistical tests for significance and normal distribution. Thus, the assumed higher influence of setae on the intraspecific variability might be attributed to difficulties in performing accurate measurements. After Maddocks (2000), the proportions of the podomeres are more important for intraspecific variability than those of the setae. However, no statistical analysis has been applied to test this hypothesis thus far. Our discrimination analyses of limb traits from *C. ilosvayi* support this hypothesis since they revealed a much higher influence of podomere ratios than of setae. In particular, the ratios of the fused podomere EIV and EIII of the antennula (A1EIV/EIII), as well as the ratio between EIV and EII of the antenna (A2EIV/EII) predominantly affects the explained variability (table V). To a minor degree, the ratios between limb pairs also affect the variability, i.e., A2 and T2. Further, the results demonstrate that the two female groups can be well separated from each other by podomere ratios (fig. 8). Considering the limb ratios as morphological trajectories (fig. 7) it is noticeable that the larger females deviate in most cases from the previous trend (e.g., A2EIII/EIIGM, fig. 7). The larger females display either lower or higher ratios than the smaller females. Naturally, one could expect a continuation of the development. This is most obvious in the antennas (i.e., A1EV/EIII to A2EIV/EIII), while there is a higher correspondence with the other groups for the ratios between limbs (e.g., A1/A2). Contrary to the deviations between the two females, the limb ratios of males (i.e., A1EIV/EIII and A2EIV/EII) are closer to that of the smaller females. This implies that the males might be the counterparts of the smaller females. The relatively large variation of ratios of the second antenna of the males could hint to a separation into two types of males (fig. 7). However, individual analyses are too small.

There is a discrepancy between the discrimination function coefficients and the correlation coefficients regarding limb ratios. Thus, high function coefficients of the proportions between limbs (e.g., A2/T2, table IV) indicate a high contribution to the discrimination function. Opposite, the correlation coefficients of the structure matrix (table V) reveal a strongly reduced number of variables correlating with the discrimination functions. The high number of function coefficients in

TABLE V

Structure matrices of three discrimination analyses (limb ratios, shells, limb ratios and shells) representing common correlations of a variable and a discrimination function within groups

	Function			
	1	2	3	4
Limb ratios				
A1EIV/EIII	*0.323*	0.197	−0.031	−0.061
A2EIV/EII	*−0.091*	0.035	0.011	0.065
A1EV/EIII	−0.063	0.287	0.001	−0.166
A2EIV/EIII	−0.229	*0.261*	0.048	−0.063
A2/T2	−0.010	*−0.248*	0.237	−0.021
A1/T1	−0.028	−0.031	*0.610*	0.033
A1/T2	0.040	−0.191	*0.473*	0.009
T1/T2	−0.098	−0.162	*0.408*	0.082
T1_EIII/EII	0.032	−0.059	*0.299*	−0.017
T2EIII/EII	0.067	0.047	*0.211*	0.055
A1/A2	0.137	0.114	0.240	*0.473*
A2EIII/EIIIGM	0.236	−0.081	0.045	*0.392*
A2EIII/EII	0.125	−0.127	0.242	*0.360*
Shells				
Length	*0.858*	0.265	0.439	−0.009
Length sulcus	0.230	*0.869*	−0.279	0.336
Length/height	−0.002	0.289	*0.904*	−0.316
Height	0.156	−0.106	0.445	0.876
Limb ratios and shells				
Length	*−0.610*	−0.297	0.207	0.088
Length sulcus	−0.160	*−0.360*	−0.067	0.345
A2EIV/EIII	−0.012	*−0.240*	0.065	0.094
A1EIV/EIII	−0.036	*0.205*	0.183	0.089
A2EIII/EIIIGM	0.000	*0.197*	0.108	−0.194
A2EIV/EII	0.002	*−0.080*	−0.001	−0.029
A2/T2	0.036	*0.046*	−0.010	0.029
Length/height	0.003	−0.122	*0.307*	−0.094
A1/T1	0.034	−0.028	*0.279*	0.172
A1/A2	−0.004	0.070	*0.273*	−0.135
Height	−0.112	0.013	*0.183*	0.054
A1/T2	0.039	0.066	*0.154*	0.107
A2EIII/EII	0.019	0.118	*0.141*	−0.131
T1EIII/EII	0.019	0.033	*0.127*	0.094
T2EIII/EII	0.013	0.057	*0.126*	0.044
T1/T2	0.040	−0.048	*0.098*	0.045
A1EV/EIII	−0.025	−0.115	0.099	0.150

Italicised values indicate correlation coefficients that represent highest absolute correlations between a variable and a discrimination function.

analysis 1 (limb ratios) is probably related to relatively high degree of interdependency between the variables. There are some highly correlated ratios, especially those which contribute one to another. For instance, T1EIII/EII is well correlated with T1/T2 ($r = 0.77$). The correlation matrix (table II) shows, however, that these ratios which have high correlation coefficients in the structure matrix do not show high interdependencies with other ratios. Since these coefficients are 'partial', they reflect an association after the influence of all other variables has been removed, and therefore it seems inappropriate to assess the importance of the limb ratio variables by the size of their coefficients (cf. Williams, 1983).

Considering both limb traits and shell features the length of the shells appears to be the most important variable for discriminating groups. Further, the position of the sulcus is responsible for a certain amount of variability and differentiation. According to low function coefficients and canonical correlations, the length of setae do not have an influence on the intraspecific variability of *C. ilosvayi*. Ratios related to thoracopods, especially ratios among limbs (e.g., A1/T1, A2/T2) have several high discrimination function coefficients. Their low absolute correlations with each discrimination function appear predominantly at function 3 (table V), indicating that thoracopods or their podomere ratios are stable and do not permit discrimination of specimens and instars, respectively.

As observed in other ostracod species, our data reveal no sexual dimorphism in the limbs (cf. Cohen & Morin, 1990) between males and smaller females. However, the majority of limb ratios of the larger females deviate from that of the males (table I).

So far, the low number of studies from different families and/or genera as well as the differences in chaetotaxy of the families prevents a definite assessment of the importance of setae and podomere ratios for the intraspecific variability of individual taxa. Comparisons with literature data are limited due to the use of different limbs, limb traits and/or setae and segments. Since measurements on the complete appendages including all setae would be extremely time consuming, a standard procedure would be needed that allows relatively fast processing. Limb traits that we used were selected according to available literature data and the manageability (easy to recognise and measure). Although terminal claws of the walking legs are found to display large variability (Yin et al., 1999; Iepure et al., 2008) we excluded them from the analyses because these are often slightly distorted and cannot be aligned evenly on the slide. The same applies for setae, which are often affected by ontogenetic changes (presence, style, position; Broodbakker & Danielopol, 1982). Thus, comparison between individuals of the last development stages is hindered. Accordingly, we choose setae that are present in instars down to A-3. Although it is reported that taxa of the subfamiliy Timiriaseviinae show higher variability in the third walking leg (Martens, 2000), we excluded this from our data since the bended position in *C. ilosvayi* precluded accurate measurements of that limb (T3).

Cytheridella morphotypes: cryptic species or mixed reproduction?

Two types of females that are predominantly differentiated by size (length) with the presence of only one type of males suggest either cryptic taxa or sympatric occurrence of mixed reproduction modes (i.e., sexual and parthenogenetic females).

Purper (1974) reported dimensions of female carapaces of *C. ilosvayi* with a length range of 957 to 1137 μm. This covers the ranges of our data set (Female 1: 1110-1140 μm length; Female 2: 921-992 μm length). Although Purper (1974) examined >1500 specimens, two (size) separated females were not mentioned. Carapace dimensions from a Guatemalan population were much smaller (females: 770-980 μm length; males: 634-787 μm length; Peréz, 2010), but the occurrences of two female morphotypes is also not reported. Beside size there are also various limb characters which differentiate the two females (fig. 7). For instance, the apical claw of the second antenna is proportionally larger than in the smaller females. The question remains, however, why there is only one type of male. Although there are hints of separation of males (see fig. 7, e.g., A1/A2) it could be possible that the morphological differences of the males are even smaller than in females (limbs) and could not have been differentiated by this kind of morphological analyses. The morphologies of the hemipenis did not reveal any differences or allowed a separation into groups. Assuming the possibility that potential male morphotypes have a strong morphological similarity, one could assign the same to the juveniles which also did not allow any separation. Because *C. ilosvayi* does not exhibits sexual dimorphism in T3 we could not differentiate A-1 in males or females. The low number of available *C. ilosvayi* specimens, however, precludes extensive analyses to test statistical significance of possible differentiation of males and A-1.

Although the knowledge of the existence of cryptic species exists for nearly 300 years, the discovery of cryptic species has increased exponentially over the past two decades because of the availability of DNA sequence data. Molecular analyses revealed exceptional cryptic diversity in some non-marine ostracods, with more than 40 cryptic taxa for *Eucypris virens* (Jurine, 1820) (e.g., Bode et al., 2010). Knowledge about cryptic ostracod species is predominantly restricted to information about genetic divergence (Adolfsson et al., 2009; Bode et al., 2010). Information about presence, extent, and style of morphological differences between cryptic species (including adults as well as juveniles) is missing.

Ostracods are known for their exceptionally wide range of different reproductive modes. Reproduction can be either sexual (species are fully sexual) or asexual (which are known as all-female population, e.g., Darwinulidae). Some of these parthenogenetic species can also reproduce sexually (e.g., *Limnocythere inopinata*; Horne & Martens, 1999). Often these species have a geographically restricted distribution which is termed geographic parthenogenesis (Martens et al., 1998b). Additionally, some ostracod species can coexist as sexually and asexually reproducing forms (e.g., Bellavere et al., 2002). The presence of two female morphotypes and only one type of males could point to a mixed reproduction population in which males and both sexual and asexual females occur sympatrically. Although it is known that most Cyprididae and a high number of candonid and cytheroid non-marine species have mixed reproduction (Martens et al., 1998b) this phenomenon is thought to be highly underestimated (Chaplin, 1993). Ostracod species groups exhibiting mixed reproduction include *Candonocypris novaezelandiae* (Baird, 1843) (cf. Chaplin, 1992, 1993), *Heterocypris incongruens* (Ramdohr, 1808) (cf. Bellavere et al., 2002) and *Eucypris virens* (e.g., Schön et al., 2000). In *H. incongruens*, coexisting female morphotypes are differentiated by a lobe-like expansion of the selvage on the posterior margin of the left valve (Bellavere et al., 2002). Sexual and parthenogenetic females of *E. virens* do not show consistent morphological differences of the limbs (Martins et al., 2009). *Candonocypris novaezelandiae* reveals two different morphs differentiated by size and colour (Chaplin, 1992, 1993). Thus, sexual/asexual reproduction might be expressed more often in shell characters. Neither parthenogenetic nor mixed reproduction populations of *C. ilosvayi* have been documented so far (Löffler, 1961; Purper, 1974; Peréz et al., 2010a, b). A reason for the lack of two female morphotypes could be the geographically limited distribution of these mixed reproduction populations (cf. Martens et al., 1998b). Many European species are represented throughout their geographical range by parthenogenetic populations while sexual populations are either unknown or restricted to a few, usually southern (circum-Mediterranean) localities, e.g., *Eucypris virens* and *Heterocypris incongruens* (cf. Horne & Martens, 1999). Based on repeated samplings Chaplin (1992) reported successive displacement of a sexual lineage by a closely related parthenogenetic form in a spatially heterogeneous and biotically diverse environment which could be related to human disturbance. This would underline the importance of ecological and climatological impacts on the reproduction mode and possibly on sympatric speciation. So far, most extant representatives of the subfamily Timiriaseviinae are known to be sexual except the parthenogenetic genus *Kovalevskiella* and sexual and/or asexual *Frambocythere* (Carbonel et al., 1986; Colin, 2011). The fossil (Miocene) species *Cytheridella danielopoli* Purper, 1979 found close to our sampling localities in the Brazilian Amazon are thought to be sexual (Gross et al., 2013). Thus, the pre-existence of parthenogenetic *Cytheridella* lineages is not documented.

Benthic non-marine ostracods likely recognise their mate through direct contact by their limbs. Copulation has been observed between males and parthenogenetic females but with limited success (Horne et al., 1998). Thus, the reason for the larger size of Female 1 could be attributed to mate behaviour, and thus copulation of the larger parthenogenetic female with males might be precluded. However, although males can build diverse structures in their copulatory organs, this does not necessarily correspond to female genitalia, although they may reproduce successfully (Broodbakker, 1982). Bellavere et al. (2002) interpreted the different body size of sexual and asexual females according to niche differentiation and assuming differences in the ecological adaptation of the morphotypes. Another reason could be the enlargement of the brood pouch of parthenogenetic females, which likely increases the number of eggs and/or juveniles and improving the survival of offspring. Nine to twelve eggs were counted in both female types, and there can only be speculations about the meaning of the size differences between the two female groups until more data are collected.

Based on morphological coherence and analogous to *E. virens* (Martens et al., 1998b), we consider both females as morphotypes of *Cytheridella*. Final clarification of whether these morphotypes represent cryptic species or mixed reproduction mode has to be determined by molecular biological analyses.

CONCLUSION

The morphological approach of this study provides quantitative information on hard- and soft-part variability of the ostracod *Cytheridella*.

Analyses of limbs and shells revealed that variability of soft part morphology is generally low. Highest variability is found in the antennula and second antenna. Thoracopods and setae have only minor influence on variability. This pattern is also shown by juveniles A-3 up to A-1. Further, males as well as females are symmetrical (left and right) at least for the antennula, second antenna as well as both thoracopods T1 and T2. This study documents that proportions between podomeres can be easily obtained and provide important information about variability.

Discrimination analyses that maximise the separation between groups indicate that shell parameters appeared to be more important for differentiating groups. In particular, shell length and the position of the transversal sulcus account for the majority of explained variability of the specimens, even in combination with limb ratios. Important limb traits discriminating the specimens are (pen)-ultimate podomeres of the antennas. Although there is some indication that proportions between limbs affect the variability, high interdependency of limbs and traits may exceed the actual influence.

Moreover, morphological analyses revealed two types of females that are separated primarily by size (length) and to a minor degree by podomere ratios of the antennas. Thus, the larger females ($\geqslant$1000 μm) were characterized by smaller A1EIV/EIII, A2EIII/EIIGM and larger A2EIV/EII ratios, respectively, than small females ($\leqslant$1000 μm) and males. The presence of two morphologically similar females and only one type of male may indicate either cryptic species in which the males cannot be identified or a mixed reproduction population in which parthenogenetic and sexual reproduction coexists.

ACKNOWLEDGEMENTS

Maria Ines Feijó Ramos (Museu Paraense Emílio Goeldi, Belém, Brazil) is thanked for organisation and kind assistance during fieldwork. Special thanks go to Liseth Peréz (Universidad Nacional Autónoma de México, Mexico City) for supply of photographic material and discussion. We are also grateful to Dan L. Danielopol for helpful comments. This study was funded by the Austrian Science Fund (P21748 and P26554). We also thank Finn Viehberg and two anonymous reviewers for critical comments on the manuscript.

REFERENCES

Adolfsson, S., Y. Michalakis, D. Paczesniak, S. N. S. Bode, R. K. Butlin, D. K. Lamatsch, M. J. F. Martins, O. Schmit, J. Vandekerkhove & J. Jokela, 2009. Evaluation of elevated ploidy and asexual reproduction as alternative explanations for geographic parthenogenesis in *Eucypris virens*, ostracods. Evolution, **64**: 986-997.

Alvarez Zarikian, C. A., P. K. Swart, J. A. Gifford & P. L. Blackwelder, 2005. Holocene paleohydrology of Little Salt Spring, Florida, based on ostracod assemblages and stable isotopes. Palaeogeogr. Palaeocl., **225**: 134-156.

Bellavere, C., G. Benassi, M. Calzolari, C. Meisch, K. Glencoe Mckenzie & V. Rossi, 2002. Heterocypris (Crustacea, Ostracoda) from the Isole Pelagie (Sicily, Italy): the coexistence of different morphotypes. Ital. J. Zool., **69**: 53-57.

Bode, S. N. S., S. Adolfsson, D. Lamatsch, M. J. F. Martins, O. Schmit, J. Vandekerkhove, F. Mezquita, T. Namiotko, G. Rossetti, I. Schön, R. K. Butlin & K. Martens, 2010. Exceptional cryptic diversity and multiple origins of parthenogenesis in a freshwater ostracod. Mol. Phylogenet. Evol., **54**: 542-552.

Broodbakker, N. W., 1982. The genus *Heterocypris* (Crustacea, Ostracoda) in the West Indies. Part 1. Taxonomic characters. Bijdr. Dierkd., **52**: 207-227.

Broodbakker, N. W. & D. L. Danielopol, 1982. The chaetotaxy of Cypridacea (Crustacea, Ostracoda) limbs: proposals for a descriptive model. Bijdr. Dierkd., **52**: 102-120.

Carbonel, P., D. L. Danielopol & L. Londeix, 1986. *Kovalevskiella* (Ostracoda, Timitriaseviinae) genre á mode de vie benthique l'Oligocène, son adaptation á la vie interstitielle. Geobios, **19**: 677-687.

Chaplin, J., 1992. Variation in the mode of reproduction among individuals of the ostracod *Candonocypris novaezelandiae*. Heredity, **68**: 411-424.

— —, 1993. The local displacement of a sexually reproducing ostracod by a conspecific parthenogen. Heredity, **71**: 259-268.

COHEN, A. C. & J. G. MORIN, 1990. Patterns of reproduction in ostracodes: a review. J. Crustacean Biol., **10**: 184-211.

COLIN, J. P., 2011. From light to darkness: from *Frambocythere* Colin, 1980 to *Kovalevskiella* Klein, 1963 (Limnocytheridae, Timiriaseviinae). Joannea Geol. Paläont., **11**: 44-47.

COLIN, J. P., Y. TAMBAREAU & V. A. KRASHENINNIKOV, 1997. An early record of the genus *Cytheridella* Daday, 1905 (Ostracoda, Limnocytheridae, Timiriaseviinae) from the Upper Cretaceous of Mali, West Africa: palaeobiogeographical and palaeoecological considerations. J. Micropalaeontol., **16**: 91-95.

DANIELOPOL, D. L., 1978. Über Herkunft und Morphologie des Süsswasser-hypogaischen Candoninae (Crustacea, Ostracoda). Sitzungsberichte der Österreichischen Akademie der Wissenschaften, mathematisch-naturwissenschaftliche Klasse, Abteilung, **1**: 44-47.

DE DECKKER, P. & R. M. FORESTER, 1988. The use of ostracods to reconstruct continental palaeoenvironmental records. In: P. DE DECKKER, J. P. COLIN & J. P. PEYPOUQUET (eds.), Ostracoda in the earth sciences: 175-199. (Elsevier, Amsterdam).

GABRIEL, J. J., E. G. REINHARDT, M. C. PEROS, D. E. DAVIDSON, P. J. HENGSTUM & P. A. BEDDOWS, 2009. Palaeoenvironmental evolution of Cenote Aktun Ha (Carwash) on the Yucatan Peninsula, Mexico and its response to Holocene sea-level rise. J. Paleolimnol., **42**: 199-213.

GROSS, M., M. RAMOS, M. CAPORALETTI & W. E. PILLER, 2013. From the edge of the Pebas System — ostracods (Crustacea) and their palaeoenvironmental implication (Late Miocene; Solimões Formation; Western Amazonia/Brazil). J. S. Am. Earth Sci., **42**: 216-241.

HIGUTI, J., L. F. M. VELHO, F. A. LANSAC-TÔHA & K. MARTENS, 2007. Pleuston communities are buffered from regional flood pulses: the example of ostracods in the Paraná River floodplain, Brazil. Freshwater Biol., **52**: 1930-1943.

HORNE, D. J., 2007. A mutual temperature range method for Quaternary palaeoclimatic analysis using European nonmarine Ostracoda. Quaternary Sci. Rev., **26**: 1398-1415.

HORNE, D. J., D. L. DANIELOPOL & K. MARTENS, 1998. Reproductive behaviour. In: K. MARTENS (ed.), Sex and parthenogenesis — evolutionary ecology of reproductive modes in non-marine ostracods: 157-195. (Backhuys Publishers, Leiden).

HORNE, D. J. & K. MARTENS, 1999. Geographical parthenogenesis in European non-marine ostracods: post-glacial invasion or Holocene stability? Hydrobiologia, **391**: 1-7.

IBM, 2010. IBM SPSS Statistics for Windows, Version 19.0. (IBM, Armonk, NY).

IEPURE, S., T. NAMIOTKO & D. L. DANIELOPOL, 2007. Evolutionary and taxonomic aspects within the species group *Pseudocandona eremita* (Vejdovsky) (Ostracoda, Candonidae). Hydrobiologia, **585**: 159-180.

— —, — — & — —, 2008. Morphological diversity and microevolutionary aspects of the lineage *Cryptocandona vavrai* Kaufmann, 1900 (Ostracoda, Candoninae). Ann. Limnol.-Int. J. Lim., **44**: 151-166.

KARANOVIC, I., 2012. Recent freshwater ostracods of the world. Crustacea, Ostracoda, Podocopida. (Springer, Berlin).

LÖFFLER, H., 1961. Zur Systematik und Ökologie der chilenischen Süßwasserentomostraken. Beitr. neotrop. Fauna, **2**: 196-206.

MADDOCKS, R. F., 2000. The antennule in podocopid Ostracoda: chaetotaxy, ontogeny, and morphometrics. Micropaleontology, **46**: i-iv + 1-72.

MARTENS, K., 1998. Sex and ostracods: A new synthesis. In: K. MARTENS (ed.), Sex and parthenogenesis — evolutionary ecology of reproductive modes in non-marine ostracods: 295-321. (Backhuys Publishers, Leiden).

— —, 2000. Factors affecting the divergence of mate recognition systems in the Limnocytherinae (Crustacea, Ostracoda). Hydrobiologia, **419**: 83-101.

MARTENS, K., G. ROSSETTI & W. GEIGER, 1998a. Intraspecific morphological variability of limbs. In: K. MARTENS (ed.), Sex and parthenogenesis — evolutionary ecology of reproductive modes in non-marine ostracods: 143-155. (Backhuys Publishers, Leiden).

MARTENS, K., G. ROSSETTI & A. BALTANÁS, 1998b. Reprodutive modes and taxonomy. In: K. MARTENS (ed.), Sex and parthenogenesis — evolutionary ecology of reproductive modes in non-marine ostracods: 197-214. (Backhuys Publishers, Leiden).

MARTINS, M. J. F., J. VANDEKERKHOVE, F. MEZQUITA, O. SCHMIT, J. RUEDA, G. ROSSETTI & T. NAMIOTKO, 2009. Dynamics of sexual and parthenogenetic populations of *Eucypris virens* (Crustacea: Ostracoda) in three temporary ponds. Hydrobiologia, **636**: 219-232.

MEISCH, C., 2000. Freshwater ostracods of western and central Europe. (Spektrum Akademischer Verlag, Heidelberg, Berlin).

MORMUL, R. P., S. M. THOMAZ, J. HIGUTI & K. MARTENS, 2010. Ostracod (Crustacea) colonization of a native and a non-native macrophyte species of Hydrocharitaceae in the Upper Paraná floodplain (Brazil): an experimental evaluation. Hydrobiologia, **644**: 185-193.

NEIL, J. V., 2000. Factors influencing intraspecific variation and polymorphism in marine podocopid Ostracoda, with particular reference to Tertiary species from southeastern Australia. Hydrobiologia, **419**: 161-180.

PEREZ-GELABERT, D., 2008. Zootaxa, arthropods of Hispaniola (Dominican Republic and Haiti). Zootaxa, **1831**: 1-530.

PERÉZ, L., 2010. Non-marine ostracodes from the Yucatán Peninsula as late Quaternary paleoenvironmental indicators. (Ph.D. Thesis, Technische Universität Braunschweig, Braunschweig).

PERÉZ, L., R. BUGJA, J. MASSAFERRO, P. STEEB, R. VAN GELDERN, P. FRENZEL, M. BRENNER, B. SCHARF & A. SCHWALB, 2010a. Post-Columbian environmental history of Lago Petén Itzá, Guatemala. Revista Mex. Cienc. Geol., **27**: 490-507.

PERÉZ, L., J. LORENSCHAT, R. BUGJA, M. BRENNER, B. SCHARF & A. SCHWALB, 2010b. Distribution, diversity and ecology of modern freshwater ostracods (Crustacea), and hydrochemical characteristics of Lago Petén Itzá, Guatemala. J. Limnol., **69**: 146-159.

PURPER, I., 1974. *Cytheridella boldii* Purper, sp. nov. (Ostracoda) from Venezuela and a Revision of the genus *Cytheridella* Daday, 1905. An. Acad. Bras. Ciênc., **46**: 635-662.

RODRIGUEZ-LAZARO, J. & F. RUIZ-MUÑOZ, 2012. A general introduction to ostracods: morphology, distribution, fossil record and applications. In: D. J. HORNE, J. HOLMES, J. RODRIGUEZ-LAZARO & F. VIEHBERG (eds.), Ostracoda as proxies for Quaternary climate change. Developments in Quaternary Science: 1-14. (Elsevier Science, Amsterdam).

SCHÖN, I., R. BUTLIN, H. I. GRIFFITHS & K. MARTENS, 1998. Slow molecular evolution in an ancient asexual ostracod. Proc. Roy. Soc. Lond. B: Biol. Sci., **265**: 235-242.

SCHÖN, I., A. GANDOLFI, E. DI MASSO, V. ROSSI, H. I. GRIFFITHS, K. MARTENS & R. BUTLIN, 2000. Persistence of asexuality through mixed reproduction in *Eucypris virens* (Crustacea, Ostracoda). Heredity, **84**: 161-169.

SCHÖN, I. & K. MARTENS, 2012. Molecular analyses of ostracod flocks from Lake Baikal and Lake Tanganyika. Hydrobiologia, **682**: 91-110.

SMITH, R. & T. KAMIYA, 2003. The ontogeny of *Loxoconcha japonica* Ishizaki, 1968 (Cytheroidea, Ostracoda, Crustacea). Hydrobiologia, **490**: 31-52.

— — & — —, 2005. The ontogeny of the entocytherid ostracod *Uncinocythere occidentalis* (Kozloff & Whitman, 1954) Hart, 1962 (Crustacea). Hydrobiologia, **538**: 217-229.

VON DADAY, E., 1905. Untersuchungen über die Süsswasser-Mikrofauna Paraguays. Zoologica, **44**-IV Ostracoda: 243-270.

WILLARD, D. A., C. E. BERNHARDT, G. R. BROOKS, T. M. CRONIN, T. EDGAR & R. LARSON, 2007. Deglacial climate variability in central Florida, USA. Palaeogeogr. Palaeocl., **251**: 366-382.

WILLIAMS, B., 1983. Some observations of the use of discriminant analysis in ecology. Ecology, **64**: 1283-1291.

WROZYNA, C., P. FRENZEL, P. STEEB, L. ZHU, R. VAN GELDERN, A. MACKENSEN & A. SCHWALB, 2010. Stable isotope and ostracode species assemblage evidence for lake level changes of Nam Co, southern Tibet, during the past 600 years. Quatern. Int., **212**: 2-13.

YIN, Y., W. GEIGER & K. MARTENS, 1999. Effects of genotype and environment on phenotypic variability in *Limnocythere inopinata* (Crustacea: Ostracoda). Hydrobiologia, **400**: 85-114.

First received 30 November 2012.
Final version accepted 14 May 2014.

[When citing this chapter, refer to Crustaceana 87 (2014) 1072–1094]

This paper is dedicated to Prof. Dr. Eugen Kempf on the occasion of his 80th birthday (16 April 2012)

THE OSTRACOD SPRINGTAIL — CAMERA RECORDINGS OF A PREVIOUSLY UNDESCRIBED HIGH-SPEED ESCAPE JUMP IN THE GENUS *TANYCYPRIS* (OSTRACODA, CYPRIDOIDEA)

BY

RENATE MATZKE-KARASZ[1,4]), CHRISTINA NAGLER[2]) and SIMON HOFMANN[3])

[1]) Department of Earth and Environmental Sciences, Palaeontology and Geobiology, GeoBio-Center[LMU], Ludwig-Maximilians-Universität München, Richard-Wagner-Strasse 10, D-80333 Munich, Germany

[2]) Center of Life and Food Sciences Weihenstephan, Technische Universität München, Alte Akademie 8, D-85354 Freising, Germany

[3]) Faculty of Electrical Engineering, University of Applied Sciences Regensburg, Seybothstraße 2, D-93053 Regensburg, Germany

ABSTRACT

During an ostracod sampling campaign in the city of Munich (Germany) samples were taken from containers in a greenhouse of the Munich Botanical Garden. Beside the ubiquitous species *Cypridopsis vidua* (O. F. Müller, 1776), the samples contained four alien species, i.e., *Chlamydotheca arcuata* (Sars, 1901), *Strandesia bicuspis* (Claus, 1892), *Tanycypris centa* Chang, Lee & Smith, 2012, and *Tanycypris alfonsi* Nagler, Geist & Matzke-Karasz, 2014. While sorting the living *Tanycypris* specimens, a yet undescribed usage of the caudal rami was observed. Freshwater ostracods usually move on or in the sediment by using their first and second antennae, walking legs and — if not reduced — their caudal rami. During (non-swimming) locomotion of most freshwater ostracods with well-developed caudal rami, they help pushing the body forward by being used as a lever. This movement can be fast, but has never been reported to include sudden jumps. In contrast, both investigated *Tanycypris* species show an extraordinarily fast movement, especially when disturbed. Recordings with a high-speed camera were made, shooting horizontally into a 1.5-mm-thick micro-aquarium. The fast movement could be identified as a powerful jump, much resembling the movement of a catapult, propelled by a very rapid repulsion of the caudal rami from the ground. Although sized only around 1 mm, the observed specimens reached top speeds of up to 0.75 ms^{-1}. Anatomically, this speed is obtained by the exceptional length of the caudal rami in *Tanycypris*, combined with a well-developed musculature, which stretches from a broadened posterior end of soft body along the so-called 'caudal rami attachment'. The jump itself resembles that of springtails or fleas, where the jump is powered by the energy previously stored in an elastic proteinaceous material; however, in *Tanycypris* no such mechanism could be detected and thus the energy for the catapult-like jump must be considered muscular, possibly aided by tendon-like structures and/or a mechanism involving a muscular pre-tension by a click-joint as recorded for Squillids.

[4]) Corresponding author; e-mail: matzke-karasz@lmu.de

Key words. — Caudal rami, caudal rami attachment, Triebel loop, fast jump, tomography

ZUSAMMENFASSUNG

Im Verlauf einer Beprobung von Gewässern der Stadt München (Deutschland) in Hinblick auf Ostracodenvorkommen wurden Proben aus Becken eines Gewächshauses des botanischen Gartens München entnommen. Neben der weltweit vorkommenden Art *Cypridopsis vidua* (O. F. Müller, 1776), beinhalteten die Proben vier exotische Arten, zwar *Chlamydotheca arcuata* (Sars, 1901), *Strandesia bicuspis* (Claus, 1892), *Tanycypris centa* Chang, Lee & Smith, 2012, und *Tanycypris alfonsi* Nagler, Geist & Matzke-Karasz, 2014. Während lebende Exemplare von *Tanycypris* aus den Proben entnommen wurden, fiel eine bisher noch nicht beschriebene Nutzung der caudalen Rami auf. Süsswasser-Ostracoden bewegen sich normalerweise mit den ersten und zweiten Antennen, dem Laufbeinpaar und — wenn nicht reduziert — ihren caudalen Rami auf oder im Sediment fort. Bei der Laufbewegung der meisten Süsswasser-Ostracoden mit gut entwickelten caudalen Rami helfen diese, den Körper vorwärts zu schieben, indem sie wie ein Hebel eingesetzt werden. Diese Bewegung kann schnell sein, doch wurde nie zuvor von plötzlichen Sprüngen berichtet. Im Gegensatz dazu zeigen beide untersuchten *Tanycypris*-Arten eine außerordentlich schnelle Bewegung, insbesondere, wenn sie gestört werden. Aufnahmen mit einer Hochgeschwindigkeitskamera wurden erstellt, indem horizontal in ein 1.5 mm dickes Mikroaquarium gefilmt wurde. Die schnelle Bewegung konnte als ein kraftvoller Sprung identifiziert werden, der einer Katapultbewegung ähnelt. Er wird vorangetrieben, indem sich die caudalen Rami sehr schnell vom Untergrund abstoßen. Obwohl die Tiere nur ca. 1 mm lang sind, erreichen sie Höchstgeschwindigkeiten von bis zu 0.75 ms^{-1}. Anatomisch wird diese Geschwindigkeit durch die außergewöhnliche Länge der caudalen Rami bei der Gattung *Tanycypris* erreicht, kombiniert mit einer gut entwickelten Muskulatur, die sich vom verbreiterten Hinterende des Weichkörpers entlang des sogenannten 'caudal rami attachment' zieht. Der Sprung selbst ähnelt dem von Springschwänzen oder Flöhen, bei denen der Sprung von der zuvor in einem elastischen Protein gespeicherten Energie angetrieben wird; bei *Tanycypris* konnte jedoch kein solcher Mechanismus nachgewiesen werden und daher muss die Energie für diesen katapultartigen Sprung muskulären Ursprungs sein, möglicherweise unterstützt durch Bänder und/oder durch einen Mechanismus, der auf einer muskulären Vorspannung und einem Klick-Gelenk basiert, so, wie er für Squillidae beschrieben wurde.

INTRODUCTION

During an ostracod sampling campaign in the city of Munich, Germany, in 2011, samples were taken from containers in a 20°C greenhouse of the Munich Botanical Garden. The ca. 1 m^2 large containers accommodate non-native hydrophytes growing on partly sandy and muddy sediment in 1-10 cm depths of water.

Beside the ubiquitous species *Cypridopsis vidua* (O. F. Müller, 1776), the samples contained four alien species, i.e., *Chlamydotheca arcuata* (Sars, 1901), *Strandesia bicuspis* (Claus, 1892), *Tanycypris centa* Chang, Lee & Smith, 2012, and *Tanycypris alfonsi* Nagler, Geist & Matzke-Karasz, 2014 (fig. 1). A subsequent image search on the internet, targeting pages of fanciers and breeders of exotic

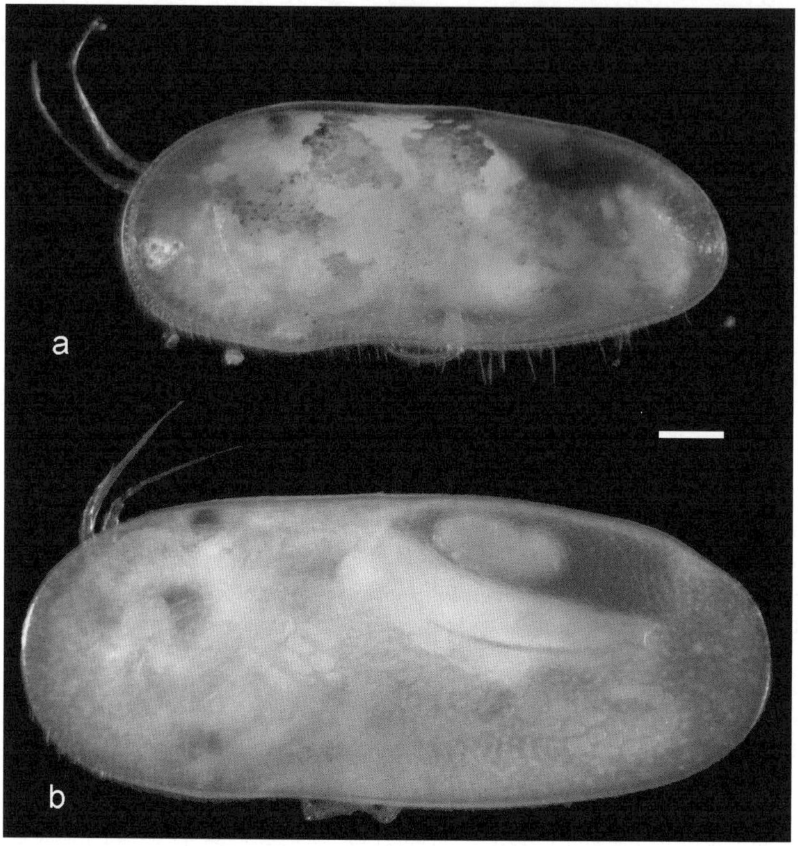

Fig. 1. *Tanycypris centa* Chang, Lee & Smith, 2012 (above) and *Tanycypris alfonsi* Nagler, Geist & Matzke-Karasz, 2014 (below), anterior to left. In the latter, right below a feces pellet accidentally left behind in the carapace lumen, the huge white muscles stretching along the caudal rami attachment are visible. Scale bar = 100 μm. This figure is published in colour in the online edition of this journal, which can be accessed via http://booksandjournals.brillonline.com/content/journals/15685403.

shrimps showed that *C. arcuata*, *S. bicuspis* and *Tanycypris* species are frequent invaders of private aquaria in the German-speaking area and are assumed to be transported attached to imported plants, either as adult animals or as eggs. However, their geographic origins cannot be traced back. Natural occurrences of the genus *Tanycypris* Triebel, 1959 have so far been reported from Asia (i.e., Korea, Thailand, Philippines, Indonesia and Japan), South America (Chile, Brazil) and Madagascar.

Species of the genus *Tanycypris* are characterized by an elongated ellipsoidal (torpedo-like) carapace shape and exceptionally long caudal rami. The distal tips of the longest claws of this posterior-most pair of appendages reach anteriorly to the mouth when folded underneath the ventrum in the resting position (fig. 2). The terminal claws are long and set ventrally with dozens of long spines.

Freshwater ostracods usually move on or in the sediment by using their first and second antennae, walking legs and — if not reduced — their caudal rami. It has been supposed that the caudal rami show a "stronger development" in those species that predominantly burrow within the sediment (Hartmann, 1967: 222). Irrespective of this hypothesis, freshwater ostracods with well-developed caudal rami use them to help push the body forward during (non-swimming) locomotion. The rami are thus acting as a lever (Alm, 1916). This movement can be fast (i.e., in *Fabaeformiscandona* species) but has never been reported to include sudden jumps of extraordinarily high speed.

However, while trying to sort living *Tanycypris* specimens from the samples, their sudden and obviously very fast movements let them escape from being sucked into the pipette. Since this movement was too fast to be resolved by the naked eye, recordings with a high-speed camera were made, shooting horizontally into micro-aquarium. By doing so, a yet undescribed usage of the caudal rami was observed and will be presented here.

A tentative record of a jumping behavior is that of specimens of *Prionocypris longiforma* Dobbin, 1941, which have been found to jump up to the water surface, when transferred into a container (Külköylüoğlu, 1999; unpublished Ph.D. thesis). This record will be discussed in the light of the results presented here.

MATERIAL AND METHODS

Movements of specimens of the species *Tanycypris centa* and of *Tanycypris alfonsi* were monitored. Sediment and water samples were collected from containers from a greenhouse of the Botanical Garden in Munich. Here, non-native hydrophytes are grown on sandy to muddy sediment in 1-10 cm depths of water. Ostracods were sorted under a Wild Stereoscopic Microscope M3Z from the samples, using pipettes.

Jumps were observed in both species; however, detailed analyses were only carried out in *Tanycypris centa*. The average body size of *Tanycypris centa* is 0.95 mm (length) and 0.46 mm (height).

Camera-recordings

Animals were placed in a micro-aquarium (1.5 mm wide) built from standard dissection glass slides, using silicone rubber. A small amount of quartz grains and some wood particles (broken off tooth-picks) were used to provide a bottom surface. Video recording was carried out using an Olympus i-SPEED 3 high-speed camera, shooting horizontally into the micro-aquarium.

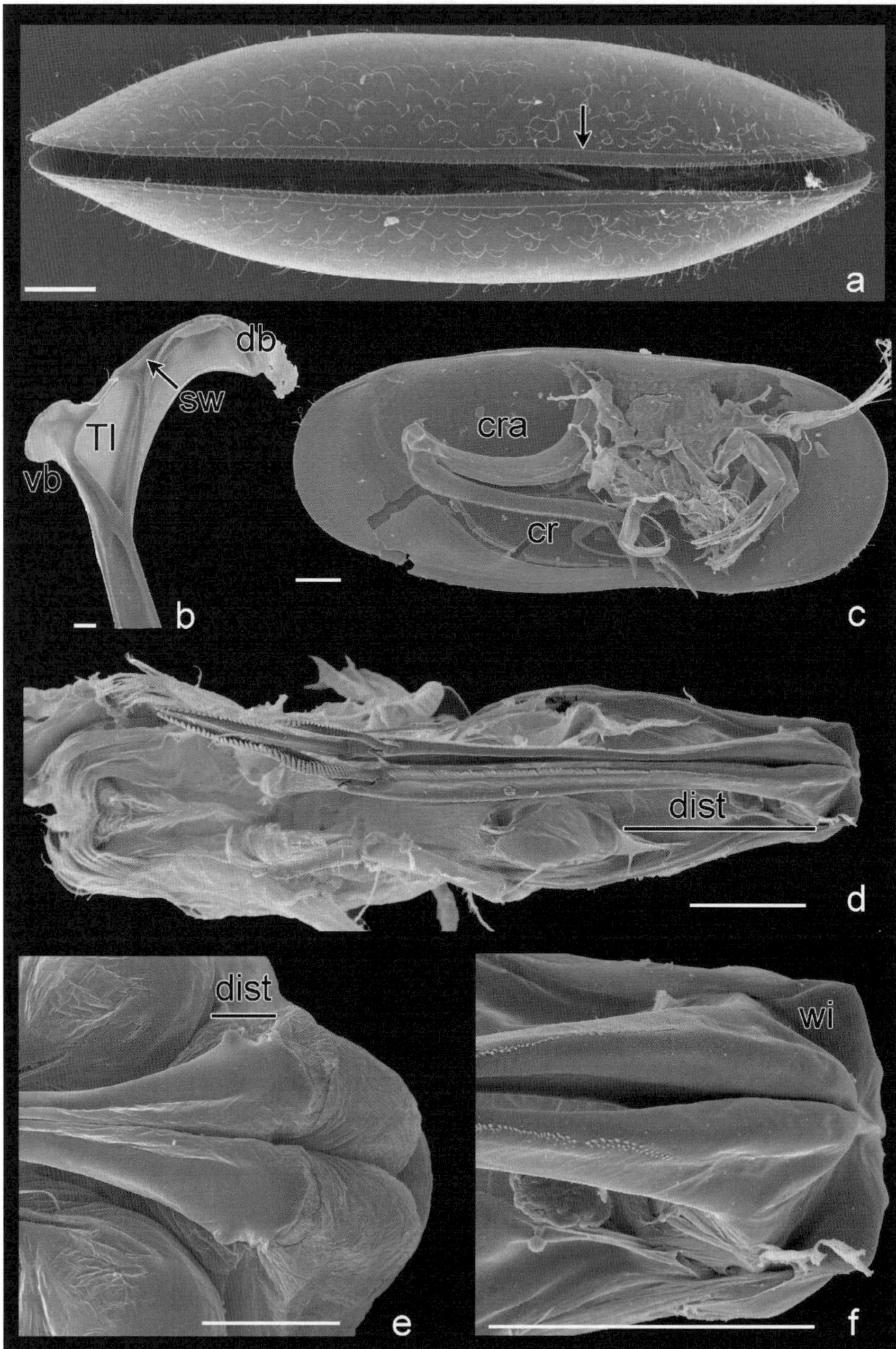
a
db
sw
Tl
vb
b
cra
cr
c
dist
d
dist
e
wi
f

Twenty-eight jumps were recorded at 2000 frames per second. The jumps were either apparently spontaneous or they were provoked by touching the ostracod with a fine tungsten wire. For illumination, four 25-50 W halogen lamps with gooseneck light guides were used. The experiments were carried out at 20°C.

The image analysis was made using the i-SPEED camera software (Olympus) and the free software ImageJ (NIH Image). Of twenty-eight recordings, twelve provided sufficient image quality to be further analysed, especially regarding a good focus over the entire jump. Of these 12 jumps of the species *Tanycypris centa*, detailed analyses of the movement of the caudal rami were made and covered distances of each jump were measured. Speed and exerted force were calculated from short video sequences, analysing every 2nd to 10th image, depending on the image quality. Since recordings were carried out at 2000 images per second continuously, the respective time period between two analysed images could be calculated, just as velocity in the particular phase of the jump. In periods of high velocity, images closer to one another were picked than in periods of low velocity. For analysis of the caudal rami extension, a 'jump' was defined by the movement cycle of the caudal rami: it started, when the rami became visible outside the carapace and was finished when they were entirely moved back into the carapace to their 'resting position' underneath the ventrum. Analysis of velocity and force over time started when the caudal rami touched the ground to push the body off.

Extension angles of the caudal rami were measured with the ventral margin of the carapace representing the zero-bottom line. Extension time and angle measurement started, when the rami appeared from the carapace. Retraction time started when the extension angle started to decrease and ended, when the rami were completely retracted into the carapace.

For scanning electron microscope (SEM) observations, ostracods were fixed and dissected in 70% ethanol, dehydrated, critical point dried (BalTec CPD 030), coated in gold (Polaron Sputter Coater), and finally studied using a Leo 1430VP.

Fig. 2. *Tanycypris alfonsi* Nagler, Geist & Matzke-Karasz, 2014 (a-d, f) and *Eucypis virens* (Jurine, 1820) (e). a, ventral view, anterior to right, arrow indicating the distal tips of the caudal rami reaching to the mouth region. b, anterior region of caudal rami attachment, with ventral branch (vb), dorsal branch (db), Triebel loop (Tl) and second window (sw). c, view of entire animal with right valve removed, caudal rami (cr) and caudal rami attachment (cra), the latter obscured by musculature. d, ventral view without carapace (anterior to left), with black line showing the long distance (dist) from elevated female reproductive area to middle of the caudal rami point of origin. e, posterior ventral view of female *E. virens*, with black line showing the short distance (dist) from elevated female reproductive area to middle of the caudal rami point of origin. Posterior end of soft body without special features. f, posterior ventral view of *Tanycypris alfonsi* female, showing the point where caudal rami fuse with caudal rami attachment, here forming wing-like processes (wi). Scale bars = 10 μm for b, 100 μm for a, c-f.

Tomographic studies result from experiment ec-945 at the European Sychrotron Radiation Facility (ID19). A pink beam of 19 keV energy (insertion device U17.6, gap 20) was used for scans with 1500 projections over 180 degrees at 0.2 s per projection. A FreLoN 2K14 ccd camera was mounted on a visible light microscope coupled to a LSO 8.8 μm scintillator. Propagation distance was 20 mm, pixel size 0.56 μm. Data reconstruction was done with a single distance phase retrieval algorithm of Paganin ring correction in post-processing on reconstructed slices, followed by 16 bits tif stack conversion. For data visualization, the Amira® software (vsg) was used.

Calculations were carried out on the basis of Kiørboe et al. (2010), who analysed escape jumps in three copepod species. By doing so, comparison of the efficiency of fast movements in both groups is possible, as far as their different locomotion techniques allow.

Abbreviations and symbols used herein are: α, aspect ratio of ostracod body; b, height of ostracod (mm); C_{D}, drag coefficient; D, drag (N); F, average force over time (μN); L, length of ostracod carapace (mm); m, mass of ostracod (kg); m_{f}, mass of water displaced by ostracod (kg); ν, kinematic viscosity (m^2 s^{-1}); ρ, ostracod density (kg m^{-3}); ρ_{f}, water density (kg m^{-3}); r, correlation coefficient; R, radius of model ostracod (mm); Re, Reynolds number; $\mathrm{Re}_{\mathrm{max}}$, Reynolds number based on maximum velocity; τ, Stoke's time scale (ms); T, temperature (°C); t, time (ms); V, ostracod velocity (mm s^{-1}).

In accordance with Kiørboe et al. (2010), we approximate the ostracod body as a prolate spheroid, with mass m (=volume) being:

$$m = \frac{4}{3}\pi\rho R^2 \tag{1}$$

With the carapace length L representing the major axis and its height b describing the equatorial diameter of the spheroid, the aspect ratio α of the ostracod is:

$$\alpha = \frac{b}{L} \tag{2}$$

to allow for the calculation of radius R:

$$R = \alpha^{\frac{2}{3}}\left(\frac{L}{2}\right) \tag{3}$$

The mean force F (averaged over jump duration; not the peak force) exerted by the ostracod was calculated using the equation for motion:

$$\left(m + \frac{1}{2}m_{\mathrm{f}}\right)\frac{\mathrm{d}V}{\mathrm{d}t} = F - D \tag{4}$$

with drag D being expressed as:

$$D = \frac{1}{2}\pi C_{\mathrm{D}}\rho_{\mathrm{f}}R^2V^2 \tag{5}$$

The drag-coefficient C_D depends on the Reynolds number Re describing the ratio of inertial forces to viscous forces in a fluid, as a function of the ostracod body radius R and its velocity V:

$$\mathrm{Re} = \frac{2RV}{\nu} \tag{6}$$

with the kinematic viscosity ν being

$$\nu = 1.004 \times 10^{-6}\ \mathrm{m^2 s^{-1}} \quad (\text{for } T = 20°\mathrm{C}).$$

Now the drag coefficient C_D can be calculated using:

$$C_D = \frac{24}{\mathrm{Re}} + \frac{5}{\sqrt{\mathrm{Re}}} + \frac{2}{5} \tag{7}$$

To calculate the propulsion force, the following equation was differentiated with respect to t:

$$V(t) = \frac{V_{\max}}{(1 + \frac{\mathrm{Re}_{\max}}{60})e^{(t - \frac{t_{\max}}{\tau})} - \frac{\mathrm{Re}_{\max}}{60}} \tag{8}$$

with τ representing the Stokes time scale as:

$$\tau = \frac{R_2}{3\nu} \tag{9}$$

Differentiating eq. (8), we obtain:

$$\frac{dV}{dt} = -\frac{60(\mathrm{Re}_{\max} + 60)\nu_{\max} e^{\frac{t\max + t}{\tau}}}{\tau[\mathrm{Re}_{\max} e^{\frac{t_{\max}}{\tau}} - (\mathrm{Re}_{\max} + 60e^{\frac{t}{\tau}})]^2} \tag{10}$$

Merging eqs. (10) and (4), the average force F exerted over a single jump can be calculated. The correlation coefficients r between force, velocity and deflection angle of the caudal rami were analysed using the Pearson product-moment correlation.

RESULTS

Analysis of movement

The extension of the long caudal rami is the essential movement in the *Tanycypris* jump; working as an effective lever, they are being forcefully expelled from their resting position within few milliseconds until their two long claws touch and, with strong impetus, push off the ground. This movement results in a sudden relocation of the ostracod body, often involving a complete back somersault, as can be seen in five videos that are made available under the Creative Commons Attribution 4.0 International Public License and are available online at http://media.ostracoda.net/2013-01/.

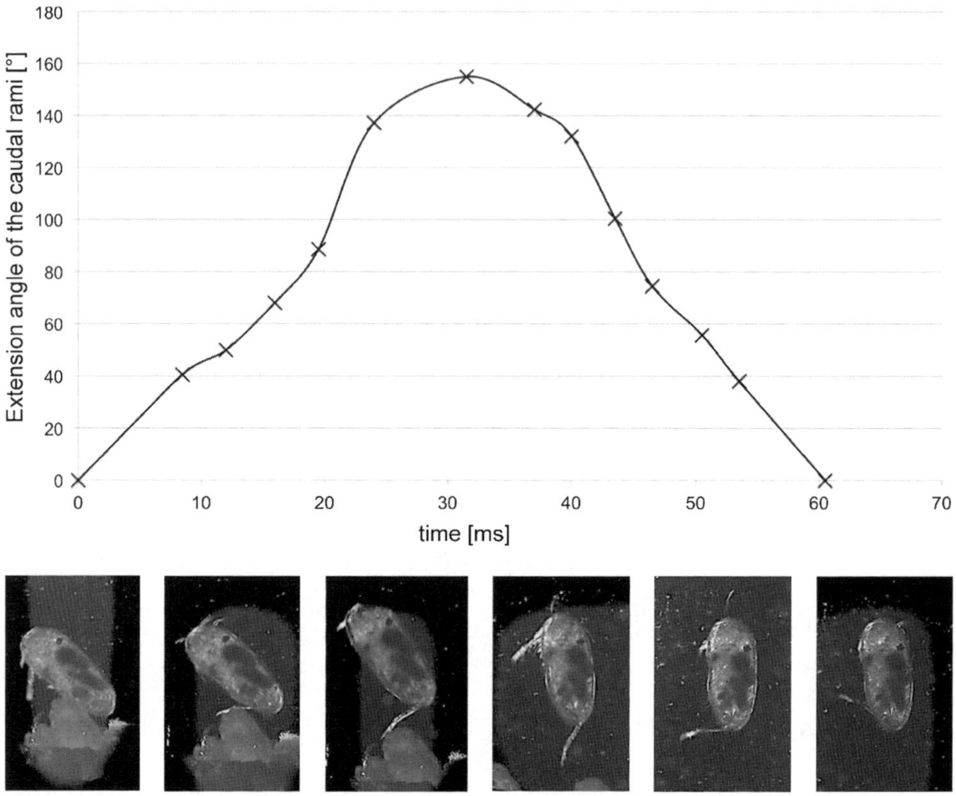

Fig. 3. *Tanycypris centa* Chang, Lee & Smith, 2012: Example of a jump, showing the extension angle of the caudal rami over the period of a single jump. Still images of the camera recordings of the respective phases are shown below the graph.

Extension angles of the caudal rami were measured in ten jumps from still images at intervals of 2-10 images, depending on caudal rami being in the image focus. Results are visualized in a sample jump illustrated with still images (fig. 3) and in a graph compiling all analysed jumps (fig. 4). Raw data are provided in table AI in the online edition of this journal, which can be accessed via http://booksandjournals.brillonline.com/content/journals/15685403.

The maximal extension amounts to 191°. In most jumps, the maximum extension is reached within the first 30 ms, while the retraction time is much more variable (between 20 and 130 ms).

Velocity was calculated from distances covered between two still images, and the lapse of time between these images. For each jump, the average and the maximum velocity were calculated. Velocity values are presented in table AI, with maximum velocity per jump covering a range from 59.30 to 749.84 mm/s. Average velocities over a single jump ranged from 51 to 238 mm/s in the different specimens. Camera recordings show that velocity increases explosively during the

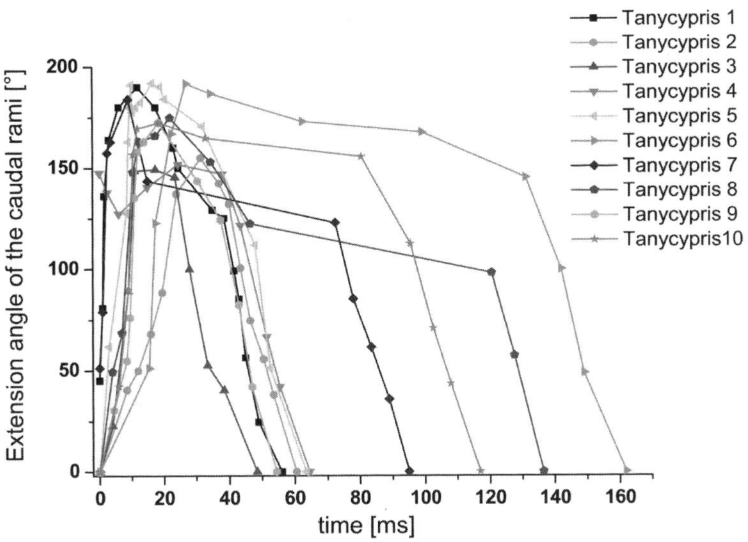

Fig. 4. *Tanycypris centa* Chang, Lee & Smith, 2012: For ten jumps (Tanycypris 1 to Tanycypris 10), the extension angles of the caudal rami have been visualized as a function of time. This figure is published in colour in the online edition of this journal, which can be accessed via http://booksandjournals.brillonline.com/content/journals/15685403.

first 2-5 ms, then decreases quickly during the next 3-5 ms, and finally goes back to almost zero within the following 20-35 ms.

Propulsion force, exerted by the ostracod to overcome the friction of the water's viscosity, was found to range from 0.27 to 7.8 μN (table AII in the online edition of this journal, which can be accessed via http://booksandjournals.brillonline.com/content/journals/15685403).

Correlations were tested between (1) velocity and propulsion force, (2) velocity and extension angle, (3) propulsion force and extension angle and (4) velocity and retracting time of caudal rami. While in (1) to (3) a positive correlation exists (values in figs. 9, 11 and 12), the velocity of a jump is not correlated to the retracting time of the rami ($r = -0.15$). Table I compiles key results on *Tanycypris centa*-jumps, here presented in the context of results on copepod fast jumps from the literature.

Anatomical prerequisites

Tanycypris species are all characterised by a small height/length ratio, and these torpedo-shaped animals can rightly be described mathematically as prolate spheroids. Another specific character of the genus *Tanycypris* is the exceptional length of the caudal rami, in resting position reaching to the mouth region with the tips of their claws (figs. 2a-d and 5). Along with the stretched pair of caudal

TABLE I

Key results of the analysis of twelve jumps in *Tanycypris centa* Chang, Lee & Smith, 2012, presented in comparison to data on Copepods

	Ostracods	Copepods		
	Tanycypris centa (this study)	*Acartia tonsa* Dana, 1849 / *Oithona davisae* Ferrari & Orsi, 1984 / *Calanus finmarchicus* (Gunnerus, 1770) (from Kiørboe et al. (2010))	*Acartia tonsa* (from Busky & Hartline (2002))	*Undinula vulgaris* (Dana, 1849) (from Lenz & Hartline (1999))
Average speed (m/s)	0.125	0.101-0.333	–	–
Maximum speed (m/s)	0.750	0.198-0.634	over 0.8	–
Maximum speed (body lengths/s)	max. ≈ 790	max. ≈ 660 (*Oithona davisae*)	–	–
Propulsion force (μN)	0.3-8.2	–	–	2.0-3.0

rami goes a long paired chitinous caudal rami attachment (fig. 2a, c). This structure reaches from the posterior body end towards anterior, embracing the hind part of the soft body laterally.

Genera of the family Cypricercinae are characterized by the so-called Triebel loop on the anterior end of the caudal rami attachment. In *Tanycypris*, the Triebel loop sits in the middle between the ventral and dorsal branches (fig. 2b) that terminate the stout attachment anteriorly. Beside the main ‘window’ addressed as Triebel loop, in both, *T. centa* and *T. alfonsi*, a second small window is present towards the dorsal branch (figs. 2b and 6). The entire anterior region of the attachment is lined by a thin, extracellular membrane. Unlike in other genera, the posterior-most edge of the soft body is shaped by the caudal rami attachment, here forming a sclerotized, rectangular supportive frame (fig. 2f).

The muscle that stretches from the posterior end of the body (where the proximal end of caudal rami meets the posterior end of caudal rami attachment) along the caudal rami attachment to its anterior end is exceptionally voluminous (fig. 5). As can be seen in a tomographic study (fig. 7), the muscle consists of four distinct bundles of different thicknesses and slightly differing orientations. The most voluminous of these four bundles is oval in a cross-section near the posterior body end (A in fig. 7), becoming more and more flattened towards anterior, until the cross section has turned into a thin line, so that the filaments can all attach to the posterior side of the dorsal branch of the caudal rami attachment (fig. 7e). Two smaller bundles (C and D in fig. 7) attach to the opposite side of this branch in the more ventral area, where the branch is slightly folded (fig. 7e). Muscle bundle B (fig. 7a, b) attaches to the main trunk of the caudal rami attachment (fig. 7c).

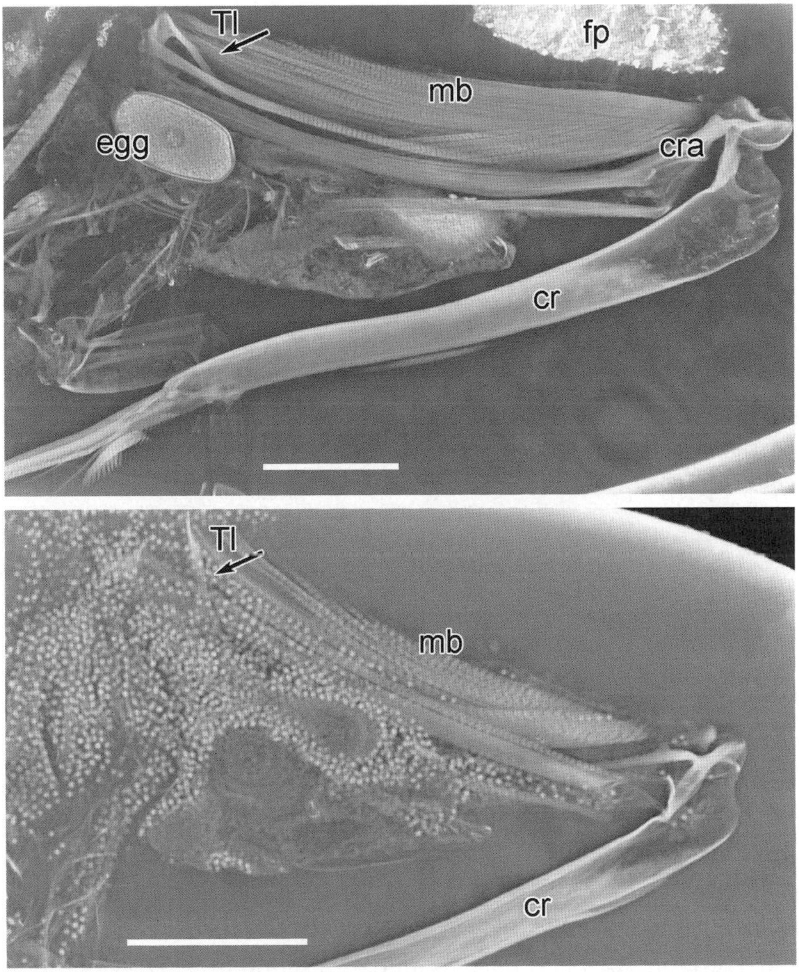

Fig. 5. Tomography of *Tanycypris alfonsi* Nagler, Geist & Matzke-Karasz, 2014 (above) and *Tanycypris centa* Chang, Lee & Smith, 2012 (below) posterior bodies (anterior to left): caudal rami (cr) and caudal rami attachment (cra) are connected in the posterior-most edge of the soft body. The caudal rami attachment embraces the hind body laterally, branching out anteriorly with a Triebel loop (Tl). It accommodates huge muscle bundles that move the caudal rami. The specimen of *Tancypris centa* is infected by microorganisms, most probably fungi (fp fecal pellet caught in the carapace lumen). Scale bars = 100 μm. This figure is published in colour in the online edition of this journal, which can be accessed via http://booksandjournals.brillonline.com/content/journals/15685403.

From the anterior side of the dorsal-most part of the dorsal branch, a thick muscle leads to the inner side of the valve (called dorsal muscle in figs. 6 and 7), thus giving additional muscular stability to the caudal rami attachment.

The joint between caudal rami and their attachments (the posteriormost part of the soft body) is a complicated and highly differentiated structure, as can be seen in virtual sections at different levels (fig. 8).

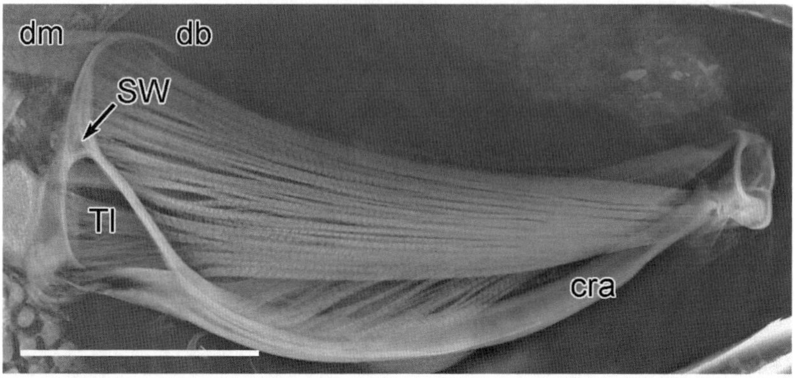

Fig. 6. Tomography of *Tanycypris alfonsi* Nagler, Geist & Matzke-Karasz, 2014 focusing on the caudal rami attachment (cra) of the left body side (anterior to left). Dorsal to the Triebel loop (Tl), a small second window (sw) leads over to the thin dorsal branch (db), which points to the body center. The dorsal muscle (dm) connects the dorsal branch with the inner carapace wall more anteriorly. Scale bar = 100 μm. This figure is published in colour in the online edition of this journal, which can be accessed via http://booksandjournals.brillonline.com/content/journals/15685403.

DISCUSSION

There are many reports available in the literature on the copepod escape jump, a sudden movement based on a very effective metachronic power stroke of several pairs of swimming legs (e.g., Alcaraz & Strickler, 1988; Lenz & Hartline, 1999; Buskey & Hartline, 2002; Lenz et al., 2004; Kiørboe et al., 2010; Kiørboe, 2011). The jumps observed here in the ostracod genus *Tanycypris* are different in that they are propelled by a very rapid repulsion of the long caudal rami from the ground. In ostracods, the caudal rami are not involved in swimming movements and the jumps can, therefore, be interpreted as an optimisation of walking abilities, not of swimming abilities. Quantifications in the present study encompass extension and retraction of the caudal rami, jump velocity, and exerted force during the jump.

The high-speed extension of the caudal rami is the key to the *Tanycypris* jump; therefore, extension and retraction time have been analysed for 10 jumps, recording angle and progress over time. A jump has been defined as the time between first appearance of the rami outside the carapace and the moment of its disappearance back into the ventral carapace space. Extension reaches a maximum of 191° (average maximum 166°); the maximum extensions were reached between 9 and 31 ms (average 22 ms) — a sudden increase in extension that is visible in the steep ascending branches of curves in fig. 4. The retraction, however, is a less consistent process, with some specimens retracting the rami as fast as they had deflected them and others folding them back to the resting position only slowly. A correlation coefficient of $r = 0.67$ proves the extension angle being positively correlated to

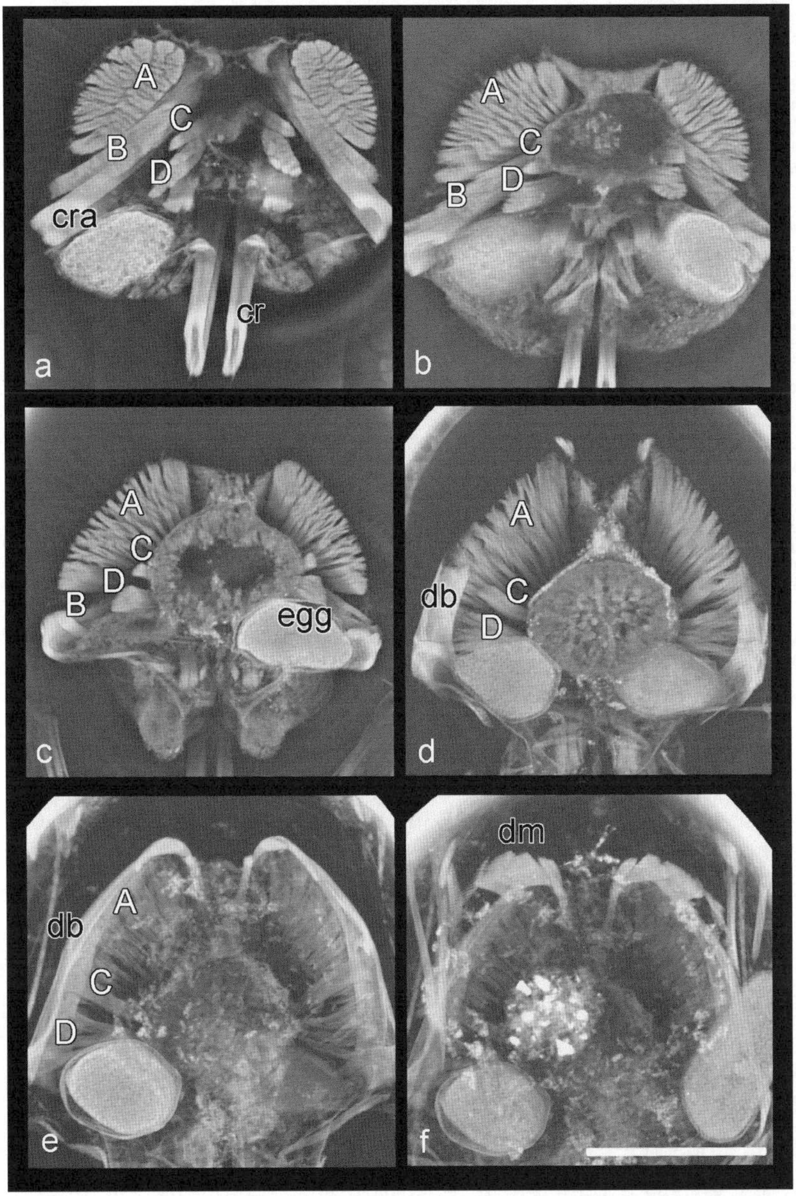

Fig. 7. Tomographic cross-sections through a *Tanycypris alfonsi* Nagler, Geist & Matzke-Karasz, 2014 specimen in a consecutive sequence from posterior to anterior. Cross-sections of the muscle bundles A-D of the main caudal rami muscle change their shape from round/oval in the posterior region to much more rectangular/linear towards anterior. Muscle bundle B attaches to the main trunk of the caudal ramus attachment (cra), while the bundles A, C and D attach to the dorsal branch (db) of the Triebel loop. A further muscle (dm) connects the anterior side of the dorsal branch with the carapace dorsally. Scale 100 μm. This figure is published in colour in the online edition of this journal, which can be accessed via http://booksandjournals.brillonline.com/content/journals/15685403.

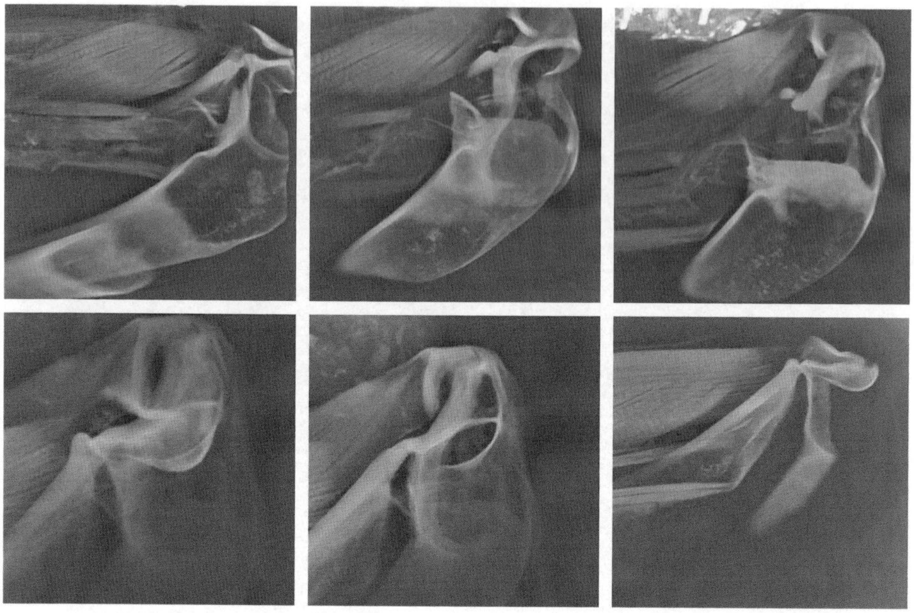

Fig. 8. Tomographies of *Tanycypris alfonsi* Nagler, Geist & Matzke-Karasz, 2014, six virtual sections of the joint between caudal rami and cauda rami attachments. Lateral views at different levels, showing a highly structured joint, possibly allowing for a click-joint-mechanism. At different scales. This figure is published in colour in the online edition of this journal, which can be accessed via http://booksandjournals.brillonline.com/content/journals/15685403.

the maximum velocity (fig. 9), while the retraction time is irrelevant for moving forward ($r = -0.15$).

This result confirms the visual impression of the ostracod jump: only the initial backstroke impulse from the ground and the subsequent stroke against water until the extension is completed accelerate the ostracod. Once the rami have reached their maximal extension point, the movement slows down and the rami are retracted.

The velocity of the *Tanycypris* jump reaches up to 750 mm/s, or, up to 790 body lengths per seconds. These values are in the same magnitude of fast movements recorded in copepods (Busky & Hartline, 2002; Kiørboe et al., 2010). When imaging velocity as a function of time, as done exemplarily for three jumps in fig. 10, it becomes clear that acceleration peaks within the first few milliseconds to then abruptly decrease again. Consequently, velocity is back to a value of 100-200 mm/s at 10 ms and near zero at around 30 to 40 ms after the jump started. Clearly, retraction time of the caudal rami (that lasted up to 130 ms after reaching the maximum extension) has no effect on the kinematics of the jump.

A positive correlation has been identified between the maximal extension angle of the caudal rami of a single jump and the calculated average force exerted over

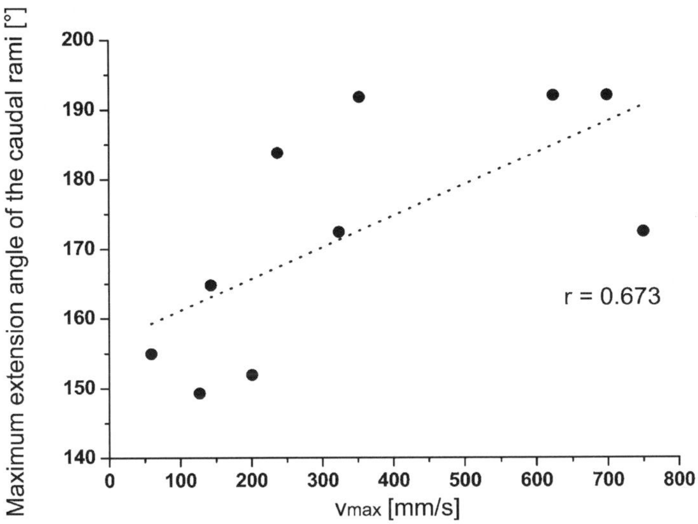

Fig. 9. The maximum extension measured for a specific jump of *Tanycypris centa* Chang, Lee & Smith, 2012 positively correlates with the maximum velocity measured for this jump.

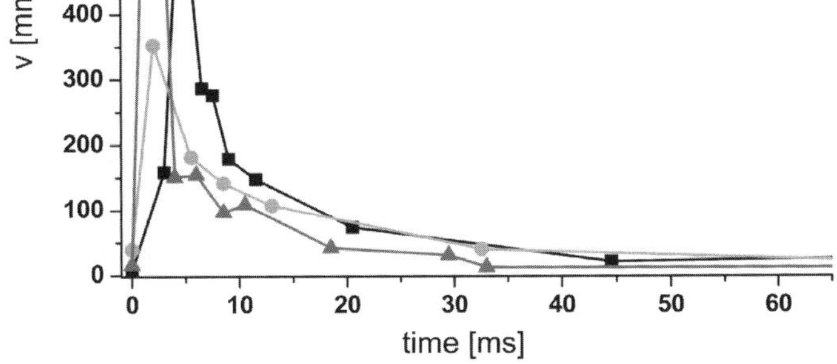

Fig. 10. Absolute velocity measured in periods of milliseconds as a function of time; here done exemplarily for three jumps of *Tanycypris centa* Chang, Lee & Smith, 2012. Acceleration peaks within the first few milliseconds to then abruptly decrease again. This figure is published in colour in the online edition of this journal, which can be accessed via http://booksandjournals.brillonline.com/content/journals/15685403.

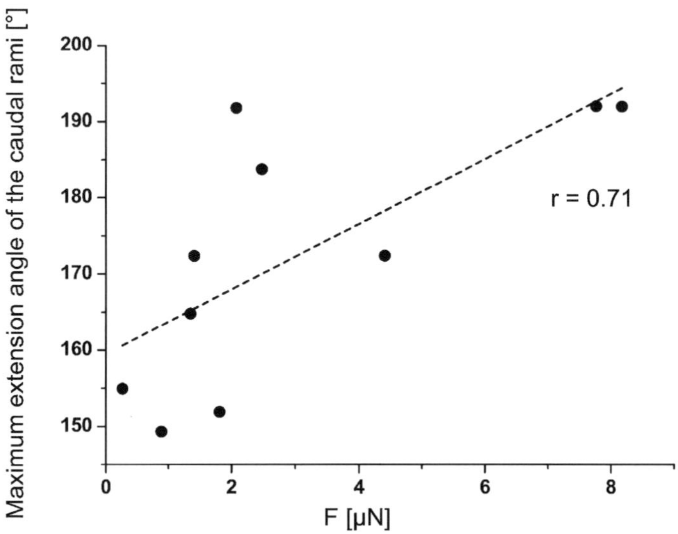

Fig. 11. *Tanycypris centa* Chang, Lee & Smith, 2012: the maximum extension measured for a specific jump positively correlates with the calculated exerted force over the entire jump.

time for this jump ($r = 0.71$) (fig. 11). At the same time, maximal velocities of jumps are positively correlated to the exerted force ($r = 0.86$) (fig. 12).

It can thus be concluded that the exerted force in the *Tanycypris* jump is linearly translated into the degree of caudal rami extension, and extension, for its part of the process, regulates maximal and average velocity of a jump.

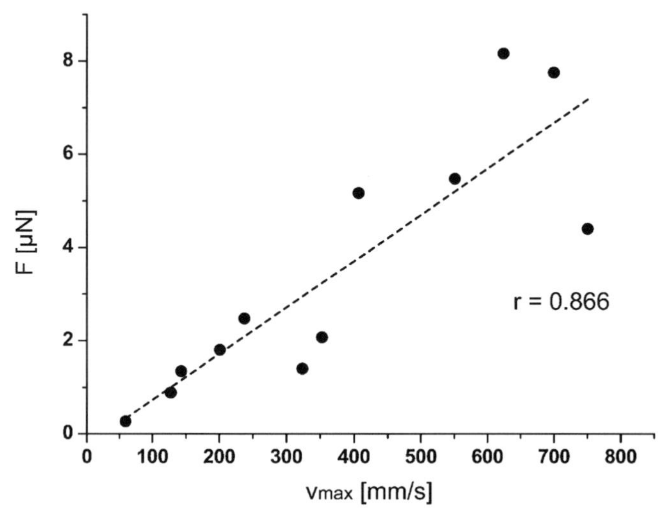

Fig. 12. *Tanycypris centa* Chang, Lee & Smith, 2012: maximum velocities reached during a specific jump positively correlates with the calculated exerted force over the entire jump.

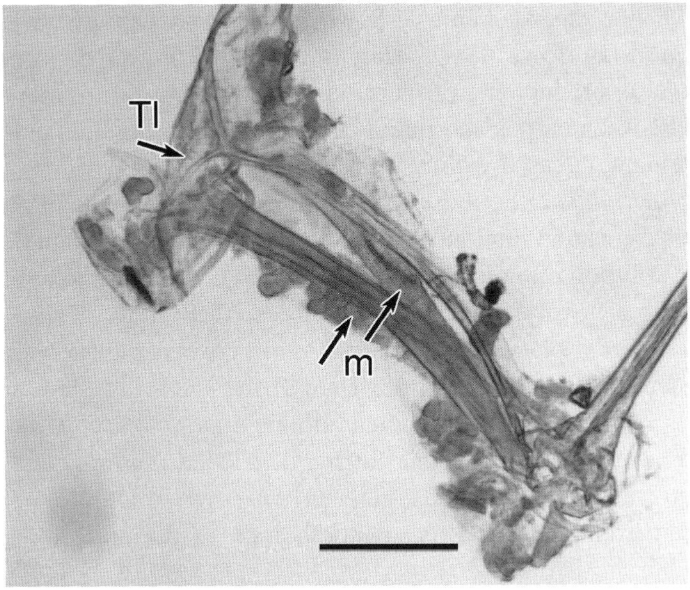

Fig. 13. *Bradleystrandesia fuscata* (Jurine, 1820): caudal ramus attachment with musculature. This species — also from the family Cypricercinae — shows much weaker developed muscles (m) in this area than the genus *Tanycypris*. The Triebel loop (Tl) has a different orientation, stabilizing the ventral and not the dorsal branch of the attachment. Scale 100 μm.

The anatomical and physiological prerequisites allowing the high velocity in a *Tanycypris* jump cannot be detailed in the frame of the present study. A combination of characters present in both *Tanycypris* species analysed here might be responsible for the powerful jump. One relevant character is the extraordinarily well-developed musculature, as shown in figs. 1, 2 and 5-7. As a comparison, the 'normally developed' musculature in *Bradleystrandesia fuscata* (Jurine, 1820) from the same family Cypricercinae is illustrated in fig. 13. Although the Triebel loop is present here as well, it is located between the main trunk and the ventral branch — not the dorsal branch as in *Tanycypris*. The interaction of muscles and caudal rami attachment must therefore be considerably different in the two groups.

This musculature is combined with an exceptional length of the caudal rami, here being used as a lever of high efficiency (fig. 2). Although the jump itself resembles that of springtails or fleas, where the jump is powered by the energy previously stored in an elastic proteinaceous material (e.g., resilin pad in fleas); no such mechanism could be detected in *Tanycypris*. However, there is a possibility of energy-storage in tendon-like structures, which have been reported to connect caudal rami muscles to the posterior end of the caudal rami attachment (Rome, 1969). Finally, the highly differentiated joint between caudal rami and caudal rami attachment, allows for hypothesizing a mechanism that produces a pre-tension, much like the mechanism shown for the prey capture strike in *Squilla* and *Hemisquilla* (Burrows, 1969). Here, a co-contraction of flexor and extensor muscles results in producing a pre-tension prior to the fast movement. Helped by a click joint between carpus and merus, this pre-tension can be fixed and, to start the fast movement, released within milliseconds.

In *Tanycypris alfonsi*, the musculature, running more or less in parallel to the main trunk of the caudal rami attachment, is composed of four bundles, as shown in fig. 7. The splitting into distinct muscle bundles could point to a diversification of tasks, i.e., to a flexor-extensor-system as described above. These bundles are hooked into different places of the paired joint between caudal rami and caudal rami attachments at the posterior-most edge of the soft body. Here are wing-like processes that increase space, thus potentially facilitating muscular attachment, or stabilizing the construction (fig. 2f). Such a developed posterior end is normally not present in freshwater ostracods; here, *Eucypris virens* (Jurine, 1820) (fig. 2e) is given as an example of the usual morphology of the posterior body end in Cyprididae.

Beside these wing-like processes, several structural components are present in this area (fig. 8), which can potentially be involved in a click-joint-mechanism stabilising a possible muscular pre-tension, thus facilitating the main muscular outburst at the moment of jump. However, at present this functional interpretation from pure morphological evidence must remain speculative.

On the opposite end of the caudal rami attachment, the rather thick main trunk forms the Triebel loop and branches out long dorsally and much more stubby ventrally. In the observed species, the Triebel loop is accompanied by a second, small window towards the dorsal branch, separated from the main Triebel loop by a chitinous strut (figs. 2b, 6, 14).

The Triebel loop characterises the subfamily Cypricercinae within the Cyprididae. Its size ranges from small (e.g., in the genera *Bradleystrandesia* and *Spirocypris*) over medium (e.g., genus *Bradleytriebella*) to big (e.g., genus *Astenocypris* and *Tanycypris*) (Savatenalinton & Martens, 2009). No hypothesis has been phrased so far about the function of the Triebel loop. In drawings and photographies, presented in literature, the caudal rami attachment has a two-dimensional aspect with all parts lying in the same plane. However, *Tanycypris alfonsi* tomographies visualise that the dorsal branch of the caudal rami attachment reaches towards the body centre, thus forming a three-dimensional framework together with the main trunk (fig. 14). While the filaments of muscle bundle B attach to the main trunk, muscle bundles A, C and D are accommodated by the dorsal branch (fig. 7). Thus, during caudal rami movements, a strong tension will be executed on this part of the chitinous construction. To prevent the dorsal branch from folding down during phases of high muscular tension, a supportive bracket is interlaced, leading

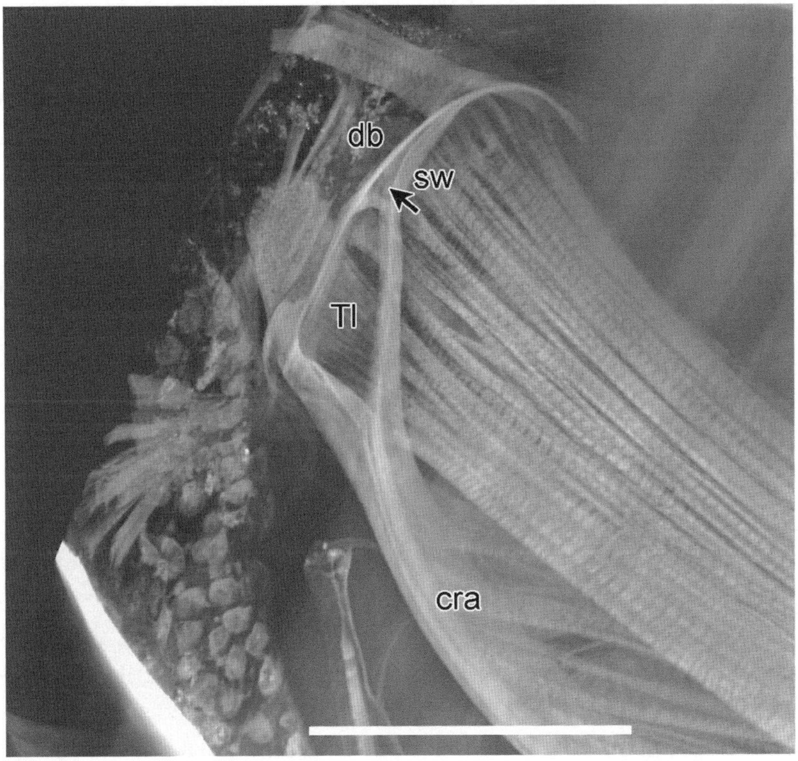

Fig. 14. Tomography of *Tanycypris alfonsi* Nagler, Geist & Matzke-Karasz, 2014 with anterior part of caudal rami attachment (cra) of the left body side. The dorsal branch reaches over towards the center of the body, while the light-weight construction of Triebel loop (Tl) and second window (sw) prevent the dorsal branch from folding down during muscle tension. Scale bar = 100 μm. This figure is published in colour in the online edition of this journal, which can be accessed via http://booksandjournals.brillonline.com/content/journals/15685403.

from the main trunk to the dorsal branch, thus forming the Triebel loop. This stabilization is even optimized in *Tanycypris* species by the small strut forming the second window (figs. 6 and 14). We therefore here interpret the Triebel loop as an integral part of a light-weight construction to provide a high mechanical stability of the caudal rami attachment at a low material expenditure.

Prionocypris longiforma has been described to carry out jumps when taken from their spring habitat into a sampling container (Külköylüoğlu, 1999). A preliminary investigation of the species' caudal rami attachment made clear that it possesses a Triebel loop at the distal end of the caudal rami attachment. Beside the taxonomical relevance of this fact (the Triebel loop is a distinctive character of Cypricercinae and thus a revision of the generic allocation of this species would be necessary) it supports the hypothesis that the Triebel loop is relevant for the development of a bigger and perhaps more specialised muscle.

The carapace of *Tanycypris* species is visibly torpedo-shaped — a shape that is favourable per se for fast movements in water. Less visibly, the soft body has also undergone a shaping, compared to species with shorter caudal rami. In both *Tanycypris centa* and *T. alfonsi* the posterior part of the body is stretched more than the anterior half. This can clearly be identified in the distance of the paired, elevated female genital regions, where gonoporus and vagina are situated. The large distance between this area and posterior-most point of the soft body is evident and might have evolved to better accommodate the long rami, their attachments, and of course the voluminous muscles.

CONCLUSION

The present study is only a superficial investigation of factors that may influence high-speed movement by small organisms in water. In the frame of this project, neither peak forces during the jumps could be analysed, nor the relationship between force and real body mass or characteristics of the muscular motor performance. However, we could showcase and partially analyse an as yet undescribed high speed movement in an ostracod genus, and tentatively describe the anatomical adaptations underlying it. In the investigated *Tanycypris* species walking abilities are brought to perfection by developing caudal rami that can be used as a kind of catapult. The length of caudal rami is present in other genera of the family, too, and possibly the here described ability of high-speed movements has been overlooked not only in congeners of our model species.

A maximum velocity of 0.75 ms^{-1} has been measured for *Tanycypris centa*; in other words this animal moves forward at ca. 790 body lengths per second. Among similar sized water animals, only copepods are known to perform movements at a similar speed. However, the mechanics propelling these high-speed moves differ considerably in these two crustacean taxa. Nonetheless, the statement that "escape velocities in pelagic copepods are about an order of magnitude higher than in other, similarly sized aquatic invertebrates" (Kiørboe, 2011: 679) is herewith challenged by the jumping performance of *Tanycypris centa*.

The common view of ostracods is that of moderately fast organisms, either crawling more or less slowly on or in the sediment or swimming unhurriedly in the water column. On this behalf, this record of exceptionally fast movements in ostracods throws a new light on these crustaceans. In the context of predator-prey-relationships, the complete carapace closure possible in freshwater ostracods has been looked upon as the most effective direct defense towards predators — based upon records of ostracods having passed unharmed the intestine of predators (e.g., Lopez et al., 2002). Further, migration to open water and alteration of swimming activity has been considered as responses to predator presence (Vandekerkhove, 2012). However, the potential to actively escape from being picked as prey by performing a 'last-second escape jump' has never been considered relevant. Our study shows that flight behaviour of ostracods needs to be evaluated much more differentially than done so far and should obviously be taken into account in future studies on predator-prey-relationships involving ostracods.

Beside escaping predators, the fast movement could as well serve other purposes, such as quickly leaving areas of unfavourable water conditions (e.g., areas of low oxygenation) or avoiding being buried in incidents of high sedimentation. No information is available on behavior of *Tanycypris* species in their natural habitats and, thus, no ultimate conclusion can be drawn here about the reason for the evolution of high-speed jumps in this group.

ACKNOWLEDGEMENTS

Foremost, we thank the lab group of Christian Laforsch and particularly Jessica Fischer (Biozentrum LMU Martinsried) for giving us access to the high-speed camera and for sharing their know-how. Eva Schmidbauer and Andreas Richter (Munich Botanical Garden) are thanked for giving us access to the greenhouse any time. We are grateful to Ludwig Nagler for general technical support. Tomographic studies were carried out at and funded by the European Synchrotron Radiation Facility in Grenoble, France (experiment ec-945). We are especially grateful to Paul Tafforeau (ESRF, ID19) for helping us to achieve incredible results from this experiment. Fruitful discussions with colleagues during the IGOM 2012 conference in Cologne and the Workshop II of the Research Initiatives in Northern Hemisphere Quaternary Non-Marine Ostracoda (Charlotte, NC, 2012) helped us a lot and we particularly acknowledge comments by Dietmar Keyser and Todd Oakley. During the ISO17 conference in Rome we had a valuable discussion with Okan Külköylüoğlu about his observation of jumping *Prionocypris* (?) *longiforma*. He is very much acknowledged for sending specimens of this species for comparison. R.M.-K. was supported by project MA 2118/3-1 of the German Research Foundation. C.N. was supported by a Deutschlandstipendium of the Bundesministerium für Bildung und Forschung.

REFERENCES

ALCARAZ, M. & J. R. STRICKLER, 1988. Locomotion in copepods: pattern of movements and energetics of *Cyclops*. Hydrobiologia, **167-168**: 409-414.

ALM, G., 1916. Monographie der schwedischen Süsswasser-Ostracoden nebst systematischen Besprechungen der Tribus Podocopida. Zoologiska Bidrag från Uppsala, **4**: 1-247.

BURROWS, M., 1969. The mechanics and neural control of the prey capture strike in the mantid shrimps *Squilla* and *Hemisquilla*. Zeitschrift für vergleichende Physiologie, **62**: 361-381.

BUSKEY, E. & D. K. HARTLINE, 2003. High-speed video analysis of the escape responses of the Copepod *Acartia tonsa* to shadows. Biological Bulletin, **204**: 28-37.

CHANG, C. Y., J. LEE & R. J. SMITH, 2012. Nonmarine ostracods (Crustacea) from South Korea, including a description of a new species of *Tanycypris* Triebel (Cyprididae, Cypricercinae). Zootaxa, **3162**: 1-19.

HARTMANN, G., 1967. Ostracoda. In: H.-E. GRUNER (ed.), Dr. H.G. Bronns Klassen und Ordnungen des Tierreichs, Band **5** (Arthropoda), Abteilung I (Crustacea), Buch 2, Teil IV (Ostracoda), Lieferung 2: 217-408. (Geest & Portig, Leipzig).

KIØRBOE, T., 2011. What makes pelagic copepods so successful? Journal of Plankton Research, **33**: 677-685.

KIØRBOE, T., A. ANDERSEN, V. J. LANGLOIS & H. H. JAKOBSEN, 2010. Unsteady motion: escape jumps in planktonic copepods, their kinematics and energetics. Journal of the Royal Society Interface, **7**: 1591-1602.

KÜLKÖYLÜOĞLU, O., 1999. Taxonomy, ecology and biogeographic distribution of spring water ostracoda (Crustacea) in Nevada. (Ph.D. Dissertation, Program in Ecology, Evolution and Conservation Biology, University of Nevada, Reno, NV).

LENZ, P. H. & D. K. HARTLINE, 1999. Reaction times and force production during escape behavior of a calanoid copepod, *Undinula vulgaris*. Marine Biology, **133**: 249-258.

LENZ, P. H., A. E. HOWER & D. K. HARTLINE, 2004. Force production during pereiopod power strokes in *Calanus finmarchicus*. Journal of Marine Systems, **49**: 133-144.

LOPEZ, L. C. S., D. A. GONÇALVES, A. MANTOVANI & R. I. RIOS, 2002. Bromeliad ostracods pass through amphibian (*Scinaxax perpusillus*) and mammalian guts alive. Hydrobiologia, **485**: 209-211.

NAGLER, C., J. GEIST & R. MATZKE-KARASZ, 2014. Revision of the genus *Tanycypris* (Ostracoda, Cypricercinae) with the description of *Tanycypris alfonsi* n. sp., and an identification key to the genus. Zootaxa, **3821**: 401-424.

ROME, D. R., 1969. Morphologie de l'attache de la furca chez les Cyprinidae et son utilisation en systématique. In: J. W. NEALE (ed.), The taxonomy, morphology and ecology of Recent Ostracoda: 168-193. (Oliver & Boyd, Edinburgh).

SAVATENALINTON, S. & K. MARTENS, 2009. Generic revision of Cypricercinae McKenzie, 1971 (Crustacea, Ostracoda), with the description of three new genera and one new species and a phylogenetic analysis of the subfamily. Hydrobiologia, **632**: 1-48.

TRIEBEL, E., 1959. Zur Kenntnis der Ostracoden-Gattungen *Isocypris* und *Dolerocypris*. Senckenbergiana biologia, **40**: 155-170.

VANDEKERKHOVE, J., T. NAMIOTKO, E. HALLMANN & K. MARTENS, 2012. Predation by macroinvertebrates on *Heterocypris incongruens* (Ostracoda) in temporary ponds: impacts and responses. Fundamental and Applied Limnology, **181**: 39-47.

First received 8 February 2013.
Final version accepted 15 December 2013.

[When citing this chapter, refer to Crustaceana 87 (2014) 1095–1123]

This paper is dedicated to Prof. Dr. Eugen Kempf on the occasion of his 80th birthday (16 April 2012)

LIFE HISTORIES AND DISTRIBUTION OF OSTRACODS WITH DEPTH IN WESTERN LAKE GENEVA (PETIT-LAC), SWITZERLAND: A RECONNAISSANCE STUDY

BY

LAURENT DECROUY[1]) and TORSTEN W. VENNEMANN
Faculty of Geosciences and Environment, Institute of Earth Surface Dynamics, University of Lausanne, Géopolis, 1015-Lausanne, Switzerland

ABSTRACT

Because environmental conditions within a given basin are different for each season and at different water depth, knowledge of the life history and depth distribution of target species is important for environmental and palaeoenvironmental interpretations based on ostracod species assemblages and/or the geochemical compositions of their valves. In order to determine the distribution of species with depth as well as the life history of species from Lake Geneva, a one-year sampling campaign of living ostracods was conducted at five sites (2, 5, 13, 33 and 70 m water depth) on a monthly basis in the Petit-Lac (western basin of Lake Geneva, Switzerland). Based on the results, the different species can be classified into three groups. Littoral taxa are found at 2 and 5 m water depth and include, in decreasing numbers of individuals, *Cypridopsis vidua* (O. F. Müller, 1776), *Pseudocandona compressa* (Koch, 1838), *Limnocythere inopinata* (Baird, 1843), *Herpetocypris reptans* (Baird, 1835), *Potamocypris smaragdina* (Vávra, 1891), *Potamocypris similis* (G. W. Müller, 1912), *Plesiocypridopsis newtoni* (Brady & Robertson, 1870), *Prionocypris zenkeri* (Chyzer & Toth, 1858) and *Ilyocypris* sp. Brady & Norman, 1889. Sublittoral species are found in a majority at 13 m water depth and to a lesser extend at 33 m water depth and include, in decreasing numbers of individuals, *Fabaeformiscandona caudata* (Kaufmann, 1900), *Limnocytherina sanctipatricii*, *Candona candida* (O. F. Müller, 1776) and *Isocypris beauchampi* (Paris, 1920). Profundal species are found equally at 13, 33 and 70 m water depth and includes, in decreasing numbers of individuals, *Cytherissa lacustris* (Sars, 1863), *Candona neglecta* Sars, 1887 and *Cypria lacustris* Lilljeborg, 1890. The occurrence of *Limnocytherina sanctipatricii* (Brady & Robertson, 1869) is restricted from late winter to late spring when temperatures are low, while *C. vidua*, *L. inopinata*, *P. smaragdina*, *P. similis*, *P. newtoni* and *Ilyocypris* sp. occur predominantly from spring to early autumn when temperatures are high. Individuals of *C. neglecta*, *C. candida*, *F. caudata*, *P. compressa*, *C. lacustris*, *H. reptans* and *Cp. lacustris* occur throughout the year with juveniles and adults occurring during the same period (*C. neglecta* at 70 m, *C. lacustris* at 13, 33 and 70 m, and *H. reptans* at 2, 5 and 13 m water depth) or with juveniles occurring during a different period of the year than adults (*C. neglecta* at 13 and 33 m and *C. candida*, *F. caudata*

[1]) Corresponding author; e-mail: laurent.decrouy@unil.ch

and *P. compressa* at their respective depth of occurrence). Among the environmental parameters investigated, an estimate of the relationship between ostracod autoecology and environmental parameters suggests that in the Petit-Lac: (i) water temperature and substrate characteristics are important factors controlling the distribution of species with depth, (ii) water temperature is also important for determining the timing of species development and, hence, its specific life history, and (iii) water oxygen and sedimentary organic matter content is less important compared to the other environmental parameter monitored.

RÉSUMÉ

Une bonne connaissance préalable du cycle de vie et de la distribution en fonction de la profondeur des différentes espèces d'ostracodes est nécessaire lorsque des études environnementales ou paléo-environnementales basées sur les assemblages d'espèces ou sur la géochimie des carapaces sont envisagées, et cela d'autant plus en milieu continental tempéré où les conditions au sein d'un même bassin peuvent varier de manière saisonnière mais aussi en fonction de la profondeur d'eau.

Afin de déterminer ces paramètres pour les différentes espèces peuplant les fonds du Petit-Lac (partie occidentale du Lac Léman, Suisse), une campagne d'échantillonnages mensuels effectuée sur cinq sites de profondeur croissante (2, 5, 13, 33 et 70 m de profondeur d'eau) a été entreprise sur une période d'une année. Les résultats montrent que les différentes espèces collectées peuvent être classifiées au sein de trois groupes. Les espèces dites littorales sont trouvées à 2 et 5 m de profondeur et comportent, selon un ordre décroissant du nombre d'individus: *Cypridopsis vidua*, *Pseudocandona compressa*, *Limnocythere inopinata*, *Herpetocypris reptans*, *Potamocypris smaragdina*, *Potamocypris similis*, *Plesiocypridopsis newtoni, Prionocypris zenkeri* et *Ilyocypris* sp. Les espèces dites sub-littorales sont principalement trouvées à 13 m mais aussi dans une moindre proportion à 33 m et comportent, selon un ordre décroissant du nombre d'individus: *Fabaeformiscandona caudata*, *Limnocytherina sanctipatricii*, *Candona candida* et *Isocypris beauchampi*. Les espèces dites profondes se retrouvent autant à 13, 33 et 70 m de profondeur et comportent, selon un ordre décroissant du nombre d'individus: *Cytherissa lacustris*, *Candona neglecta*, et *Cypria lacustris*. *Limnocytherina sanctipatricii* n'a pu être collectée que de la mi-hiver à la mi-printemps, alors que les températures de l'eau étaient les plus basses. *Cypridopsis vidua*, *L. inopinata*, *P. smaragdina*, *P. similis*, *P. newtoni* et *Ilyocypris* ont été, pour leur part, principalement récoltés du début du printemps à la mi-automne alors que les températures de l'eau étaient élevées. Juvéniles et adultes ont pu être récoltés simultanément, et ceci tout au long de l'année, pour les populations de *C. neglecta*, *C. candida*, *F. caudata*, *P. compressa*, *C. lacustris*, *H. reptans* et *Cp. lacustris* à 70 m de profondeur ainsi que pour les populations de *C. lacustris* à 13, 33 et 70 m de profondeur et les populations de *H. reptans* à 2, 5 et 13 m de profondeur. Par contre, pour les populations de *C. neglecta* à 13 et 33 m et les populations de *C. candida*, *F. caudata* et *P. compressa*, les juvéniles et les adultes n'apparaissent pas conjointement mais sont récoltés à des périodes différentes de l'année. Une analyse sommaire des relations entre autoécologie des espèces et paramètres environnementaux investigués suggère que pour le cas du Petit-Lac (i) La température de l'eau et les caractéristiques du substrat sont des facteurs importants contrôlant la distribution des espèces en fonction de la profondeur d'eau, (ii) La température de l'eau est aussi un paramètre important contrôlant le 'timing' du développement des individus et ainsi le cycle de vie des différentes espèces et (iii) La teneur en oxygène dissous ainsi que la teneur en matière organique dans les sédiments semblent être des paramètres moins importants comparé aux autres paramètres environnementaux examinés.

Ce papier est dédié au Professeur Docteur Eugen Kempf pour l'anniversaire de ses 80 ans (16 avril 2012).

INTRODUCTION

Studies of ostracod communities, their fossil assemblages or valve geochemistry provide valuable information about modern, as well as past environmental conditions (e.g., Chivas et al., 1986; Von Grafenstein et al., 1999; Ricketts et al., 2001; Schwalb, 2003; Decrouy et al., 2011a, b; Horne et al., 2012). As ostracods grow by successive ecdysis, the developmental stage of each individual can be assessed based on the valve size and soft part morphology. For (paleo)environmental studies, adult valves are commonly used and juvenile individuals down to stage A-4 can be determined to species level and are often more numerous compared to adults, providing more material for geochemical analyses and higher individual counts for statistical interpretations of (palaeo)ecological studies (e.g. McKenzie & Swain, 1967; Von Grafenstein et al., 1999). In addition, many species have a seasonal life-cycle, with adults and juveniles occurring only during a limited time period of the year or adults being found during a different part of the year compared to juveniles (e.g., Meisch, 2000). If seasonal variations occur and adults and juveniles calcify at different periods, the environmental conditions at the time of valve calcification can be substantially different for adults compared to juveniles (e.g., Decrouy, 2012). It is, therefore, essential to know the timing of moulting for adults and the development stages of these ostracods before the results can be interpreted.

Two main types of life-cycles have been described, eurychronal and stenochronal. The latter is generally subdivided into early year, summer, and autumnal forms. For most common species, the general life-cycle is known and abundance of noteworthy observations has been published since the pioneering work of early ostracodologists (e.g., Müller, 1900; Wolf, 1920). Yet, it is often difficult to get the period of occurrence for each development stage from the literature. In addition, the variability of life-cycles for a given species as well as the factors controlling the timing of moulting are often unknown, complicating extrapolation of data from previous to present studies.

The depth distribution and life history of 16 taxa collected at monthly intervals during one year at different depth in the Petit-Lac (western basin of Lake Geneva, Switzerland) are described below. A modified graphical model is used to illustrate ostracod development throughout the year. This allows the life history of a target species to be compared at different depth and with other, closely related species and also allows for a comparison with measured environmental parameters. In addition, the data presented will provide useful information about the life-cycles of species frequently employed in environmental and palaeoenvironmental studies but for which timing of moulting is not well known.

MATERIAL AND METHODS

Study site

Lake Geneva straddles the border between Switzerland and France (fig. 1A) and is, in terms of volume, the largest freshwater lake in Central Europe with a surface area of about 580 km^2 and a total volume of 89 km^3. It is divided into two contrasting sub-basins: the Grand-Lac (eastern part) and the Petit-Lac (western part, fig. 1B). The Petit-Lac is a medium-sized basin, 23.3 km long, 4.7 km wide, with a mean depth of 41 m and a maximum depth of 76 m. It is a mesotrophic, soft water lacustrine system with a pH of approximately 8. It is calcium-bicarbonate

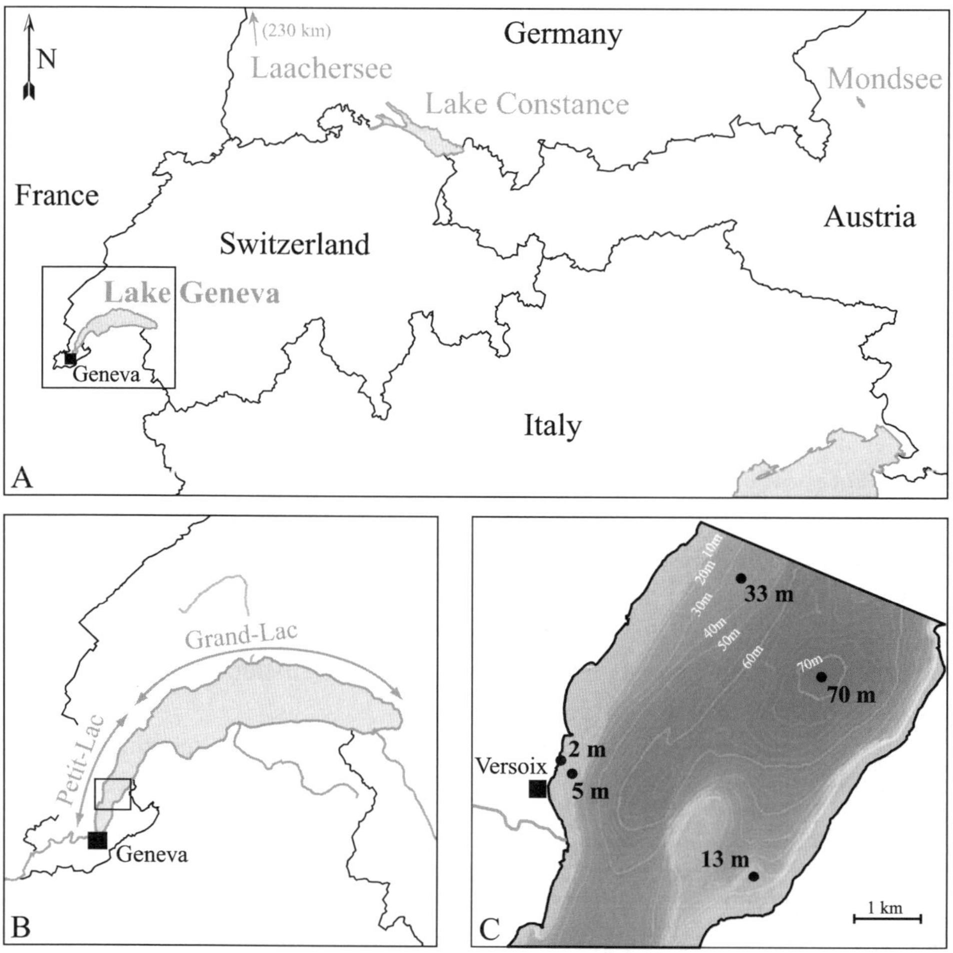

Fig. 1. Geographic setting of Lake Geneva and sampling sites of this study. Locations in the Swiss coordinate system are 3 m (502,085/126,280), 5 m (502,310/126,280), 13 m (505,055/125,100), 33 m (504,860/129,555), and 70 m (506,000/128,055).

dominated (about 104 mg/l HCO_3^- and 43 mg/l Ca^{2+}) with a low salt content (294 mg/l TDS, 46.8 mg/l SO_4^{2-}, 8.7 mg/l Cl^-, 6.1 mg/l Na^+, 5.9 mg/l Mg^{2+}, 1.4 mg/l K^+, 1.3 mg/l SiO_2) (CIPEL, 1984). It is monomictic and the deep water is well-oxygenated throughout the year because of an annual overturn in winter and mixing induced by wind action. From April 2006 to May 2007, water temperature had a range between 4.4°C and 26.6°C and oxygen concentration between 7.9 mg/l and 14.7 mg/l (data from "Service de l'Écologie de l'Eau", SECOE). This basin accounts for only 4% of the total volume of water of the entire lake. Because of its good oxygen supply and relatively quiet autochthonous-dominated sedimentation, the Petit-Lac was chosen for this study. Five sampling sites situated along gently sloping declines have been chosen at different depth (2, 5, 13, 33 and 70 m depth; fig. 1C).

Measurements of environmental parameters

The temperature of lake water was measured continuously every three hours from January 2006 to July 2007 with data-loggers (Vemco Minilog) installed at 2, 5, 13 and 33 m depth. Water temperature was also measured in situ at the five sites with a temperature probe at the time of sampling. Water temperatures from 23 January 2006 to 20 March 2006 and on 21 May 2007, as well as monthly oxygen saturation levels of the water column were measured by the SECOE. Hence, continuous water temperature record is available for the whole ostracod collection period for sites at 2, 5, 13 and 33 m water depth while only punctual water temperature measurements are available for the 70 m water depth site. Samples for analyses of the isotope composition of dissolved inorganic carbon ($\delta^{13}C_{DIC}$) were also taken at the same time (procedure described in Decrouy et al., 2011a). Sediment texture, macroscopic vegetation and fauna were determined for each sampling site on the basis of Dussart (1966).

Sampling and separation of ostracods

Living ostracods were sampled at monthly intervals from April 2006 to May 2007 (table I). At 13, 33 and 70 m water depth, sediments consist of silty sand or clayey silt. A single-tube gravity corer, such as a Kajak corer, is an effective way to sample ostracods in such sediments (Danielopol et al., 2002). Yet, the sampling cross-section of gravity corers is generally modest and numerous replicate cores must be collected to increase sampling size and hence get representative samples (e.g., Viehberg, 2006). As ostracod sediment penetration depth were also determined, a method that could collect bottom samples without disturbing the sediment structure was required (Decrouy et al., 2012). Also, preliminary results suggested that adult/juvenile ratios may be biased and the population density underestimated when grab or box corers are used. Hence, although the sampling cross-section of

TABLE I
Dates of ostracod and water sampling as well as physico-chemical measurements of the bottom water

13 and 70 m		2, 5 and 33 m	
Sampling occasion	Date	Sampling occasion	Date
A	04.07.06	A	04.07.06
B	04.19.06	C	04.25.06
D	05.10.06	E	05.24.06
F	06.12.06	G	06.19.06
H	07.11.06	I	07.25.06
J	08.10.06	K	08.31.06
L	09.12.06	M	10.04.06
N	10.10.06	O	10.25.06
P	11.15.06	Q	11.28.06
R	12.12.06	S	01.09.07
T	01.16.07	U	02.15.07
V	02.20.07	W	03.12.07
X	03.27.07	Y	04.10.07
Z	04.25.07	aa	05.01.07

the chosen gravity corers (either a ETHZ or a UWITEC using plastic tubes of 5.8 cm in diameter, hence a sample cross-section of 26.4 cm^2) is rather small, this approach with several cores taken at the same locality to increase the representativeness of the results was taken to be the best option to sample ostracods at 13, 33 and 70 m depth. For each core, the top six centimetres of sediment were sampled using a piston inserted in the lower end of the tube to collect living ostracods. Given the time available for each sampling session (one day) and the number of sites visited (2-3, see table I), four to five replicate cores (n) per site and per sampling session were normally scheduled. Nevertheless, depending on meteorological conditions, it was occasionally practically impossible to collect four to five replicate cores for each month. Hence, for several samples, n is sometimes lower with only two or three cores per site.

In contrast, sediment at 2 and 5 m depth consists of sand beds and pebbles. At these sites, a Van Veen-type sediment grab was used to recover pebbles (at 2 m), sand (at 5 m) and, when present, algae (at 2 and 5 m). Pebbles, sand, and algae were directly washed on board with lake water over a 200 μm mesh size sieve. Sediments and live organisms were transferred with lake water into flasks closed with a pierced lid, stored during transport to the laboratory in a cold box, and kept refrigerated at about 6 to 8°C until further processing.

For subsequent geochemical analyses of shells, individual ostracods needed to be sampled alive and should not have moulted between sampling and separation (see Decrouy et al., 2011b). As species were to be separated within 2-3 days following sampling, washing sediment samples through a 200 μm mesh sieve was found to be the best way to separate living ostracods within this period without losing all juvenile stages. Once separated, living ostracods were first killed in 30% alcohol and then stored in pure ethanol.

Taxonomy, ontogeny and individual abundance

Taxonomy is based on Meisch (2000). Danielopol (1969) was used for the identification of *Candona neglecta* Sars, 1887. For *Cypria lacustris* Lilljeborg, 1890, the classification of Meisch (2000), suggesting that this species and *Cypria ophtalmica* (Jurine, 1820) are two extreme forms of one single species, has not been followed. As a matter of fact, the same author stated during the 17th International Symposium on Ostracoda held in 2013 in Roma that he now considers the two forms to be classified as two different species (personal communication). Martens & Savatenalinton (2011) also consider the two forms as two distinct species. Identification of development stages are based on the general knowledge on larval development for Cypridoidea Baird, 1845 and Cytheroidea Baird, 1850 (Henderson, 1900; Meisch, 2000). Identification of development stages of *Cytherissa lacustris* (Sars, 1863) is based on Danielopol & Tetart (1990). Micro-dissections of soft parts of the ostracods were carried out following Danielopol (1982). Identification of development stages was also based on valve size. For each sample, the number of adults and juveniles was determined. When possible, gender was also determined for adults and A-1 instars. Mean monthly number of individuals per sample was calculated for each development stage. The monthly relative abundance of each development stage to total number of individuals was also calculated. Because the sampling area is known for samples taken at 13, 33 and 70 m depth, the mean number of individuals per surface area can be roughly estimated. Hence, monthly mean number of individuals was calculated for sites at 13, 33 and 70 m depth (juveniles and adults summed up). Annual mean number of individuals corresponds to the average of monthly mean numbers of individuals.

RESULTS

Environmental parameters

Water temperature at 2, 5 and, with a certain delay, at 13 m depth parallels seasonal variations (fig. 2). Temperature at 33 and 70 m depth increased continuously during 2006 and decreased only slightly during the winter of 2006-2007. In the epilimnion, O_2 saturation varies seasonally with higher O_2 saturation level during summer and autumn (up to 150% oxygen saturation at 2.5 and 5 m depth). The O_2 saturation also decreases with increasing depth (between 85-120% at 15 m and between 65-105% at 70 m; SECOE; CIPEL, 2011). $\delta^{13}C_{DIC}$ values at 2, 5 and 13 m depth are positively correlated with water temperature ($r^2 = 0.62$, $n = 42$),

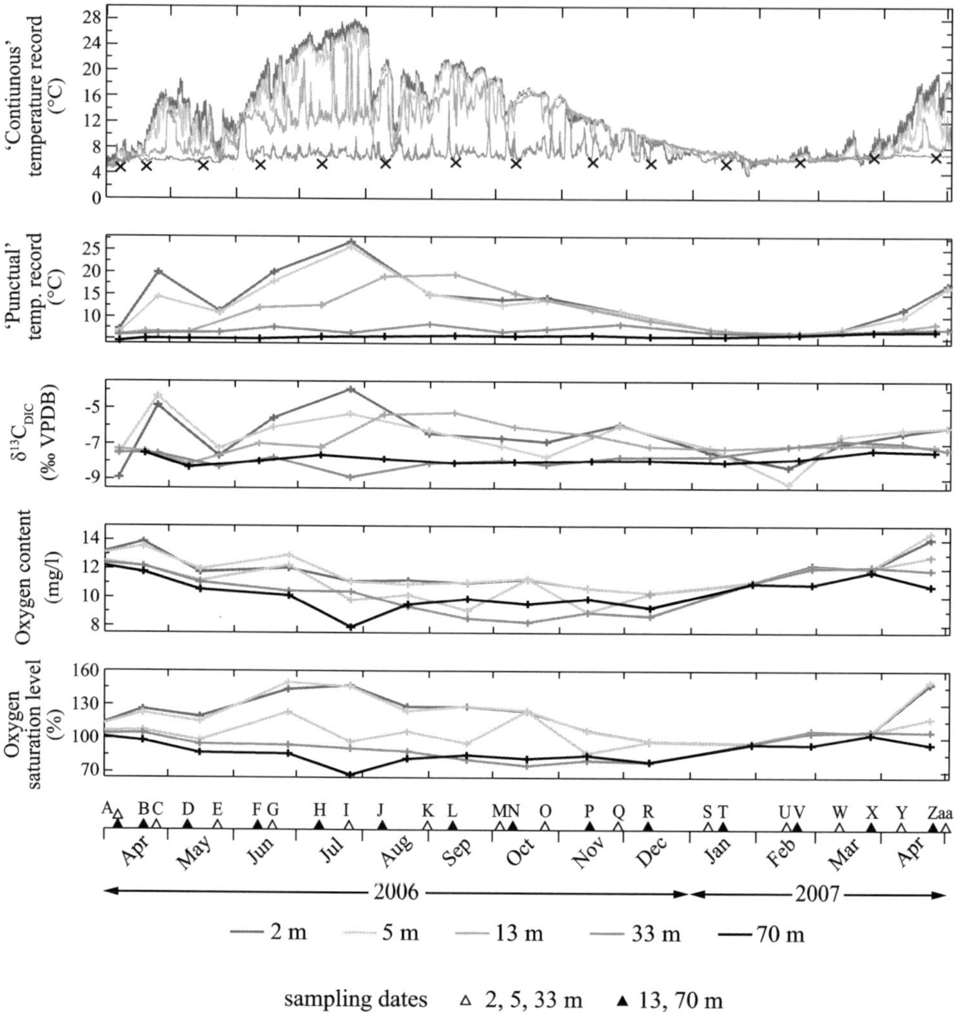

Fig. 2. Continuous record of water temperature at 2, 5, 13 and 33 m water depth and monthly variations of major environmental parameters at 2, 5, 13, 33 and 70 m water depth. 'Punctual' temp. record is the water temperature measured in situ at the time of ostracod collection, $\delta^{13}C_{DIC}$ is the carbon isotope composition of dissolved inorganic carbon, dissolved oxygen content is given both in mg/l and in saturation level. Temperature and isotopic data are from Decrouy et al. (2011a). Dissolved oxygen content was measured within the water column at 2.5, 5, 15, 30 and 70 m water depth at the 70 m water depth site shown in fig. 1C (data from "Service de l'Écologie de l'Eau", SECOE) and was related to the sites of 2, 5, 13, 33 and 70 m depth to allow for comparison. This figure is published in colour in the online edition of this journal, which can be accessed via http://booksandjournals.brillonline.com/content/journals/15685403.

whereas the values remain constant at 33 and 70 m depth. Variations in $\delta^{13}C_{DIC}$ at 2, 5 and 13 m depth are interpreted as resulting from preferential uptake of ^{12}C

in organic matter during photosynthesis, which can be used to track variations in primary productivity (see Decrouy et al., 2011a).

Substratum characteristics, sediment textures, as well as results on the geochemical composition of surface sediment at 13, 33 and 70 m water depth from Decrouy et al. (2011a) are summarised in table II.

Ostracod systematics, depth distribution, and number of individuals

In total 16 taxa were recovered alive during the sampling year (table III). Fig. 3A represents a simplified composite cross-section of the studied zone, while fig. 3B displays the species distribution with depth as well as approximate abundances. The species can be regrouped as follows: (1) littoral taxa living at 2 and 5 m depth (*Pseudocandona compressa* (Koch, 1838), *Ilyocypris* sp. Brady & Norman, 1889, *Prionocypris zenkeri* (Chyzer & Toth, 1858), *Herpetocypris reptans* (Baird, 1835), *Cypridopsis vidua* (O. F. Müller, 1776), *Plesiocypridopsis newtoni* (Brady & Robertson, 1870), *Potamocypris similis* (G. W. Müller, 1912), *Potamocypris smaragdina* (Vávra, 1891), and *Limnocythere inopinata* (Baird, 1843)), of which three taxa (*H. reptans* and *L. inopinata* and to a lesser extent *Ilyocypris* sp.) also colonise the sublittoral zone; (2) sublittoral species living at 13 m depth (*Candona candida* (O. F. Müller, 1776), *Fabaeformiscandona caudata* (Kaufmann, 1900), *Isocypris beauchampi* (Paris, 1920) and *Limnocytherina sanctipatricii* (Brady & Robertson, 1869)), and to a lesser extent at 33 m depth (*C. candida* and *I. beauchampi*); and (3) profundal species that are found from 13 m to 70 m depth (*C. neglecta*, *Cypria lacustris* and *Cytherissa lacustris*).

At 2 and 5 m depth, the results indicate that *C. vidua* and *P. compressa* were most abundant, *H. reptans* and *L. inopinata* were moderately abundant, and all the other species were less abundant. Table IV summarises the monthly mean number of individuals for 13, 33 and 70 m water depth as well as the number of replicate cores. Mean annual numbers of individuals calculated on the monthly mean number of individuals are given in table V and illustrated in fig. 3C. The number of individuals is higher at 13 m compared to 33 and 70 m depth. *Cytherissa lacustris* dominates the population at the three deeper sites followed by *C. neglecta*. *Cypria lacustris* is also abundant at 13 and 70 m depth.

Species phenology

For most species, illustrations of instar-frequency distributions shown in figs. 4, 5 and 6 allow the phenology of species to be followed throughout the year. However, for *C. neglecta* at 70 m and *C. lacustris* at 33 and 70 m depth, the development of individuals occurs throughout the year, making it difficult to distinguish life histories from these populations. For *Cp. lacustris* and, to a lesser

TABLE II

Sediment, vegetation, and fauna found at the five sampling sites

Site	Sediment type and texture	Grain size	H_2O (wt%)	$CaCO_3$ (dry wt%)	TOC	Vegetation	Fauna
2 m	Large banks of pebbles and few beds of coarse sand situated in depressions	Pebbles and coarse sand	–	–	–	Macroalgae and macrophytes from end spring to end autumn	Large number of zebra mussels, isopods, some gastropods, *Pisidium*, *Limnaea*, *Planaria* and Trichoptera larvae
5 m	Medium sized sand punctuated by isolated pebbles sometime grouped in banks	Middle sized sand and pebbles	–	–	–	Macroalgae and macrophytes from end spring to end autumn	Large number of zebra mussels, several isopods, and gastropods (of which *Bithynia tentaculata* (L., 1758)), some *Pisidium*, *Limnaea*, *Planaria* and Trichoptera larvae
13 m	Intercalation of large beds of zebra mussel and beds of sand	Silty sand	85	62	1.4	Little or no vegetation, apart from detritus of alga and plant	Large beds of zebra mussels; in the sand: some chironomids and *Pisidium* sp. and few isopods and gastropods
33 m	Very fine sediment, very rich in water with a high organic matter content, having a "jelly-like" texture	Clayey silt	93	45	2.9	No vegetation	Large number of copepods and many chironomids
70 m	Very fine sediment, very rich in water with a high organic matter content, having a "jelly-like" texture	Clayey silt	95	44	3.3	No vegetation	Large number of copepods and many chironomids

H_2O (wt%) represents the water content of surface sediments (mean of seasonal samples), $CaCO_3$ and TOC give the content of carbonate and total organic mater, respectively. Both are expressed in dry weight percent. H_2O, $CaCO_3$ and TOC correspond to average of seasonal sample values.

TABLE III
Ostracod systematics

Genera	Species
Candona s. str.	*candida* (O. F. Müller, 1776)
	neglecta Sars, 1887
Fabaeformiscandona	*caudata* (Kaufmann, 1900)
	cf. *protzi* (Hartwig, 1898)*
Pseudocandona	*compressa* (Koch, 1838)
Cryptocandona	*reducta* ? (Alm, 1914)*
Cypria	*lacustris* Lilljeborg, 1890
Ilyocypris	*Ilyocypris* sp.
Prionocypris	*zenkeri* (Chyzer & Toth, 1858)
Herpetocypris	*reptans* (Baird, 1835)
Isocypris	*beauchampi* (Paris, 1920)
Cypridopsis	*vidua* (O. F. Müller, 1776)
Plesiocypridopsis	*newtoni* (Brady & Robertson, 1870)
Potamocypris	*similis* G. W. Müller, 1912
	smaragdina (Vávra, 1891)
Limnocythere s. str.	*inopinata* (Baird, 1843)
Limnocytherina	*sanctipatricii* (Brady & Robertson, 1869)
Leucocythere	*mirabilis* Kaufmann, 1892*
Cytherissa	*lacustris* (Sars, 1863)

* Species found only as sub-fossil valves in core sediment.

extent, *H. reptans*, no clear pattern in the development is observed and instar abundances seem to vary randomly.

DISCUSSION

Monthly and annual mean number of individuals

Table IV indicates that for monthly samples the variance (S^2), in general, is several times larger than the mean of the total number of individuals (table IV), suggesting highly clumped populations (Ludwig & Reynolds, 1988). Spatial distribution of benthic invertebrates is known to be complex with individuals that tend to agglomerate on the lake floor, complicating studies of benthic organism distributions (e.g., Elliot, 1977). Yet, because of the small cross-sectional area of the sampling device (24.6 cm^2) and the low number of replicates per sample, potential outliers might bias the data and calculated means are less precise. The monthly results should, therefore, be interpreted with caution. Enlarging the sampling cross-sectional area or alternatively sampling several cores at the same site would certainly improve the precision and statistical representativeness of different sites. However, given the logistic restrictions, the number of sites sampled, and the number of species studied, it was neither possible to increase the

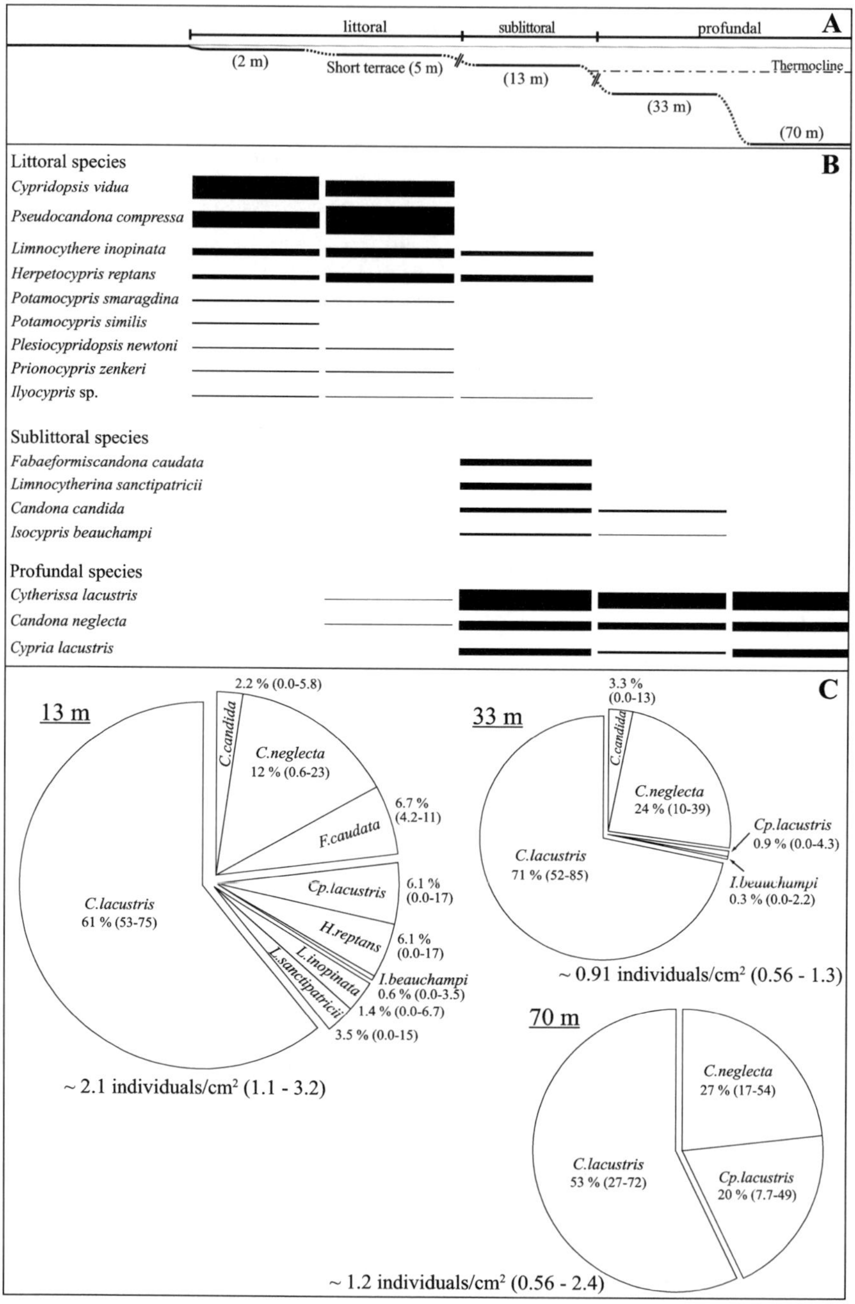

Fig. 3. (A) Simplified composite cross-section of the studied zone; (B) estimated number of individuals for each species at the different sampling sites, the thicker the line, the higher the number of individuals; (C) relative individual abundances of species for sites at 13, 33 and 70 m depth, based on the annual mean number of individuals (minimal and maximal values of monthly samples are given in parentheses).

TABLE IV
Monthly mean number of individuals per sample (26.4 cm^2)

Depth													
13 m													
Sample	B	D	F	H	J	L	N	P	R	T	V	X	Z
C. candida	0.0	0	2.7	3.5	2.3	2.0	2.0	2.3	1.0	0.5	0.0	0.25	0.7
C. neglecta	4.0	1	0.3	11	13	12	13	4.0	12	14	4.5	4.5	1.3
F. caudata	2.5	5.5	3.7	3.5	3.5	3.0	6.7	2.3	3.3	5.8	2.75	2.75	1.3
Cp. lacustris	0.0	3.5	4.7	1.0	5.5	2.0	3.0	2.0	2.0	0.8	2.75	5	5
H. reptans	5.0	0.5	3.3	9.5	4.0	2.0	2.7	3.0	2.0	2.8	2.5	0.25	2.3
I. beauchampi	0.0	0	0.3	0.5	0.0	0.0	0.3	0.3	0.0	0.0	0.5	1.25	0
L. inopinata	0.0	0	1.7	3.0	5.5	0.3	2.7	0.0	0.3	0.3	0.0	0	0
L. sanctipatricii	5.5	2	4.7	1.5	0.0	0.0	0.0	0.0	0.7	0.5	1.8	1.5	1.3
C. lacustris	20	37	32	51	48	48	51	26	34	36	19	21	17
All species	37	50	54	84	82	69	81	40	55	61	34	36	29
s^2	–	–	137	–	1182	1720	772	576	492	137	202	266	41
n	2	2	3	2	4	3	3	3	3	4	4	4	3
33 m													
Sample	C	E	G	I	K	M	O	Q	S	U	W	Y	aa
C. candida	0.0	–	0.3	2.0	1.0	1.0	1.2	0.8	1.0	0.8	0.3	0.3	0.3
C. neglecta	5.0	–	6.0	3.3	3.7	3.3	2.4	4.3	7.3	10.0	8.5	8.5	7.0
Cp. lacustris	0.3	–	0.7	0.0	0.0	0.0	0.0	0.3	0.0	0.3	0.5	0.3	0.0
I. beauchampi	0.0	–	0.3	0.0	0.0	0.0	0.0	0.0	0.5	0.0	0.0	0.0	0.0
C. lacustris	9.5	–	8.0	11	19	17	20	19	26	21	23	17	16
All species	15	–	15	16	24	22	23	25	34	32	33	26	24
s^2	77	–	21	53	6.3	91	50	115	85	195	110	63	33
n	4	–	3	4	3	4	5	4	4	4	4	4	4
70 m													
Sample	B	D	F	H	J	L	N	P	R	T	V	X	Z
C. neglecta	6.8	8.0	8.0	6.8	5.0	9.3	8.0	3.3	10	9.3	8.0	7.5	7.5
Cp. lacustris	6.0	16	13	1.3	3.0	1.3	5.0	2.8	11	8.8	4.3	11	4.8
C. lacustris	15	9.0	27	6.8	8.0	6.7	28	11	24	16	20	17	32
All species	27	33	47	15	16	17	41	17	45	34	33	35	44
s^2	88	–	–	345	–	30	1303	57	74	169	312	327	184
n	4	2	2	4	2	3	4	4	4	4	4	4	4

core size nor the number of replicate cores beyond that sampled here. In contrast to the monthly data, the mean annual numbers of individuals per sample in table V correspond to the average of 12 to 13 monthly values and represent a total of 40, 47 and 44 replicates for sites at 13, 33 and 70 m depth, respectively. Hence, although the variances calculated for these figures are high and the ranges (min-max) wide, these estimates are considered to be approximations of the populations inhabiting these sites.

TABLE V
Mean annual number of individuals per sample (26.4 cm^2)

Depth	Species	Mean	S^2	Min	Max
13 m ($n = 13$)	All species	55	379	29	84
	C. candida	1.3	1.4	0.0	3.5
	C. neglecta	7.2	27	0.3	14
	F. caudata	3.6	2.3	1.3	6.7
	Cp. lacustris	2.9	3.2	0.0	5.5
	H. reptans	3.1	5.3	0.3	9.5
	I. beauchampi	0.3	0.1	0.0	1.3
	L. inopinata	1.1	2.9	0.0	5.5
	L. sanctipatricii	1.5	3.1	0.0	5.5
	C. lacustris	34	161	17	51
33 m ($n = 12$)	All species	24	44	15	34
	C. candida	0.7	0.3	0.0	2.0
	C. neglecta	5.8	6.2	2.4	10
	Cp. lacustris	0.2	0.1	0.0	0.7
	I. beauchampi	0.1	0.0	0.0	0.5
	C. lacustris	17	29	8.0	26
70 m ($n = 13$)	All species	31	136	15	47
	C. neglecta	7.5	3.3	3.3	10
	Cp. lacustris	6.7	22	1.3	16
	C. lacustris	17	73	6.7	32

For the calculation the monthly mean numbers of individuals per sample from table IV were used.

Depth distribution of species

The depth distribution of species in Lake Geneva is similar to that of Lake Constance (Löffler, 1969), Laachersee (Kempf & Scharf, 1980) and Mondsee (Yin & Geiger, 1995), whose geographical locations are shown in fig. 1A. The number of taxa is high in the littoral zone (9 taxa), as well as in the sublittoral zone (10 taxa), but decreases with depth (5 and 3 species at 33 and 70 m depth, respectively). The number of individuals presents a similar pattern as species abundances, being high at 2, 5 and 13 m, and low at 33 and 70 m depth (fig. 3B, tables IV and V). According to Yin & Geiger (1995), this pattern is typical for mesotrophic lakes: "optimal oxygen abundance and prevailing food conditions in parallel with habitat richness characterise littoral areas, while beneath the thermocline harsher conditions prevail due to changes in physical condition and/or supply of organic matter as nutriments and/or predation increase. At even greater depth, conditions often change for the better with higher amounts and/or better preservation of organic matter accumulating in zones with lower oxygen content". The next

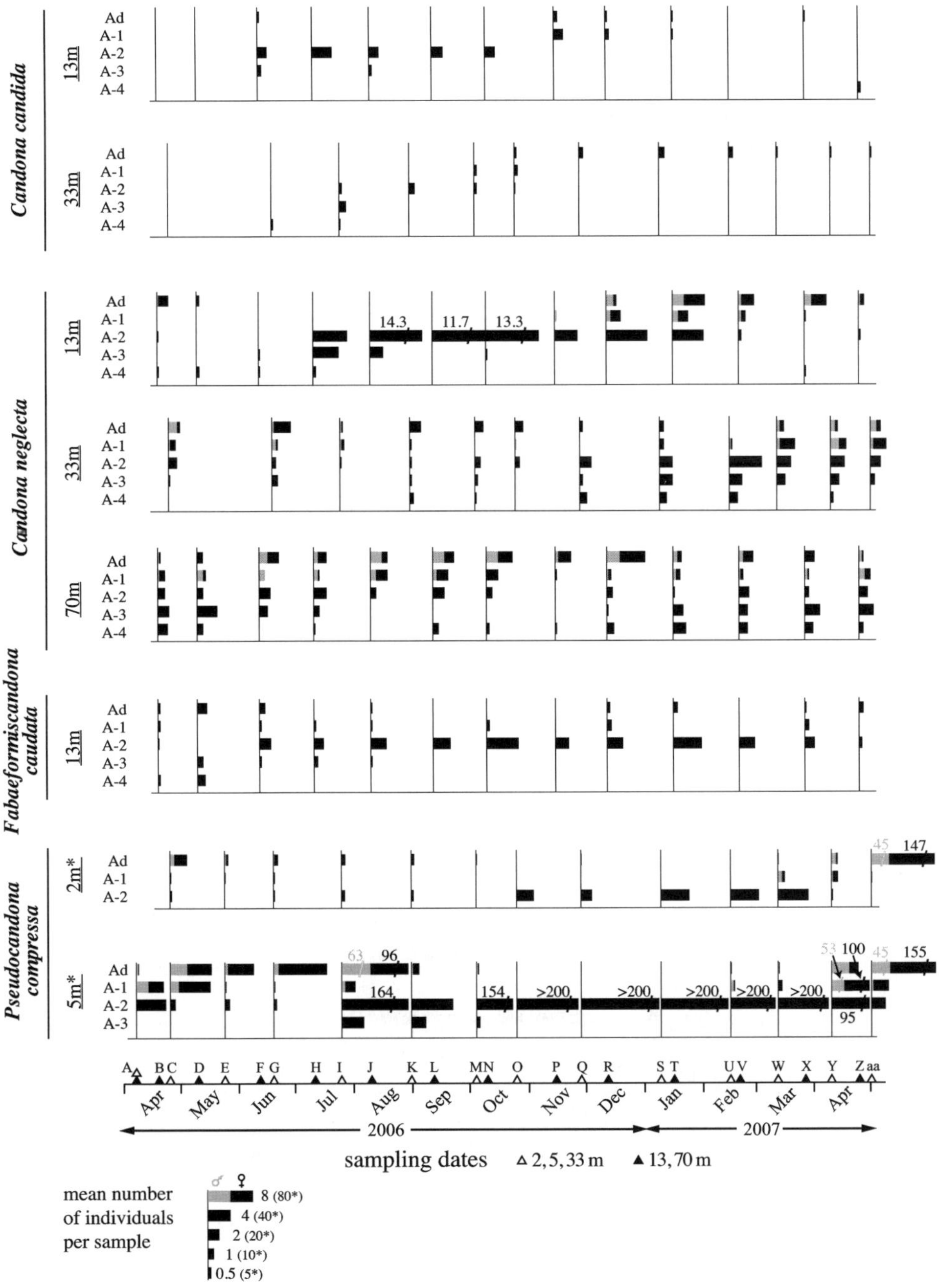

Fig. 4. Instar-frequency distributions based on monthly number of individuals per sample for *Candona candida* (O. F. Müller, 1776), *Candona neglecta* Sars, 1887, *Fabaeformiscandona caudata* (Kaufmann, 1900) and *Pseudocandona compressa* (Koch, 1838) at different depth in Lake Geneva. The location of the *y*-axes corresponds to the date of sampling given at the bottom of the figure.

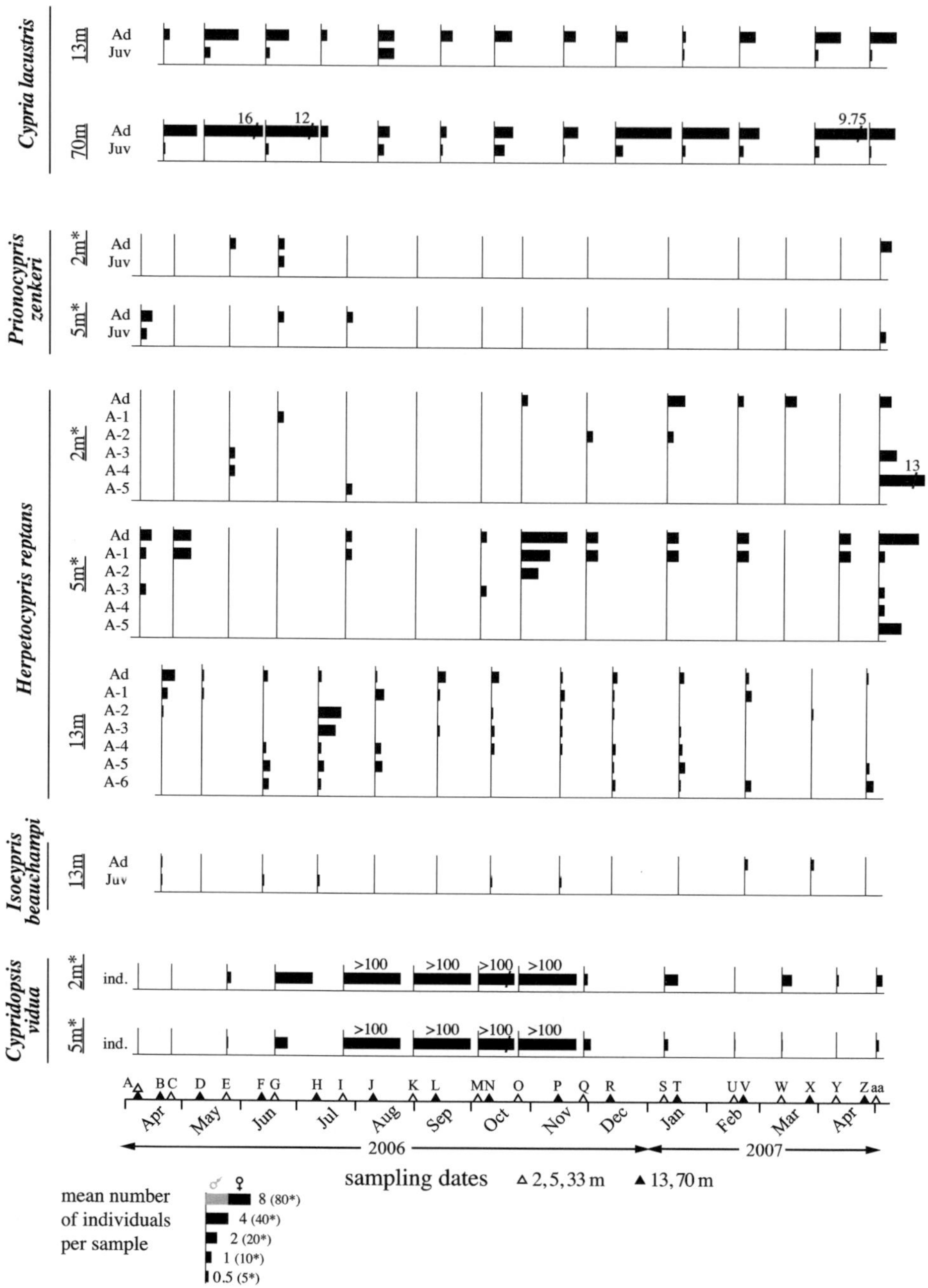

Fig. 5. Instar-frequency distributions based on monthly number of individuals per sample for *Cypria lacustris* Lilljeborg, 1890, *Prionocypris zenkeri* (Chyzer & Toth, 1858), *Herpetocypris reptans* (Baird, 1835), *Isocypris beauchampi* (Paris, 1920) and *Cypridopsis vidua* (O. F. Müller, 1776) at different depth in Lake Geneva. The location of the *y*-axes corresponds to the date of sampling given at the bottom of the figure.

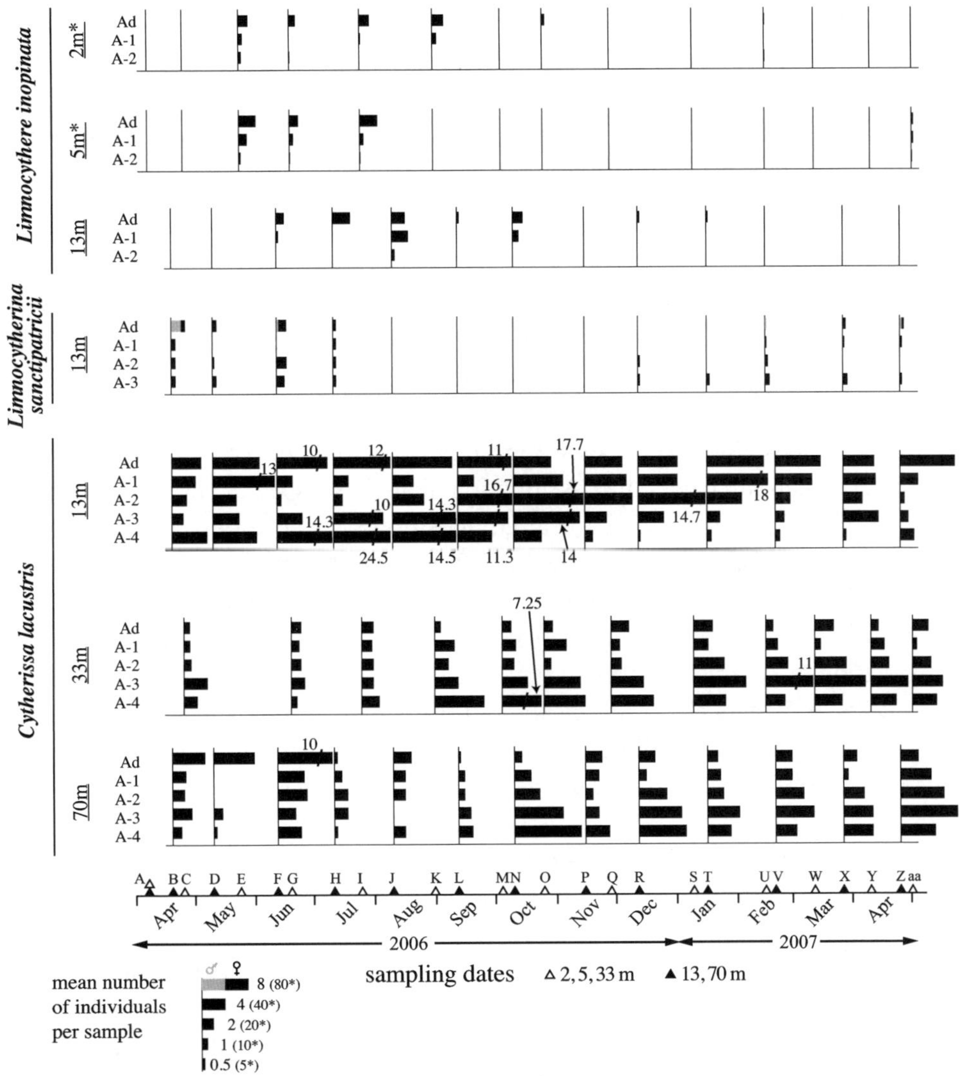

Fig. 6. Instar-frequency distributions based on monthly number of individuals sample for *Limnocythere inopinata* (Baird, 1843), *Limnocytherina sanctipatricii* (Brady & Robertson, 1869) and *Cytherissa lacustris* (Sars, 1863) at different depth in Lake Geneva. The location of the *y*-axes corresponds to the date of sampling given at the bottom of the figure.

sections aim to further examine the influence of several environmental parameters on the depth distribution of species-specific assemblages.

Oxygen saturation level

Oxygen availability is one of the potential factors controlling the bathymetric distribution of ostracods (e.g. Delorme, 1978; Geiger, 1990a; Curry, 2010). Oxygen content measured for the water column of the Petit-Lac over the past half-century has always been relatively high (minimal value of 6.04 mg/l O_2 for the 1957-2010 period; SECOE; CIPEL, 2006, 2011). As oxygen content is high at all depth throughout the year, it is suggested that oxygen content plays less of a role on controlling the species distribution in the Petit-Lac.

Food resources

Although crucial for the establishment of ostracod populations, little is known about the feeding habits of ostracods. Most of the species inhabiting sites at 13, 33 and 70 m water depth are infaunal (Decrouy et al., 2011). Yet, it is not known whether these species are able to select their source of food from the sediment at these sites or if they ingest the fine fraction of sediment without any discrimination. Under the assumption that these species are not available to select their food, the geochemical composition of sedimentary organic matter provides some information about the food resources. Results from a previous study indicate that the carbon/nitrogen ratio of sedimentary organic matter (C/N) at 13, 33 and 70 m depth have a range between 7.1 and 7.8 and that the carbon isotopic composition of sedimentary organic matter ($\delta^{13}C_{OM}$) has a range between −26.7 and −27.9‰ VPDB (Decrouy et al., 2011a). These results suggest (i) that the principal component of sedimentary organic matter consists mostly of freshwater algae produced in the epilimnion (Meyers, 1994) and (ii) that the organic matter contained in the surface sediment is similar at 13, 33, and 70 m depth (Decrouy et al., 2011a). Results also indicate that the sediment TOC increases with depth (table II), suggesting that organic matter is better preserved in sediments in deeper sites. Hagermann (1967), Reyment (1975), Heip (1976a), Jungwirth (1979), as well as Rieradevall & Roca (1995) showed that high number of individuals corresponds to sediments with high organic matter content. Yet, the number of individuals and total number of species are higher at 13 m depth, where TOC is low, suggesting that TOC within the sediments is less of a restrictive parameter for the ostracod populations in Lake Geneva. Nevertheless, the carbon isotope compositions of dissolved inorganic carbon ($\delta^{13}C_{DIC}$) are higher at 13 m compared to 33 and 70 m depth and increase by about 2‰ during summer and autumn (fig. 2). Given that $\delta^{13}C_{DIC}$ values reflect the importance of photosynthetic activity compared to aerobic respiration, these results indicate active photosynthesis at 13 m depth and a decrease of primary productivity with depth. Thus, higher availability of food directly derived from the primary productivity or via the food chain might favour ostracod populations at 13 m water depth. Benzie (1989) showed that *H. reptans* has a preferential diet for degraded macrophytes and old macroalgae before faecal material, fresh macrophytes and chitin detritus. It is, therefore, also possible that slight differences in the type of organic matter, as well as the relative abundance may influence ostracod populations (see also below).

Water temperature

Species found in the littoral zones, except *P. compressa*, *P. zenkeri* and *H. reptans*, are mesothermophilic to warm stenothermal forms (Meisch, 2000). *Limnocythere inopinata* can also be found during late spring to early autumn at the site of 13 m depth, suggesting that higher tolerance to low temperatures allows this species to live at this depth. Hence, water temperature may well be one of the factors controlling bathymetric distribution of *C. vidua*, *P. smaragdina*, *P. similis* and *P. newtoni*. Concerning *P. compressa*, an increase in water temperature during spring triggers resumption of individual development and final maturity (see below). Scarce data for *P. zenkeri* preclude any conclusions to be drawn for this species, whereas the occurrence of individuals of *H. reptans* seem not to be constrained by the temperature of water, adult and juveniles being found during the whole year at temperature between 6 and 24°C. Other parameters, such as food resource and substrate characteristics, may also restrict littoral species to the shallow zones of lakes (see below).

Species collected in the sublittoral and profundal zones are considered in Western and Central Europe as cold water stenothermal to oligothermophilic forms (*C. candida*, *C. neglecta*, *F. caudata*, *Cp. lacustris* and *L. sanctipatricii*) or eurythermal forms with preference for cold water (*C. lacustris*) (Alm, 1915; Wolf, 1920; Nüchterlein, 1969; Hiller, 1972; Roca & Baltanás, 1993; Rieradevall & Roca, 1995; Meisch, 2000; Mezquita et al., 2005; Viehberg, 2006). Hence, tolerance for cold water may allow *C. candida*, *C. neglecta*, *F. caudata*, *L. sanctipatricii* and *Cp. lacustris* to colonize deep zones of lakes but other factors, such as food resources or substrate characteristics, may explain their absence in littoral zones, and the highest species abundance and the larger number of individuals at 13 m depth.

Substrate characteristics

According to previous studies, the type of substrate and sediment (pebbles, algae, sand beds, sediment grain size, texture, etc.) may control the spatial distribution of ostracod species (e.g., Krutak, 1971; Powell, 1976; Carbonnel, 1980; Benzie, 1989; Rieradevall & Roca, 1995). A comparison of the species water depth distribution and substrate characteristics (fig. 3, table II) suggests that the littoral zone (2 and 5 m depth) is inhabited by epifaunal species (*P. compressa*, *H. reptans*, *C. vidua*, *P. newtoni*, *Potamocypris* sp. and *L. inopinata*), but no infaunal species, whereas the profundal zones (33 and 70 m depth) are home to the infaunal species (*C. candida*, *C. neglecta* and *C. lacustris*). At the site of 13 m depth, in contrast, both epi- and infaunal species occur. Because the predominant wind direction is SW-NE, trending the long axis of this elongated basin of the Petit-Lac, the effective fetch is important along the shore lines (Girardclos, 2001) and only coarse sediments are found within littoral zones. Although macrophytes, pebbles, and coarse sand may host food sources, provide a grazing surface and also provide shelter for epifaunal species (Benzie, 1989; Mbahinzireki et al., 1991; Roca & Danielopol, 1991; Roca et al., 1993), such zones may not be favourable for infaunal species. Some ostracods may need to burrow in order to survive within a shelter, avoid predation (Mbahinzireki et al., 1991), search for food (Roca & Danielopol, 1991), or adjust the oxygenation of their tissue (Corbari et al., 2004). Lack of fine sediment in littoral zones might, therefore, explain the restriction of infaunal species (*C. candida*, *C. neglecta*, *F. caudata* and *C. lacustris*) to the sublittoral and profundal zones, as well as the absence of common infaunal littoral species such as *Darwinula stevensoni* (Brady & Robertson, 1870).

Life histories of the species

While significant variations of individual abundance for the different development stages can be observed in instar-frequency distributions illustrated in figs. 4, 5 and 6, it remains difficult to grasp the species life history for each population. Graphical illustrations based on the Ostracod Watch Model (OWM) presented by Okan Külköylüoğlu in 1998 were, therefore, used to represent juvenile development throughout the year. Fig. 7 explains how these graphical illustrations are to be read. These illustrations are based on a watch dial, each hour corresponding to a month of the year and 1 January being placed at noon. The watch dial is subdivided into a succession of concentric rings. Each of these rings stands for a development stage, the youngest stage being in the centre of the dial and the adult being shown in the most external ring of the dial. The monthly relative individual abundance of each development stage is represented using a grey-scale shading of the box corresponding to the development stage and collection month. The darker the shading, the higher is the individual abundance. To indicate populations dominated by males, a black/white star is inserted in the box. In addition, the monthly mean water temperature is given outside of the dial. Figs. 8 and 9 illustrate the life history of the most important species in the Petit-Lac using these graphical illustrations.

In previous studies, the life-cycles of target species were generally determined for one single site (e.g., Ferguson, 1944; Theise, 1966; Heip, 1976; Fenwick, 1984; Mallwitz, 1984; Kamiya, 1988; Mezquita et al., 2002; Van Doninck et al., 2003; Wojtasik, 2008), or in laboratory cultures (Ikeya & Kato, 2000; Gandolfi et al.,

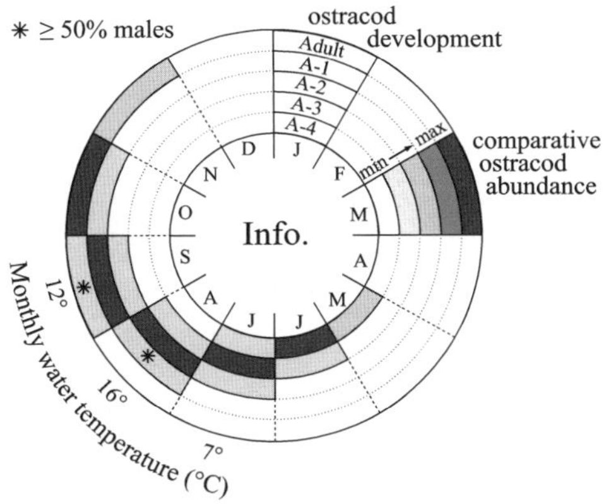

Fig. 7. Construction of the graphical illustration showing the life-cycle of dominant species from Lake Geneva. Each concentric ring corresponds to a development stage, the younger juvenile stage being in the centre of the illustration and the adult being shown in the most external ring. Individual relative abundance is represented using a grey-scale shading; the darker the shading, the higher is the individual abundance. A black/white star is inserted to indicate when the population is dominated by males. Monthly mean water temperature is given for each month around the illustration.

2001). Although numerous early studies on ostracod ecology noted that the life history of a given species can change from one site to the other (e.g., Alm, 1915; Wolf, 1920; Nüchterlein, 1969), suggesting an environmental control on ostracod development, studies that have compared the life histories of ostracods at different depth within a single site remain scarce (Savolainen & Valtonen, 1983; Geiger, 1990b; Rieradevall & Roca, 1995). Savolainen & Valtonen (1983) and Geiger (1990b) found that juvenile development is prolonged and their life span longer at deeper sites compared to shallow sites. Besides, Geiger (1990b) and Roca & Wansard (1997) showed that the periods during successive moulting decrease with increasing temperature, suggesting that water temperature might influence life history of ostracod species. In addition, Horne (1983) stated that temperature is generally regarded to be the main factor controlling periodicity in the life-cycles of ostracods.

Figs. 8 and 9 show that each species illustrated has its own life history and that the life history of certain species can vary with water depth in the Petit-Lac. Species found in the littoral zone, except *P. compressa*, *P. zenkeri* and *H. reptans*, are stenochronal and are restricted to the warm months of the year, i.e., from May to October (figs. 4, 5 and 6). As stated above, these species are mesothermophilic to warm stenothermal forms (Meisch, 2000). Hence, it is suggested that because they need to complete their total life-cycle before the return of colder conditions these species have a short life-cycle and a rapid development. *Pseudocandona compressa* is the only species with a life-cycle lasting one year and inhabiting the littoral zone. In the sublittoral zones, species adapted to colder conditions with life-

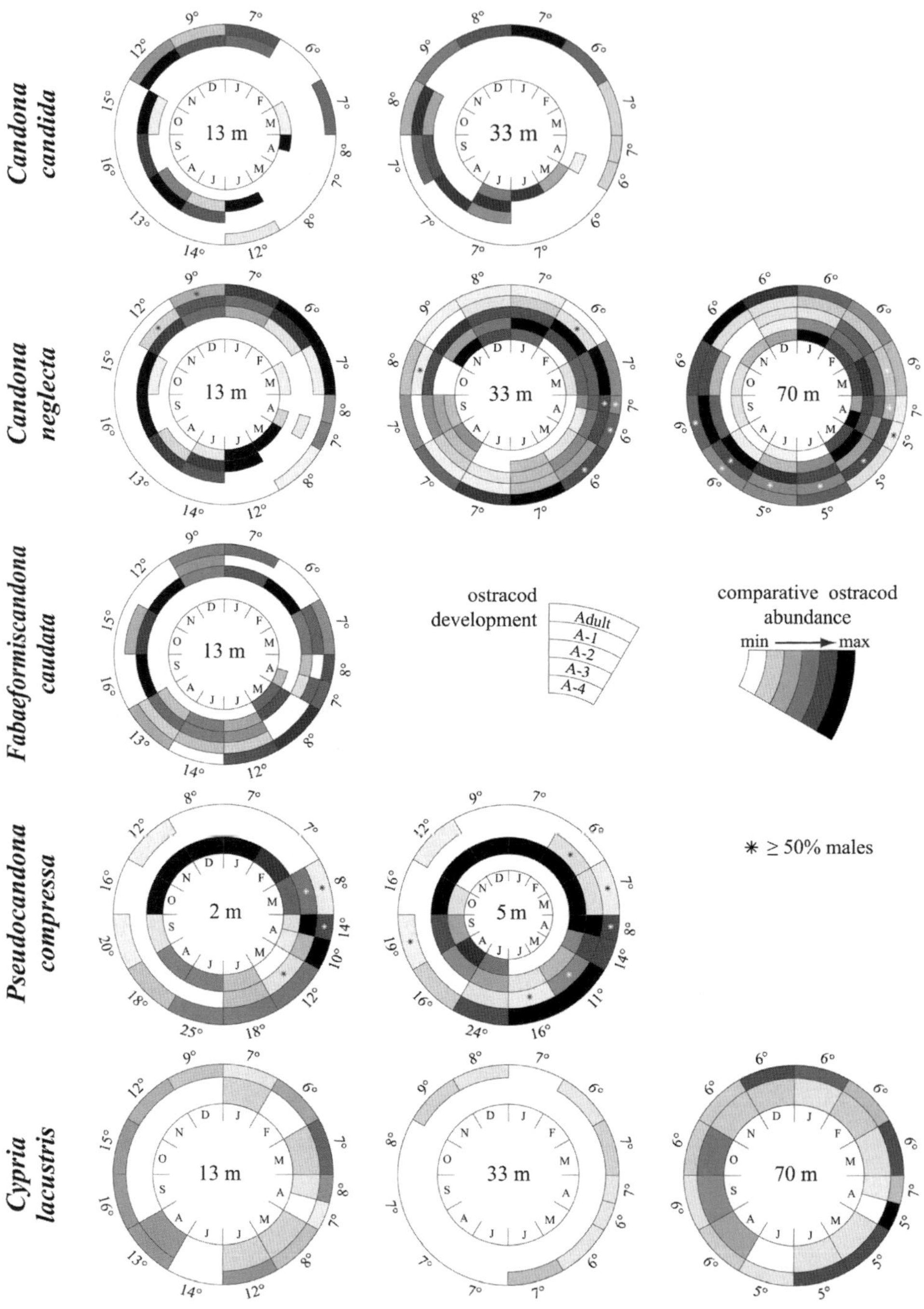

Fig. 8. Graphical illustrations of the life-cycle of species of Candonidae Kaufmann, 1900. See fig. 7 for more information on the graphical illustration of the ostracod life-cycle.

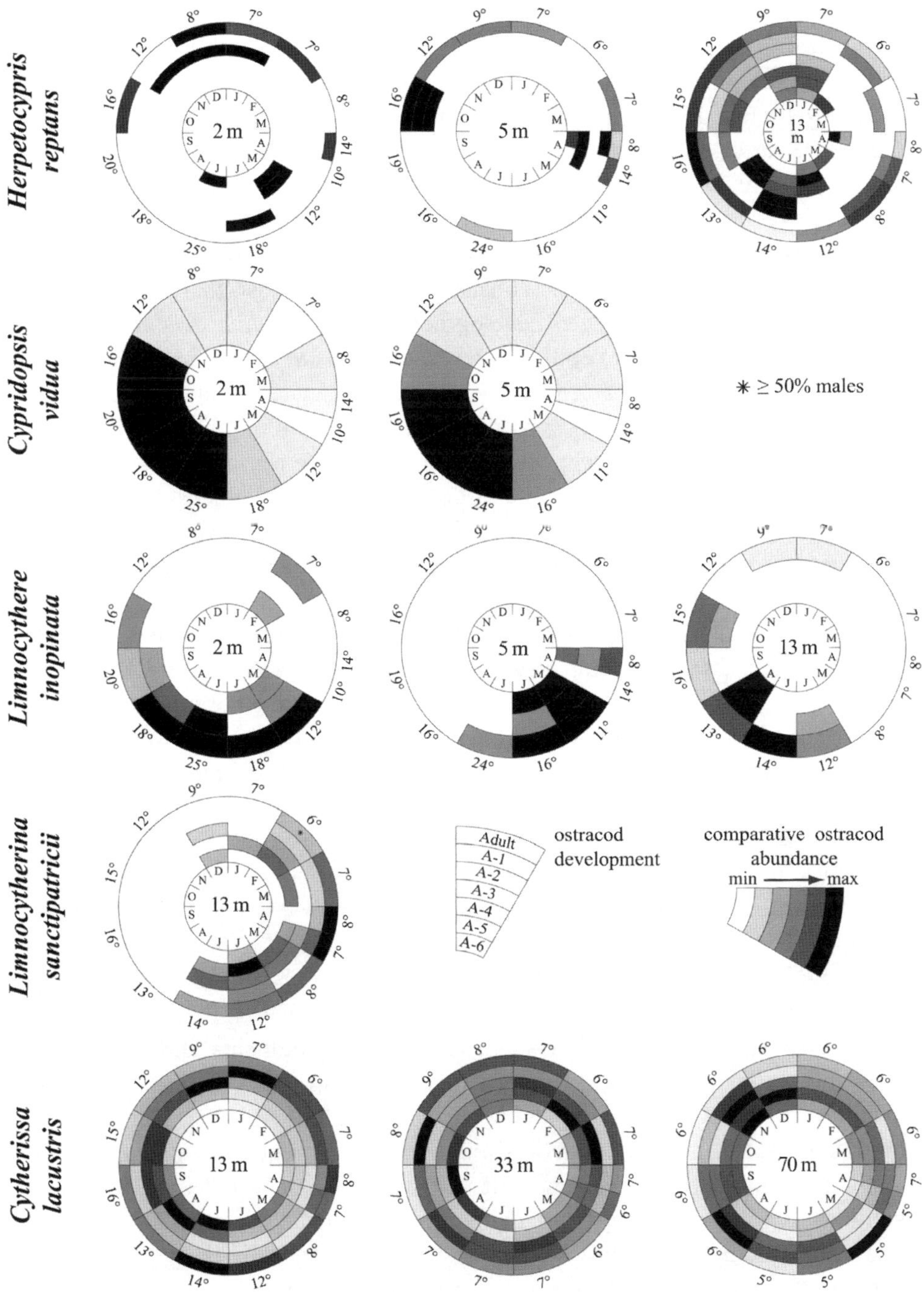

Fig. 9. Graphical illustration of life-cycle of species of Cyprididae Baird, 1845 and Cytheroidea Baird, 1850. See fig. 7 for more information about the graphical illustration of the ostracod life-cycle.

cycles of one year or more dominate. However, two species, *L. sanctipatricii* and *L. inopinata*, have short lifecycles at this depth. These species have a restricted temperature tolerance and can only develop during a short lapse of time: winter and spring for the former, summer and autumn for the latter. In deeper zones, temperature varies only slightly and species adapted to the cold with life-cycles of one year or more can develop continuously. At 33 m depth, temperature variations are still relevant and seasonality can still be observed even if the development is spread over time (*C. candida*, *C. neglecta* and in a less clear way *C. lacustris*). At 70 m depth, temperature is almost constant (fig. 2) and must, therefore, play less of a role in controlling the species development. *Cytherissa lacustris*, for example, shows no seasonality at this depth. However, *C. neglecta* and, although less clear, *Cp. lacustris*, both present a seasonal life-cycle. Other seasonal factors, such as food availability linked to primary productivity in the epilimnion and/or variations of oxygen concentrations in deep water, may control the seasonal development of these two species. Von Grafenstein and co-authors (1996) suggest, for example, that population dynamics of *Fabaeformiscandona levanderi* (Hirschmann, 1912) and *C. candida* in Ammersee and Starnbergsee is related to mixing within the lakes, relevant to sites below the thermocline.

A closer look at fig. 8 reveals that the development of Candoninae is not linear in the Petit-Lac. At 13 m depth, *Candona candida*, *C. neglecta* and *F. caudata*, remain confined to the developmental stage of A-2 for a long period, only to moult into A-1 and reach adulthood as the water temperature decreases down to 6 or 7°C at the end of the year for most adults of *C. candida* and *C. neglecta*, and 7 to 9°C in autumn and in spring for *F. caudata*. The same pattern is found at 2 and 5 m depth for *P. compressa* that spends winter as A-2 and resumes its development when temperature increases up to 9 to 14°C in spring. Viehberg (2006) estimated that in northeast Germany the optimum water temperature is 7.2 ± 1.9°C for *Candona candida* (±1.9°C being the tolerance), 7.6 ± 2.2°C for *Candona neglecta*, and 13.5 ± 5.6°C for *P. compressa*. These values correspond to the temperature at which last moulting of the three species occurs. This implies (i) that the life history is limited by the species-specific environmental requirements and (ii) juvenile stages might have different environmental optimum and tolerance compared to adults. This is supported by the work of Külköylüoğlu and Dügel (2004), who demonstrated that species with similar ecological requirements have similar phenology. This also suggests that the A-2 individuals "wait" until favourable conditions occur before they start their development and moult. As the period without any moulting can last several months (fig. 8), it is likely that the A-2 juveniles of Candoninae are able to decrease their metabolic activity to endure harsh conditions, making a kind of a diapause (high or low water temperature in the present case). This hypothesis is supported by the findings of Delorme & Donald (1969), who showed that A-2 individuals of *Candona rawsoni* Tressler, 1957 enter a torpid stage to survive ephemeral drying of their environment. Differences between environmental tolerances of juvenile stages and adults have important implication for palaeoenvironmental interpretations based on fossil assemblages or on ostracod valve geochemistry. For examples, Viehberg (2006) found that an optimum water temperature for *C. neglecta* was 7.6°C with a tolerance of 2.2°C, while Mezquita et al. (2005) obtained an optimal average annual air temperature of 10.7°C with a tolerance of 1.8°C (2.23 ± 1.6°C and 20.59 ± 1.99°C for optimal and tolerances average January and July air temperatures, respectively). In the study of Viehberg (2006), data were only used when adult and A-1 instars were collected simultaneously to make sure that samples represent the species ecological optimum, the latter being assumed to exist at the time of the final moulting and subsequent reproduction (De Deckker, 2002). In contrast, Mezquita et al. (2005) did not take phenology into account but only the presence or absence of species. Although it is difficult to compare optimal temperatures of both studies, the difference might be explained by a difference in the ecological optimum and/or tolerance between juvenile and adult stages. Care has to be taken, however, when interpreting the ecological optimum to reflect the whole development, as is the case in Mezquita et al. (2005), or only the environmental conditions required for the final moulting are considered, as in Viehberg (2006), when the fossil assemblages are to be interpreted as palaeoenvironmental indicators. For geochemical composition of the valves, one must be aware that environmental conditions might be different during each period of moulting. Decrouy (2012) has shown, for example, that a decrease of up to 1‰ for carbon isotope compositions and an increase of approximately 1‰ for oxygen isotope compositions between A-2 and A-1 instars of *P. compressa* can be explained by different times of moulting between both juvenile stages, the moulting from A-3 to A-2 instars occurring in late summer when water temperature is high (approx. 16-25°C), whereas moulting to A-1 occurs in early spring when temperature is still relatively cold (approx. 7-14°C). Knowledge of the life-cycle of target species is, therefore, important when ostracod species assemblages and/or geochemical compositions of the valves are used for environmental and palaeoenvironmental interpretations.

ACKNOWLEDGEMENTS

The research presented in this manuscript was generously funded by the Swiss National Science Foundation (SNF) projects (SNF 200021-107958 and 200020-119935). The authors would also like to express their gratitude towards the Institute F.-A. Forel for the use of the research vessel and the limnological material necessary for sediment sampling. Many colleagues and friends are thanked for their help with the fieldwork. Patrick De Deckker, Jonathan Holmes and Antje

Schwalb kindly contributed with scientific discussions while Dan Danielopol generously gave advice on ostracod identification. Many thanks are due to two anonymous reviewers and to Finn Viehberg for their critical comments and pertinent suggestions that helped to improve this manuscript. Thanks are also due to Lisa Park for language editing.

REFERENCES

ALM, G., 1915. Monographie der schwedischen Süsswasser-Ostracoden nebst systematischen Besprechungen der Tribus Podocopa. Zoologiska Bidrag fran Uppsala, **4**: 1-248.

BENZIE, J. A. H., 1989. The distribution and habitat preference of ostracods (Crustacea, Ostracoda) in a coastal sand-dune lake, Loch of Strathbeg, Northeast Scotland. Freshwater Biology, **22**: 309-321.

CARBONNEL, P., 1980. Les Ostracodes et leur Intérêt dans la Définition des Ecosystèmes Estuariens et de Plateforme Continentale. Essai d'Application à des Domaines Anciens. Mémoires de l'Institut de géologie du bassin d'Aquitaine, **11**: 1-350.

CHIVAS, A. R., P. DE DECKKER & J. M. G. SHELLEY, 1986. Magnesium and strontium in non-marine ostracod shells as indicators of palaeosalinity and palaeotemperature. Hydrobiologia, **143**: 135-142.

CIPEL, 1984. Le Léman: Synthèse 1957-1982. (CIPEL, Lausanne).

— —, 2006. Rapport Campagne 2005. (CIPEL, Lausanne).

— —, 2011. Rapport Campagne 2010. (CIPEL, Lausanne).

CORBARI, L., P. CARBONEL & J.-C. MASSABUAN, 2004. How a low tissue O_2 strategy could be conserved in early crustaceans: the example of the podocopid ostracods. The Journal of Experimental Biology, **207**: 4415-4425.

CURRY, B. B. & G. M. FILIPPELLI, 2010. Episodes of low dissolved oxygen indicated by ostracodes and sediment geochemistry at Crystal Lake, Illinois, USA. Limnology and Oceanography, **55**: 2403-2423.

DANIELOPOL, D. L., 1969. Recherches sur la morphologie de l'organe copulateur male chez quelques ostracodes du genre *Candona* Baird (fam. Cyprididae Baird). In: J. NEALE (ed.), The taxonomy, morphology and ecology of recent Ostracoda: 136-153. (Oliver & Boyd, Edinburgh).

— —, 1982. The preparation of ostracods limbs, for optical microscopy (dissection, staining and mounting procedures). In: R. F. MADDOCKS (ed.), Applications of Ostracoda to economic and scientific problems: 41-43. (University of Houston, Houston, TX).

DANIELOPOL, D. L., E. ITO, G. WANSARD, T. KAMIYA, T. M. CRONIN & A. BALTANÁS, 2002. Techniques for collection and study of Ostracoda. In: J. A. HOLMES & A. R. CHIVAS (eds.), The Ostracoda: applications in quaternary research. Geophysical monograph, **131**: 65-97. (American Geophysical Union, Washington, D.C.).

DANIELOPOL, D. L. & J. TETARD, 1990. Morphology of *Cytherissa* and *Cyprideis*: supplementary data on the appendages and the caryotype. Bulletin de l'Institut de Géologie du Bassin d'Aquitaine, **47**: 55-67.

DE DECKKER, P., 2002. Ostracod palaeoecology. In: J. A. HOLMES & A. R. CHIVAS (eds.), The Ostracoda: applications in quaternary research. Geophysical monograph, **131**: 121-134. (American Geophysical Union, Washington, D.C.).

DECROUY, L., 2012. Biological and environmental controls on isotopes in ostracod shells. In: D. J. HORNE, J. A. HOLMES, J. RODRIGUEZ-LAZARO & F. VIEHBERG, Ostracod as proxies for quaternary climate change. Developments in Quaternary Science Series, **17**: 165-181.

DECROUY, L., T. W. VENNEMANN & D. ARIZTEGUI, 2011a. Controls on ostracod shell geochemistry: Part 1. Variations of environmental parameters within ostracod (micro-)habitat. Geochimica et Cosmochimica Acta, **75**: 7364-7379.

— —, — — & — —, 2011b. Controls on ostracod shell geochemistry: Part 2. Carbon and oxygen isotopes compositions. Geochimica et Cosmochimica Acta, **75**: 7380-7399.

— —, — — & — —, 2012. Sediment penetration depth of epi- and infaunal ostracods from Lake Geneva (Switzerland). Hydrobiologia, **688**: 5-23.

DELORME, L. D., 1978. Distribution of freshwater ostracodes in Lake Erie. Journal of Great Lakes Research, **4**: 216-220.

DELORME, L. D. & D. DONALD, 1969. Torpidity of freshwater ostracodes. Canadian Journal of Zoology, **47**: 997-999.

DUSSART, B., 1966. Limnologie — L'Étude des Eaux Continentales. (Gauthier-Villars, Paris).

ELLIOTT, J. M., 1977. Some methods for the statistical analysis of aamples of benthic invertebrates. Freshwater Biological Association Scientific Publication, **25**.

FENWICK, G. D., 1984. Life history and population biology of the giant ostracod *Leuroleberis zealandica* (Baird, 1850) (Myodocopida). Journal of Experimental Biology and Ecology, **77**: 255-289.

FERGUSON, E. F., 1944. Studies on the seasonal life history of three species of freshwater ostracoda. American Midland Naturalist, **32**: 713-727.

GANDOLFI, A., E. B. A. TODESCHI, V. ROSSI & P. MENOZZI, 2001. Life history traits in *Darwinula stevensoni* (Crustacea: Ostracoda) from Southern European populations under controlled conditions and their relationships with genetic features. Journal of Limnology, **60**: 1-10.

GEIGER, W., 1990a. The role of oxygen in the disturbance and recovery of the *Cytherissa lacustris* population of Mondsee (Austria). Bulletin de l'Institut de Géologie du Bassin d'Aquitaine, **47**: 167-189.

— —, 1990b. Field and laboratory studies on the life cycle of *Cytherissa lacustris* (Sars) (Crustacea, Ostracoda) with special emphasis on the role of temperature. Bulletin de l'Institut de Géologie du Bassin d'Aquitaine, **47**: 191-208.

GIRARDCLOS, S., 2001. Sismostratigraphie et Structure Sédimentaire en 3D d'un Bassin Lacustre du Retrait Glaciaire à nos Jours (Lac Léman, Suisse). Terre et Environnement, **33**. (Institut F.-A. Forel, University of Geneva, Geneva).

HAGERMAN, L., 1967. Ostracods of the Tvärminne area, Gulf of Finland. Commentationes Biologicae Societas Scientiarum Fennicae, **30**: 1-12.

HEIP, C., 1976a. The spatial pattern of *Cyprideis torosa* (Jones, 1850) (Crustacea: Ostracoda). Journal of the Marine Biological Association of the United Kingdom, **56**: 179-189.

— —, 1976b. The life-cycle of *Cyprideis torosa* (Crustacea, Ostracoda). Oecologia, **24**: 229-245.

HENDERSON, P. A., 1990. Freshwater Ostracods. Synopses of the British Fauna (New Series), **42**. (Universal Book Services, Oegstgeest).

HILLER, D., 1972. Untersuchungen zur Biologie un zur Ökologie limnischer Ostracoden aus der Umgebung von Hamburg. Archiv für Hydrobiologie (Supplement), **40**: 400-497.

HORNE, D. J., 1983. Life-cycles of Podocopid ostracoda — a review (with particular reference to marine and brackish-water species). In: R. F. MADDOCKS (ed.), Applications of Ostracoda: 581-590. (University of Houston, Houston, TX).

HORNE, D. J., J. A. HOLMES, J. RODRIGUEZ-LAZARO & F. VIEHBERG, 2012. Ostracod as proxies for quaternary climate change. Developments in Quaternary Science Series, **17**. (Elsevier Science, Amsterdam).

IKEYA, N. & M. KATO, 2000. The life history and culturing of *Xestoleberis hanaii* (Crustacea, Ostracoda). Hydrobiologia, **419**: 149-159.

JUNGWIRTH, M., 1979. *Limnocythere inopinata* (Baird) (Cytheridae, Ostracoda): its distribution pattern and relation to the superficial sediments of Neusiedlersee. In: H. LOEFFLER (ed.), Neusiedlersee: the limnology of a shallow lake in central Europe. Monographiae Biologicae **37**: 385-388. (Dr. W. Junk, The Hague).

KAMIYA, T., 1988. Contrasting population ecology of two species of *Loxoconcha* (Ostracoda, Crustacea) in recent *Zostera* (eelgrass) beds: adaptive differences between phytal and bottom-dwelling species. Micropaleontology, **34**: 316-331.

KEMPF, E. K. & B. W. SCHARF, 1980. Lebende und fossile Muschelkrebse (Crustacea, Ostracoda) vom Laachersee. Mitteillungen der Pollichia, **68**: 205-236.

KRUTAK, P. R., 1971. The recent Ostracoda of Laguna Mandinga, Veracruz, Mexico. Micropaleontology, **17**: 1-30.

KÜLKÖYLÜOĞLU, O., 1998. Freshwater Ostracoda (Crustacea) and their quarterly occurrence in Şamlar Lake (Istanbul, Turkey). Limnologica, **28**: 229-235.

KÜLKÖYLÜOĞLU, O. & M. DÜGEL, 2004. Ecology and spatiotemporal patterns of Ostracoda (Crustacea) from Lake Gölcük (Bolu, Turkey). Archiv für Hydrobiologie, **160**: 67-83.

LÖFFLER, H., 1969. Recent and subfossil distribution of *Cytherissa lacustris* in Lake Constance. Mitteilungen. Internationale Vereiningung für Theoretische und Angewandte Limnologie, **17**: 240-251.

LUDWIG, J. A. & J. F. REYNOLDS, 1988. Statistical Ecology. (Wiley-InterScience, New York, NY).

MALLWITZ, V. J., 1984. Untersuchungen zur Ökologie litoraler Ostracoden im Schmal- und Lüttauersee (Schleswig-Holstein, BRD). Archiv für Hydrobiologie, **100**: 311-339.

MARTENS, K. & S. SAVATENALINTON, 2011. A subjective checklist of the Recent, free-living, non-marine Ostracoda (Crustacea). Zootaxa, **2855**: 1-79.

MBAHINZIREKI, G., F. UIBLEIN & H. WINKLER, 1991. Microhabitat selection of ostracods in relation to predation and food. Hydrobiologia, **222**: 115-119.

MCKENZIE, K. G. & F. M. SWAIN, 1967. Recent ostracoda from Scammon Lagoon, Baja California. Journal of Paleontology, **41**: 281-305.

MEISCH, C., 2000. Freshwater Ostracoda of western and central Europe. Süsswasser von Mitteleuropa, **8/3**. (Spektrum Akademischer Verlag, Heidelberg).

MEYERS, P. A., 1994. Preservation of elemental and isotopic source identification of sedimentary organic matter. Chemical Geology, **114**: 289-302.

MEZQUITA, F., M. D. BORONAT & M. R. MIRACLE, 2002. The life history of *Cyclocypris ovum* (Ostracoda) in a permanent karstic lake. Archiv für Hydrobiologie, **155**: 687-704.

MEZQUITA, F., J. R. ROCA, J. M. REED & G. WANSARD, 2005. Quantifying species-environment relationships in non-marine Ostracoda for ecological and palaeoecological studies: examples using Iberian data. Palaeogeography, Palaeoclimatology, Palaeoecology, **225**: 93-117.

MÜLLER, G. W., 1900. Deutschlands Süsswasser-Ostracoden. Zoologica, Original-Abhandlungen aus dem Gesamtgebiet der Zoologie, **12**(30): 1-112.

NÜCHTERLEIN, H., 1969. Süsswasserostracoden aus Franken. Ein Beitrag zur Systematik und Ökologie der Ostracoden. Internationale Revue der gesamten Hydrobiologie, **54**: 223-287.

POWELL, S., 1976. Einige Aspekte der Beziehung zwischen Sedimenteigenschaften und der Fortbewegung benthischer Süsswasser-Ostracoden, mit spezieller Berücksichtigung der *Cytherissa lacustris* (Sars). (Ph.D. Thesis, University of Vienna, Vienna).

REYMENT, R. A., 1975. Canonical correlation analysis of hemicytherinid and trachyleberinid ostracodes in the Niger Delta. In: F. M. SWAIN (ed.), Biology and paleobiology of Ostracoda. Bulletins of American Paleontology, **65**: 141-146.

RICKETTS, R. D., T. C. JOHNSON, E. T. BROWN, K. A. RASMUSSEN & V. V. ROMANOVSKY, 2001. The Holocene paleolimnology of Lake Issyk-Kul, Kyrgyzstan: trace element and stable isotope composition of ostracodes. Palaeogeography, Palaeoclimatology, Palaeoecology, **176**: 207-227.

RIERADEVALL, M. & J. R. ROCA, 1995. Distribution and population dynamics of ostracodes (Crustacea, Ostracoda) in a karstick lake: Lake Banyoles (Calalonia, Spain). Hydrobiologia, **310**: 189-196.

ROCA, J. R. & A. BALTANÁS, 1993. Ecology and distribution of Ostracoda in Pyrenean springs. Journal of Crustacean Biology, **13**: 165-174.

ROCA, J. R. & D. L. DANIELOPOL, 1991. Exploration of interstitial habitats by the phytophilous ostracod *Cypridopsis vidua* (O.F. Müller): experimental evidence. Annales de Limnologie, **27**: 243-252.

ROCA, J. R. & G. WANSARD, 1997. Temperature influence on development and calcification of *Herpetocypris brevicaudata* Kaufmann, 1900 (Crustacea: Ostracoda) under experimental conditions. Hydrobiologia, **347**: 91-95.

ROCA, J. R., A. BALTANÁS & F. UIBLEIN, 1993. Adaptative response in *Cypridopsis vidua* (Crustacea, Ostracoda) to food and shelter offered by a macrophyte (*Chara fragilis*). Hydrobiologia, **262**: 127-131.

SAVOLAINEN, I. & T. VALTONEN, 1983. Ostracods of the north-eastern Bothnian Bay and population dynamics of the principal species. Aquilo Series Zoologica, **22**: 69-76.

SCHWALB, A., 2003. Lacustrine ostracodes as stable isotope recorders of late-glacial and Holocene environmental dynamics and climate. Journal of Paleolimnology, **29**: 267-351.

THEISEN, B. F., 1966. The life history of seven species of ostracods from a Danish brackish-water locality. Meddelelser fra Danmarks Fiskeri- og Havundersøgelser, N.S., **4**: 215-269.

VAN DONINCK, K., I. SCHÖN, K. MARTENS & B. GODDEERIS, 2003. The life-cycle of the asexual ostracod *Darwinula stevensoni* (Brady & Robertson, 1870) (Crustacea, Ostracoda) in a temporate pond. Hydrobiologia, **500**: 331-340.

VIEHBERG, F. A., 2006. Freshwater ostracod assemblages and their relationship to environmental variables in waters from northeast Germany. Hydrobiologia, **571**: 213-224.

VON GRAFENSTEIN, U., H. ERLENKEUSER, A. BRAUER, J. JOUZEL & S. J. JOHNSEN, 1999. A Mid-European decadal isotope-climate record from 15,500 to 5000 years B.P. Science, **284**: 1654-1657.

VON GRAFENSTEIN, U., H. ERLENKEUSER, J. MÜLLER, P. TRIMBORN & J. ALEFS, 1996. A 200 year mid-European air temperature record preserved in lake sediments: an extension of the $\delta^{18}O_P$-aire temperature relation into the past. Geochimica et Cosmochimica Acta, **60**: 4025-4036.

WOJTASIK, B., 2008. Life cycle of *Tonnacypris glacialis* (Crustacea: Ostracoda). Polish Polar Research, **29**: 33-44.

WOLF, J. P., 1920. Die Ostracoden der Umgebung von Basel. Archiv für Naturgeschichte, Abteilung A, **85**: 1-100.

YIN, Y. & W. GEIGER, 1995. Spatial distribution of ostracods related to abiotic factors in Mooswinkl Bay, Mondsee, Austria. In: J. RIHA (ed.), Ostracoda and biostratigraphy: 381-387. (A. A. Balkema, Rotterdam).

First received 27 November 2012.
Final version accepted 14 May 2014.

[When citing this chapter, refer to Crustaceana 87 (2014) 1124–1135]

This paper is dedicated to Prof. Dr. Eugen Kempf on the occasion of his 80th birthday (16 April 2012)

LIVING OSTRACODA (CRUSTACEA) FROM THE TOWN MOAT OF BREMEN, GERMANY

BY

BURKHARD SCHARF[1,3]) and FINN A. VIEHBERG[2])
[1]) Ellhornstrasse 21, D-28195 Bremen, Germany
[2]) University of Cologne, Institute of Geology and Mineralogy, Zülpicher Strasse 49a, D-50674 Cologne, Germany

ABSTRACT

Between 2009 and 2014 we collected 23 ostracod species in the historical town moat of Bremen, Germany. Today, the moat is part of an urban park, separated into six basins by various bridges. Brackish water, pumped from the Weser River into the moat, regulates the water level and is the main factor controlling the various physical and chemical parameters of the basins. Brackish water ostracod species originate from this artificial inflow, but only a minority of these species establishes stable populations. The Bremen moat ostracod assemblages are compared with the moat of Greifswald. The different environmental setting (e.g., water current) and the number of studied localities can explain the differences in the occurrence of ostracods in these two localities.

Key words. — Ostracoda, Crustacea, town moat, Bremen, Greifswald

ZUSAMMENFASSUNG

Zwischen den Jahren 2009 und 2014 wurden im historischen Stadtgraben der Stadt Bremen 23 Ostracoden-Arten gefunden. Heute ist der Stadtgraben ein Teil eines städtischen Parks, der durch Brücken in sechs Becken unterteilt ist. Die Einleitung von Brackwasser aus der Weser in den Stadtgraben zur Regulierung des Wasserstandes ist die Hauptursache für die unterschiedlichen physikalischen und chemischen Eigenschaften und die Ostracodenbesiedlung der einzelnen Becken. Die meisten Brackwasser-Arten, die durch das Pumpwasser in den Stadtgraben eingespült wurden, konnten keine stabile Populationen aufbauen. Abschließend wurde die Besiedlung der Stadtgräben von Bremen und von Greifswald verglichen. Verschiedene Umweltbedingungen (z. B. Fließgeschwindigkeit) und die Anzahl der untersuchten Probestellen sind wahrscheinlich die Ursache für die Unterschiede in den beiden beobachteten Ostracodenassoziationen.

[3]) Corresponding author; e-mail: burkhard.w.scharf@t-online.de

INTRODUCTION

In 2009-2010, the population dynamics of three microcrustacean groups (cladocerans, calanoid copepods and ostracods) were studied in the town moat of Bremen (Brandorff et al., 2013). The authors revisited the type locality of *Leptodora kindtii* Focke, 1844 and compared the results with the historical cladoceran findings of Poppe (1889). In total 45 species were detected, including *L. kindtii*. The cladoceran species assemblages were characterized by increased salinity and vegetation. The diversity of calanoid copepods was relatively high compared with other lakes in Northern Europe. *Eudiaptomus gracilis* (G. O. Sars, 1862), *Skistodiaptomus pallidus* (Herrick, 1879) and *Eurytemora velox* (Lilljeborg, 1853) occurred constantly, whereas *Eurytemora affinis* (Poppe, 1880) and *Eudiaptomus graciloides* (Lilljeborg, 1888) were only detected sporadically.

The town moat of Bremen is located in a residence area of the city and delimited by municipal infrastructure. In urban environments, ecosystems exist in artificial settings and are often continuously managed or altered due to constructions. In addition, human activity is a constant factor that biases the resilience of the ecosystems (e.g., recreation, pollution, eutrophication, management). The Bremen town moat has undergone many uses and forms since its creation in the Middle Ages, but continued to exist through time and enabled a long-term succession of the faunal elements and development of stable populations of aquatic flora and fauna.

STUDY SITE AND METHODS

Study site

At the end of the Middle Ages, the town of Bremen was one of the best protected cities in Europe, fortified by a wall, bastions and a town moat (fig. 1). With the invention of far-reaching cannons, these defensive measures ceased to be effective. Thus, in 1802, the administration of Bremen decided to level the wall and the bastions and create a public park. The original outline of the town moat was altered and shortened to approx. 2.8 km length and about 50 m width at its widest part. Later, the construction of bridges intersected the moat into six basins that are only connected by small passages. Water is pumped from the Weser River into the moat (near Station 6 in fig. 2), controlling the water level in the moat. The water of the Weser River has a high nutrient load and an increased conductivity ranging from 1100 to 1400 μS cm^{-1}, making it a eutrophic and brackish (oligohaline) habitat. From sample sites 6 to 1, the conductivity decreases (figs. 2 and 3). In 2009 and 2010, additional groundwater with higher conductivity of about 1600 μS cm^{-1}

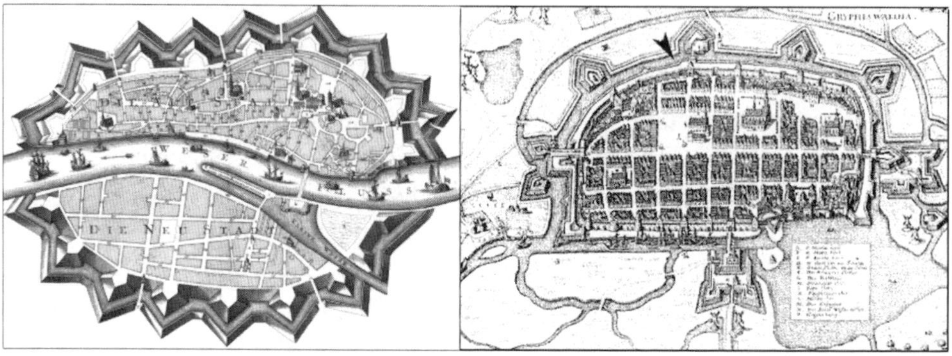

Fig. 1. City map of Bremen in the year 1739 (left) and of Greifswald in the year 1652. The arrow indicates the location of the sample from the town moat of Greifswald. Modified from Brandorff et al. (2013) and Frenzel et al. (2004).

was discharged into the basin near station 6 from a construction site (Brandorff et al., 2013).

The town moat is relatively shallow, with a maximum depth of approximately 3 m. The water is eutrophic and the flora is dominated by submerged macrophytes (Kuhbier & Kesel, 2013). Various tree species occur along the embankment.

Methods

Five ostracod samples were collected at each sampling site (2009: 8 October; 2010: 18 April, 17 June, 6 July and 6 September), using a hand-net (200 μm mesh

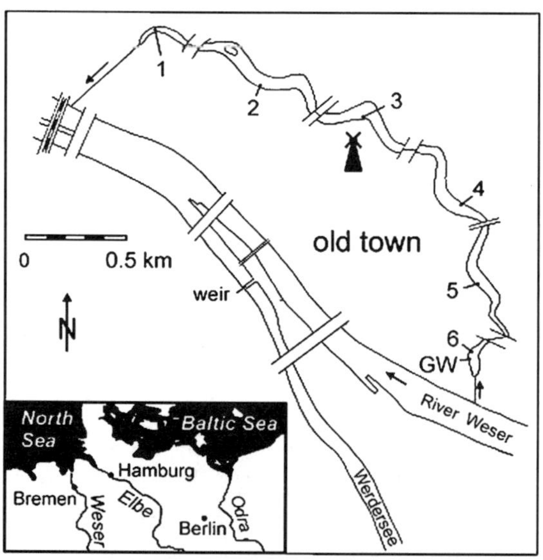

Fig. 2. Sample sites in the town moat of Bremen. GW, artificial groundwater inlet by the construction of a house.

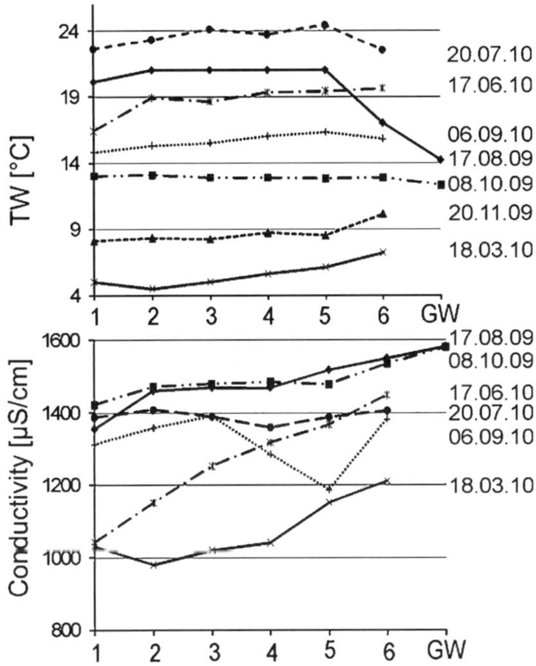

Fig. 3. Water temperature and conductivity at the sampling sites in the town moat of Bremen in 2009 and 2010. GW, groundwater of the inlet near locality 6. Modified from Brandorff et al. (2013).

size) (fig. 2). Two additional samples were taken on 5 February 2014 at site 6 and on 30 December 2011 between sites 4 and 5 for other studies (Scharf et al., 2014). The collected samples were immediately sieved with a 5-mm-mesh net in order to eliminate coarse sediment. Subsequently, the residual fine sediment was transferred to a plastic basket (6 × 12 cm) within a larger container filled with water following the description of Scharf et al. (2014). Overnight, most of the ostracods migrated from the plastic basket into the container, where they could easily be captured in clear water. Living ostracods were put in boiling water for 10 s so they would die with a slightly open carapace. For permanent slides, the soft parts were embedded in Hydromatrix® (Micro-Tech-Lab, Graz, Austria). The remaining animals were stored in 96% ethanol in the collection of B.S.

Statistics and indices

The frequency (F, in %) of a species between all sites is calculated as

$$F = a_i * 100/A_{\mathrm{a}}$$

where a_i is the number of samples sites containing a specific species i and A_{a} is the number of all sample sites. The frequency is then classified using the following intervals: ⩾71% = very constant; 70-46% = constant; 45-26% = inconstant; 25-10% = sporadic; ⩽10% = random.

The dominance (*D*, in %) of a species in one sample is calculated as:

$$D = n_i * 100/N_\mathrm{a}$$

where n_i is the number of individuals of species i and N_a the number of the individuals of all species in the sample. The classification follows Engelmann (1978) and Mühlenberg (1989): 100-32% = eudominant; 31.9-10% = dominant; 9.9-3.2% = subdominant; 3.1-1.0% = recent; <0.99% = subrecent.

The relation of the environmental parameters (water temperature and conductivity) to the abundance of the dominant species was investigated by performing a redundancy analysis (RDA) using the software Canoco 4.5 (ter Braak & Smilauer, 2002). Thus, ecological preferences are shown two-dimensionally.

RESULTS

Between 2009 and 2014, a total of 22 539 adult ostracods individuals comprising 23 species were counted from samples of the town moat of Bremen (table I). *Cypridopsis vidua* (O. F. Müller, 1776) and *Cypria ophtalmica* (Jurine, 1820) were the only eudominant species in this study (table I). Both species occurred consistently (table II). Other dominant species included *Cyclocypris ovum* (Jurine, 1820) and *Pseudocandona compressa* (Koch, 1838). The following taxa were subdominant in one or more sample sites: *Candona neglecta* Sars, 1887, *Cyclocypris ovum*, *Cypria ophtalmica* Redeke, 1936, *Cypridopsis vidua*, *Fabaeformiscandona protzi* (Hartwig, 1898), *Ilyocypris bradyi* Sars, 1890, *Ilyocypris decipiens* Masi, 1905, *Limnocythere inopinata* (Baird, 1843) and *Plesiocypridopsis newtoni* (Brady & Robertson, 1870). All other species were recent or subrecent in their dominance.

A similar species assemblage was observed on December 30, 2011, when an additional sample was taken under the bridge between sample sites 4 and 5. The sediment top at the site consisted mainly of maple and oak leaves and gravel. The following ostracod species were collected (dominance given in parentheses): *Cypridopsis vidua* (52%), *Fabaeformiscandona protzi* (15%), *Cypria ophtalmica* (13%), *Candona candida* (10%), *C. neglecta* (8%). *Cyclocypris ovum*, *Plesiocypridopsis newtoni*, *Pseudocanona compressa*, *Herpetocypris reptans* (Baird, 1835), *Darwinula stevensoni* (Brady & Robertson, 1870) and *Limnocythere inopinata* were recent.

On 5 February 2014 a further sample was taken at sample site 6 because the artificial inlet of groundwater with the high conductivity stopped and the water became clear. In addition, since 2013, the artificial input of water from the Weser

TABLE I

Sum of adult ostracods and dominance at each locality in the town moat of Bremen in 2009/2010 and 5 February 2014

	Locality						
	1	2	3	4	5	6	
Date	2009/2010	2009/2010	2009/2010	2009/2010	2009/2010	2009/2010	2014
Latitude (N)	53°05′01″	53°04′54″	53°04′51″	53°04′40″	53°04′31″	53°04′19″	
Longitude (E)	8°47′38″	8°48′07″	8°48′26″	8°48′44″	8°48′49″	8°48′51″	
Candona candida (O. F. Müller, 1776)	11 sr	12 sr	23 sr	12 r	18 sr	•	17 sr
Candona neglecta Sars, 1887	•	6 sr	•	14 r	•	3 sr	391 sd
Cyclocypris ovum (Jurine, 1820)		49 r	71 r	63 sd	477 d	1470 d	968 d
Cypria ophtalmica (Jurine, 1820)	251 d	434 d	834 d	51 sd	249 sd	156 sd	2814 ed
Cypria subsalsa Redeke, 1936						8 sr	379 sd
Cyprideis torosa (Jones, 1850)							1 sr
Cypridopsis vidua (O. F. Müller, 1776)	1135 ed	3035 ed	4598 ed	850 ed	1563 ed	114 sd	67 r
Cytheromorpha fuscata (Brady, 1869)						52 r	77 r
Darwinula stevensoni (Brady & Robertson, 1870)	6 sr	9 sr	50 sr	33 r	4 sr	22 sr	49 sr
Fabaeformiscandona protzi (Hartwig, 1898)	20 r	67 r	35 sr	2 sr	112 sd		
Herpetocypris reptans (Baird, 1835)				11 r	2 sr	13 sr	8 sr
Herpetocypris chevreuxi (Sars, 1896)							1 sr
Ilyocypris bradyi Sars, 1890	2 sr				1 sr	82 r	181 sd
Ilyocypris decipiens Masi, 1905	4 sr				•	175 sd	
Isocypris beauchampi (Paris, 1920)						13 sr	51 r
Limnocythere inopinata (Baird, 1843)	3 sr			1 sr	8 sr	150 sd	
Notodromas monacha (O. F. Müller, 1776)	3 sr	2 sr					
Physocypria kraepelini G. W. Müller, 1903	6 sr	26 sr	77 r	7 sr	43 r	84 r	
Plesiocypridopsis newtoni (Brady & Robertson, 1870)	17 r	50 r	51 sr	41 sd	102 sd	13 sr	
Potamocypris smaragdina (Vávra, 1891)	2 sr						
Potamocypris variegata (Brady & Norman, 1889)						15 sr	30 sr
Prionocypris zenkeri (Chryzer & Toth, 1858)				2 sr	•	8 sr	47 sr
Pseudocandona compressa (Koch, 1838)	29 r	25 sr	102 r	17 r	42 r	310 d	
Sum	1489	3715	5841	1104	2621	2688	5081

A circle (•) indicates that the species was found only as larvae at that locality; ed, eudominant; d, dominant; sd, subdominant; r, recent; sr, subrecent.

TABLE II

Frequency (in %) of the occurrence of ostracods in the town moat of Bremen in 2009 and 2010 (modified from Brandorff et al., 2013)

Very constant (⩾71%)		Constant (70-46%)		Inconstant (45-26%)		Sporadic (25-10%)		Random (⩽10%)	
Cypridopsis vidua	100	*Darwinula stevensoni*	58	*Pseudocandona compressa*	42	*Limnocythere inopinata*	23	*Cyprideis torosa*	3
Cypria ophtalmica	97	*Plesiocypridopsis newtoni*	52	*Physocypria kraepelini*	39	*Ilyocypris bradyi*	19	*Herpetocypris chevreuxi*	3
Cyclocypris ovum	81	*Candona candida*	48	*Fabaeformiscandona protzi*	26	*Candona neglecta*	16	*Potamocypris smaragdina*	3
						Cytheromorpha fuscata	16		
						Potamocypris variegate	16		
						Prionocypris zenkeri	16		
						Herpetocypris reptans	13		
						Ilyocypris decipiens	13		
						Cypria subsalsa	10		
						Isocypris beauchampi	10		
						Notodromas monacha	10		

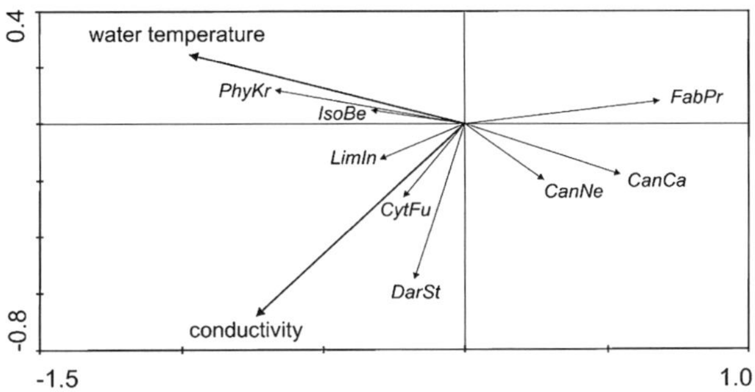

Fig. 4. Redundancy analysis of the environmental parameters water temperature and conductivity and the eight most important ostracod species (excluding larvae). *CanCa* = *Candona candida* (O. F. Müller, 1776), *CanNe* = *Candona neglecta* Sars, 1887, *CytFu* = *Cytheromorpha fuscata* (Brady, 1869), *DarSt* = *Darwinula stevensoni* (Brady & Robertson, 1870), *FabPr* = *Fabaeformiscandona protzi* (Hartwig, 1898), *IsoBe* = *Isocypris beauchampi* (Paris, 1920), *LimIn* = *Limnocythere inopinata* (Baird, 1843), *PhyKr* = *Physocypria kraepelini* G. W. Müller, 1903. Modified from Brandorff et al. (2013).

River into the town moat has increased. During sampling, a thin layer of ice covered the water surface and the conductivity was 1453 μS cm^{-1}. Two additional species were found in the town moat for the first time, *Cyprideis torosa* (Jones, 1850) and *Herpetocypris chevreuxi* (Sars, 1896) (table I). *Cypria subsalsa* Redeke, 1936 was more abundant than in 2009/2010 and many larvae of *Cytheromorpha fuscata* (Brady, 1869) were found.

The frequency of the occurrence of ostracods is shown in table II. Only *Cypridopsis vidua*, *Cypria ophtalmica* and *Cyclocypris ovum* occur consistently, while most species occur sporadically.

The species relationship to environmental parameters, i.e., water temperature and conductivity indicate a distinct pattern (fig. 4). Based on these results we note a higher abundance of *Cytheromorpha fuscata* in sites with increased salinity. *Physocypria kraepelini* G. W. Müller, 1903 is found in samples with increased water temperature, while adults of *Candona candida* and *C. neglecta* are found in cold water mainly in the winter.

In addition, a tapeworm cysticercus was identified in one female of *Candona candida* at the sampling site 5. The detection of a tapeworm cysticercus has to be discussed in a separate paper to take justice of this finding.

DISCUSSION

Occurrence of ostracods in the town moat of Bremen

Prior to 2009, the only known publication about ostracods in the Bremen town moat was by Poppe (1889), who identified only one ostracod species,

Pseudocandona compressa ('*Cypris compressa*') that still existed in the water in 2009/2014. It is thought that it is most likely that more species were present in former times.

Ostracods characteristic of the Weser River, including *Cypria subsalsa*, *Cytheromorpha fuscata*, *Herpetocypris chevreuxi* and *Prionocypris zenkeri* (Chryzer & Toth, 1857), are likely to be introduced in the town moat. *C. subsalsa* and *C. fuscata* are known from brackish water species that are widely distributed in Belgium, in the Netherlands, in the Baltic Sea, and also in the Odra lagoon (Klie, 1938; Frenzel et al., 2010; Bąk & Szlauer-Łukaszewska, 2012). Finding both species in the region of the Weser estuary closes a gap in the disjunctive geographical range between the Netherlands and the Baltic. Other species of the Weser marsh like *Cyprideis torosa*, *Herpetocypris chevreuxi*, *Ilyocypris decipiens*, *Stenocypria fischeri* (Lilljeborg, 1883) or *Leptocythere ilyophila* (Hirschmann, 1909) were not found during the survey in 2009 and 2010 in the town moat of Bremen (Scharf, unpublished). Later, at sampling site 6 in 2014, one male individual of *Cyprideis torosa* and one female individual of *Herpetocypris chevreuxi* were detected. In addition, several individuals, adults and larvae, of *Cytheromorpha fuscata* found in the latest sample imply that the population might be stable. The change of the ecosystem with the increased input of water from the Weser River is probably responsible for the differences of the results in 2014 in comparison with 2009/2010 (table I).

The occurrence of *Prionocypris zenkeri* is noteworthy in that it prefers slow flowing waters and is likely to originate from the Weser River as well. The species' lack of constant presence in the town moat suggests that this species is not able to sustain a stable population in the sampled habitat. Another notable occurrence in this study is *Isocypris beauchampi* (Paris, 1920). It was unknown to exist in Germany before 1951 (Herbst, 1951), but since then, this large conspicuous species has become widely distributed throughout Germany (Meisch, 2000). It is found also in the town moat of Bremen. In 2009 and 2010, the moat was colonized by many submersed macrophytes, in particular *Myriophyllum* and *Ceratophyllum* species (Kuhbier & Kesel, 2013). *Cypridopsis vidua* thrives among such macrophytes, and was the dominant species in 2009 and 2010. *Notodromas monacha* (O. F. Müller, 1776) lives in very calm water, often between macrophytes. The calmest water within the town moat was that at sampling sites 1 and 2, and it was here that this species occurred, though in small numbers.

The seasonality of the collected species corresponds to the phenology described in Meisch (2000). *Cytheromorpha fuscata* is not mentioned by Meisch (2000), unfortunately our observations allow no remarks on the seasonality; still, in 2014 there is a hint for a stable population of *C. fuscata* in the town moat of Bremen.

TABLE III
List of adult ostracods found in the town moats of Bremen and Greifswald and their dominance

	Bremen		Greifswald	
	Adult animals	Dominance	Valves	Dominance
Candona candida	93	sr	14	d
C. neglecta	414	sr		
Cyclocypris ovum	3098	d	15	d
Cypria ophtalmica	4789	d		
C. subsalsa	387	r		
Cyprideis torosa	1	sr		
Cypridopsis vidua	11 362	ed		
Cytheromorpha fuscata	129	sr		
Darwinula stevensoni	173	sr		
Fabaeformiscandona hyalina			8	sd
F. protzi	236	r	11	d
Herpetocypris reptans	34	sr		
H. chevreuxi	1	sr		
Ilyocypris bradyi	266	r		
I. decipiens	179	sr		
Isocypris beauchampi	64	sr		
Limnocythere inopinata	162	sr	2	r
Notodromas monacha	5	sr		
Physocypria kraepelini	243	r		
Plesiocypridopsis newtoni	274	r	7	sd
Potamocypris smaragdina	2	sr		
P. variegata	45	sr		
Prionocypris zenkeri	57	sr		
Pseudocandona compressa	525	r	27	d
Sum	22 539		84	

Abbreviations: d, dominant; sd, subdominant; r, recent; sr, subrecent.

With respect to species diversity, that in the town moat of Bremen is similar to the littoral margin of a lake, where the largest number of species are usually discovered, in contrast to the profundal zone, e.g., in the Laacher See (with 4 profundal species of altogether 13; Kempf & Scharf, 1980), the Arendsee (9-18; Scharf, 1998), or Lake Biwa (12-40; Smith et al., 2011). Finally, we identified 23 ostracod species in our study area similar to the species number of many lakes in temperate regions.

Comparison of the ostracod colonization of the town moats of Bremen and Greifswald

The city of Greifswald also had — as was typical in the Middle Ages — a town moat, bastions and a wall for protecting the city (fig. 1). In 2003, Frenzel et al. (2004) did not find living ostracods in the Greifswald town moat. However, in a single sample from old sediments (14th-18th century) of the moat there was a rich ostracod fauna including the following species: *Candona candida*, *Cyclocypris ovum*, *Fabaeformiscandona hyalina* (Brady & Robertson, 1870), *F. protzi*, *Limnocythere inopinata*, *Physocypria kraepelini* and *Pseudocandona compressa* (table III). Only *F. hyalina* was not found in the town moat of Bremen. Species typical for brackish water, such as *Cyprideis torosa*, *Cypria subsalsa* or *Cytheromorpha fuscata*, were not found, but all detected species were saline tolerant. It is surprising that *Cypria ophtalmica* and *Cypridopsis vidua* were not found at Greifswald compared with their dominance at Bremen. One difference between the moats is that the water in the Bremen moat is stagnant, while water in the Greifswald moat flows slowly. It is the difference in current velocity that is most likely the cause for the lack of *C. ophtalmica* and *C. vidua* because both taxa are swimming species and prefer stagnant waters. The ratio of crawling (i.e., without long natatory setae on antenna) to swimming (with long natatory setae) ostracod species in the Bremen moat is 9:12, and that of Greifswald 5:2, i.e., the crawling forms dominate in the town moat of Greifswald and may indicate a higher current velocity. However, many of the differences between the town moats of Bremen and Greifswald are difficult to compare, because of the different numbers of samples and diversity of sites.

ACKNOWLEDGEMENTS

We acknowledge the permission of Detlef Metzing, editor of the journal 'Abhandlungen des Naturwissenschaftlichen Vereins zu Bremen', for granting us the right to present translated parts of previously published content and figures of the paper of Brandorff et al. (2013). We like to thank Roland Sookias (University of Munich, Munich, Germany) for improving the English text of a previous version of this manuscript and Lisa Park (University of Akron, Akron, OH, U.S.A.) for final language editing. We also acknowledge the three anonymous reviewers for their critical comments, which helped to improve the manuscript.

REFERENCES

BĄK, M. & A. SZLAUER-ŁUKASZEWSKA, 2012. Bioindicative potential of diatoms and ostracods in the Odra mouth environment quality assessment. Nova Hedwigia, Beih., **141**: 463-484.

BRANDORFF, G.-O., W. HOLLWEDEL, B. SCHARF & F. A. VIEHBERG, 2013. Der Bremer Stadtgraben — die derzeitige Besiedlung mit Cladocera, calanoiden Copepoda und Ostracoda (Crustacea), mit einer Betrachtung der historischen Entwicklung des Gewässers. Abh. Naturwiss. Ver. Bremen, **47**: 93-119.

ENGELMANN, H.-D., 1978. Zur Dominanzklassifizierung von Bodenarthropoden. Pedobiologia, **18**: 378-380.

FRENZEL, P., J. ANSORGE & F. A. VIEHBERG, 2004. Der Greifswalder Stadtgraben vor der Industrialisierung — eine Paläomilieurekonstruktion vor archäologischem Hintergrund. Rostock. Meeresbiol. Beitr., **12**: 23-38.

FRENZEL, P., D. KEYSER & F. A. VIEHBERG, 2010. An illustrated key and (palaeo)ecological primer for postglacial to Recent Ostracoda (Crustacea) of the Baltic Sea. Boreas, **39**: 567-575.

HERBST, H. V., 1951. *Isocypris quadrisetosa* Rome (Crustacea, Ostracoda) aus der Eider. Anhang: *Isocypris laskaridisi* n. sp. von der Insel Milos. Abh. Naturwiss. Ver. Bremen, **32**: 403-414.

KEMPF, E. K. & B. W. SCHARF, 1980. Lebende und fossile Muschelkrebse (Crustacea: Ostracoda) vom Laacher See. Mitt. Pollichia, **68**: 205-236.

KLIE, W., 1938. Ostracoda, Muschelkrebse. In: F. DAHL (ed.), Die Tierwelt Deutschlands und der angrenzenden Meeresteile nach ihren Merkmalen und nach ihrer Lebensweise, **34**(3): 1-230. (Gustav Fischer, Jena).

KUHBIER, H. & R. KESEL, 2013. Wasserpflanzen im Bremer Stadtgraben im Sommer 2010. Abh. Naturwiss. Ver. Bremen, **47**: 121-123.

MEISCH, C., 2000. Crustacea: Ostracoda. In: J. SCHWOERBEL & P. ZWICK (eds.), Süßwasserfauna von Mitteleuropa, **8**(3): 1-522. (Gustav Fischer, Stuttgart).

MÜHLENBERG, M., 1989. Freilandökologie. (Quelle & Meyer, Heidelberg).

POPPE, S. A., 1889. Notizen zur Fauna der Süsswasser-Becken des nordwestlichen Deutschland mit besonderer Berücksichtigung der Crustaceen. Abh. Naturwiss. Ver. Bremen, **10**: 517-551.

SCHARF, B. W., 1998. Eutrophication history of Lake Arendsee (Germany). Palaeogeogr., Palaeoclimatol., Palaeoecol., **140**: 85-96.

SCHARF, B., A. ADLER & F. A. VIEHBERG, 2014. New methods for collecting Ostracoda (Crustacea) in stony sediments with remarks to the separation of ostracods from the collected sediment. Crustaceana, **87**: 1136-1147.

SMITH, R., H. JANZ & I. OKUBO, 2011. Recent Cyprididae and Ilyocyprididae (Crustacea: Ostracoda) from Lake Biwa, Japan, including a summary of the lake's fauna. Zootaxa, **2874**: 1-37.

TER BRAAK, C. J. F. & P. SMILAUER, 2002. CANOCO — Reference manual and CanoDraw for Windows user's guide: software for canonical community ordination (version 4.5). (Microcomputer Power, Ithaca, NY).

First received 19 March 2013.
Final version accepted 23 May 2014.

[When citing this chapter, refer to Crustaceana 87 (2014) 1136–1147]

This paper is dedicated to Prof. Dr. Eugen Kempf on the occasion of his 80th birthday (16 April 2012)

NEW METHODS FOR COLLECTING OSTRACODA (CRUSTACEA) IN STONY SEDIMENTS WITH METHODOLOGICAL REMARKS ON THE SEPARATION OF OSTRACODS FROM SEDIMENT

BY

BURKHARD SCHARF[1,4]), ANJA ADLER[2]) and FINN A. VIEHBERG[3])
[1]) Ellhornstrasse 21, D-28195 Bremen, Germany
[2]) University of Halle-Wittenberg Institute of Geosciences and Geography, Von-Seckendorff-Platz 3, D-06120 Halle/Saale, Germany
[3]) University of Cologne Institute of Geology and Mineralogy, Zülpicher Strasse 49a, D-50674 Cologne, Germany

ABSTRACT

New methods for sampling ostracods are described specifically for hyporheic environments and coarse sediments (i.e., pebbles and rocks). The main module is a diaphragm pump that establishes an artificial current strong enough to force individuals from their habitat. Two other methods are presented for separating living ostracods from sediment samples taking advantage of their biological and physical properties: (1) emigration out of a basket; (2) harvesting ostracods floating on the water surface.

Key words. — Ostracoda, Crustacea, method, collecting, stony sediment, hyporheal

ZUSAMMENFASSUNG

Neue Methoden zum Fang von Ostracoden werden beschrieben und evaluiert. Die Methoden eignen sich besonders zur Beprobung des hyporheischen Lebensraums und von Habitaten mit grobklastischem Sediment (Kies und Steine). Kernstück ist eine Membranpumpe, die einen starken Sog erzeugt, um Individuen von ihrem Habitat abzusaugen. Zwei weitere Methoden werden vorgestellt, um lebende Ostracoden aus Sedimentproben zu isolieren: (1) Auswandern aus einem Korb; (2) Abschöpfen von aufschwimmenden Ostracoden.

INTRODUCTION

Sampling bottom fauna, in particular the meiofauna in coarse-grained sediments, is difficult, because conventional methods are designed for fine-grained

[4]) Corresponding author; e-mail: burkhard.w.scharf@t-online.de

sediments like clay and sand. In addition, various grab-samplers are not suitable for these conditions, because they frequently do not close tight enough to avoid loss of sample material. In addition, all methods relying on natural drift or current, which are characteristic for high energy waters (summarised in, e.g., Macan, 1958 and Schwoerbel, 1994), are not appropriate for slow-flowing waters or stony littoral settings. Hence, there is a need for a device to allow for reliable sampling in coarse sediments.

In addition, a substantial amount of sediments is typically needed to sample ostracods. Separating ostracods from these sediments can be very time consuming and is often done in the lab under a stereo microscope by means of a pipette. Viehberg (2002) developed a handheld exhauster to isolate microcrustacenas in the field. Other methods for separating ostracods from the sediments include placing samples in a high density liquid to float and isolate lightweight ostracods from the denser liquid (P. De Deckker, pers. comm.) or aerating the samples to make the ostracods float (Szlauer-Łukaszewska & Radziejewska, 2013). However, effective field methods are not very prevalent and therefore two simple methods are proposed for use during sampling campaigns.

DESCRIPTION OF NEW METHODS FOR COLLECTING OSTRACODS

Sampling in stony sediments

When organisms are disturbed in stony habitats, they tend to adhere to larger particles or try to escape from the source of disturbance. A pulsed water jet seems to be most effective for sampling such organisms. Our new methods are based on the experience of sampling invertebrates in the interstitial of rivers (Bou & Rouch, 1967; Marmonier, 1988; Brunke & Gonser, 1999; Scharf & Brunke, 2013).

A net is fixed at the outflow of a diaphragm pump (fig. 1) and an aluminium pipe is connected to the inflow of a diaphragm pump by a vacuum hose. The aluminium pipe intrudes the interspaces between the stones. The diaphragm pump then creates a recurring sucking water pulse that collects ostracods and other organisms. A wooden lath, which is fixed at the pipe serves as the handling device that can be controlled from an onshore position. A second person operates the pump (figs. 1 and 3).

Pump design.— The aluminium pipe has an inner diameter of 26 mm. The top is cone-shaped and narrows to an inner diameter of 6 mm over a length of 40 mm. The pipe is robust enough to penetrate between rocks without being damaged. The aluminium pipe is fixed to a wooden lath (figs. 1 and 2). A vacuum coupling element has to be installed to connect the vacuum hose and the diaphragm pump. The vacuum hose has an inner diameter of approximately 20 mm. The self-priming

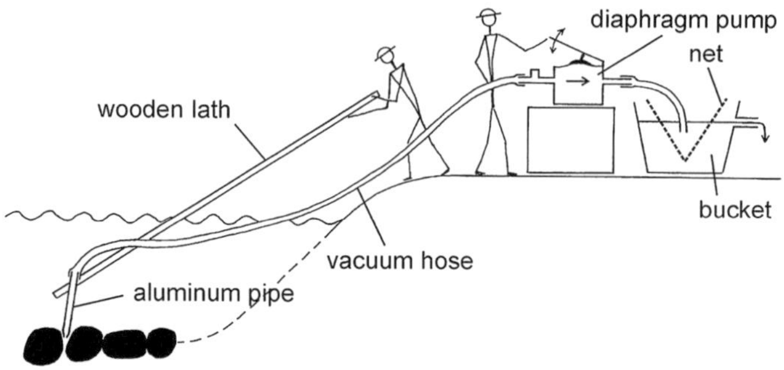

Fig. 1. Schematic illustration of ostracods recovered from stony sediment.

diaphragm pump (JOLLY MINI of Fluxinos®) is most effective when the system is filled with water. To achieve this condition, a T-piece is interconnected in front of the pump (see fig. 2, not shown in fig. 3). The system can also be filled with water by putting the entire device into the water that will be sampled. After a short time of pumping, the system should be free of air. Then, the end of the vacuum hose should be closed with a rubber stopper. This method does not require a T-piece. The diaphragm pump has a capacity of 30-40 l min^{-1}. About 400 ml are pumped with each stroke of the pump. The net, which acts as a collector, is suspended in a bucket to protect the net (figs. 1 and 3).

Proof of concept.— On 16 and 22 January 2010, the device was used in the Oker Creek near the town of Vienenburg, Germany (51°56′00″N 10°29′54″E), north of the Harz Mountains. The bottom sediment at the sampling site is coarse sand to cobbles with a maximum grain size of 15 cm. During a sampling run of 4-5 min, the device yielded 10 adult females, 6 larvae, and some empty carapaces and valves of *Candona candida* (O. F. Müller, 1776) and 7 adult females and 2 larvae of *Fabaeformiscandona latens* (Klie, 1940), as well as several individuals of *Asellus aquaticus* (Linnaeus, 1758) and Copepoda (Crustacea), some *Pisidium* sp. (Mollusca), Nematoda, larvae of Plecoptera, Ephemeroptera, Trichoptera,

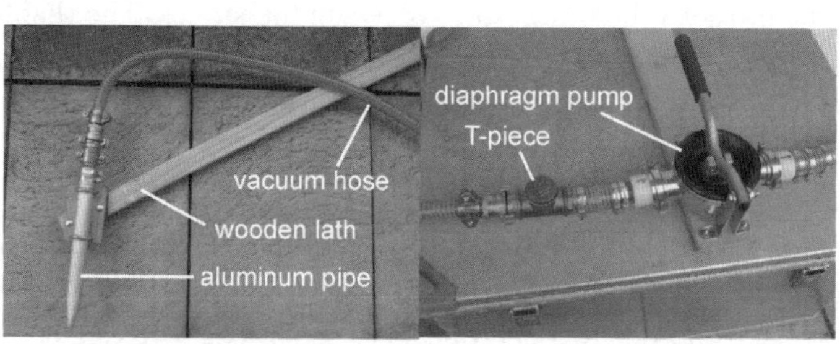

Fig. 2. The aluminium pipe, vacuum hose, T-piece and diaphragm pump. This figure is published in colour in the online edition of this journal, which can be accessed via http://booksandjournals.brillonline.com/content/journals/15685403.

Fig. 3. The diaphragm pump in action. This figure is published in colour in the online edition of this journal, which can be accessed via http://booksandjournals.brillonline.com/content/journals/15685403.

Megaloptera, Coleoptera and Diptera. All organisms passed through the valves within the pump without being damaged; the valves have a diameter of 2 cm.

A similar method for catching ostracods in the hyporheal

Sampling device description.— Fig. 4 schematically shows the device for sampling ostracods in the hyporheal. Our device is composed of a steel pipe with an inner diameter of 5 cm and a cone-shaped steel head at the top. The lower end of the steel pipe is screened with holes that are 5 mm in diameter (figs. 4 and 5). The steel pipe is pressed into the sediment, so that the groundwater with the ostracods can intrude into the pipe through the holes. A self-priming diaphragm pump (JOLLY J-MARINA of Fluxinos®) with 2 valves and a capacity of 50-60 l min^{-1} is fixed to the upper end of the pipe. The water with the animals is then transferred by the pump into a collector that contains an interior 90 μm mesh net. The organisms are collected in a removable bottle.

RESULTS

The device was successfully tested for sampling hyporheic living organisms (e.g., Bou & Rouch, 1967; Marmonier, 1988; Brunke & Gonser, 1999) and specifically for hyporheic bacteria and ostracods in a subaqueous dune within the Elbe River (Wilczek et al., 2004; Scharf & Brunke, 2013). Many species could be detected in the hyporheal of the different habitats in the Elbe River (table I). The juveniles of *Limnocythere inopinata* (Baird, 1843) and *Pseudocandona albicans* (Brady, 1864) seem to prefer to live hyporheicly. *Cypria ophtalmica* (Jurine,

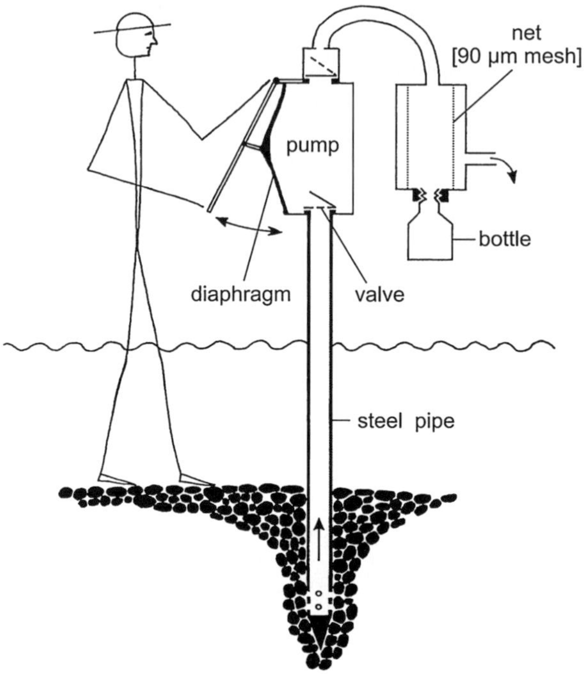

Fig. 4. Schematic illustration of the device for collecting hyporheic animals.

1820), *Herpetocypris chevreuxi* (Sars, 1896), *Ilyocypris monstrifica* and *Darwinula stevensoni* (Brady & Robertson, 1870) were not present in the hyporheal, but in the benthal (Scharf & Brunke, 2013).

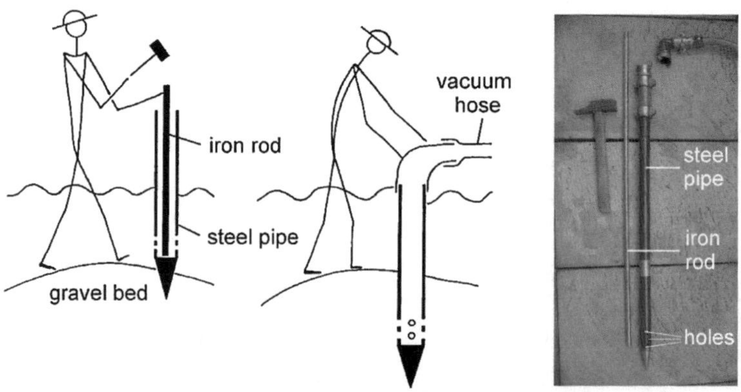

Fig. 5. Collecting of hyporheic ostracods with a modification of the method illustrated in fig. 4. This figure is published in colour in the online edition of this journal, which can be accessed via http://booksandjournals.brillonline.com/content/journals/15685403.

TABLE I

Occurrence of ostracods in the hyporheal of different habitats in the river Elbe near Coswig in 2000 (modified from Scharf & Brunke, 2013)

Species/animals found	Habitat					
	Groyne fields		Groyne heads		Subaqueous dune	
	ad	juv	ad	juv	ad	juv
Limnocythere inopinata (Baird, 1843)	47	36	4	3	1	1
Pseudocandona albicans (Brady, 1864)	18	90		25		155
Candona candida (O. F. Müller, 1776)		7		1	1	2
Cypridopsis vidua (O. F. Müller, 1776)	1	3				
Physocypria kraepelini G. W. Müller, 1903	4				1	
Potamocypris smaragdina (Vávra, 1891)	2	2				
Fabaeformiscandona wegelini (Petkovski, 1962)	5	3				
Candona neglecta Sars, 1887						2

ad, adult; juv, juvenile.

FUTURE WORK

A potential modification of the method used for collecting hyporheic ostracods is a connection of a steel pipe to the diaphragm pump using a brass arch and a vacuum hose (fig. 5). The robust steel pipe can then be hammered into the coarse sediment to sample the fauna in greater water depths (fig. 5, left), although this method has not yet been tested.

In the littoral margin of some large lakes or tundra ponds gravel and cobbles are often accumulated due to erosional or deglaciation processes. In these settings, the conventional methods are not applicable and therefore we propose an improved sampling technique that allows for sampling of coarse sand to cobble sediments, utilising a pump for greater water depths. Therefore, an interim container is installed to avoid sample material to be flushed back into the sediment (fig. 6). This becomes necessary because a diaphragm pump works in two cycles: (i) the sample gets sucked through the vacuum hose into the pump and (ii) the sample flows into the net. During the second working cycle, the current in the vacuum hose ceases and the organisms in the vacuum hose sink, especially when the hose is in a vertical position. The interim container must be controlled by one person in the boat while a second person operates the pump. The sample accumulates in the interim container and the net behind the pump. After a certain sampling time in one depth, the container is lifted up and the collected organisms can now be extracted via the hose, which is closed with a rubber plug. After the removal of the rubber plug, the interim container can be emptied in a net. Although this method samples the sediment in a qualitative way, it is a practical alternative to scuba diving.

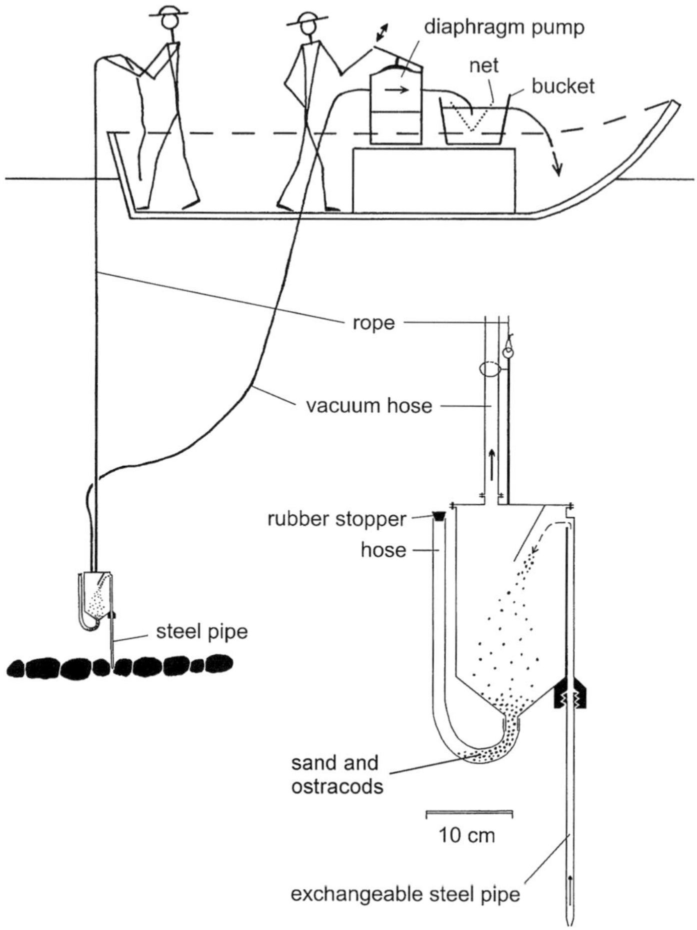

Fig. 6. Schematic illustration of the collecting of ostracods between stones in the littoral of lakes. The storage vessel is illustrated on a larger scale at the right lower corner.

DESCRIPTION OF NEW METHODS FOR ISOLATING OSTRACODS

Between 2009 and 2014 many samples were taken from the town moat of Bremen using a handnet (200 μm mesh size) (Scharf & Viehberg, 2014). The collected samples were immediately sieved with a 5-mm-mesh sieve in order to eliminate coarse sediment. Subsequently, the residual fine sediment was transferred to a plastic basket (6 cm × 12 cm) within a larger container filled with water (fig. 7). Overnight, most of the ostracods migrated from the plastic basket into the container, where they could easily be captured in clear water. We call this method as the "emigration method".

Another method for separating ostracods from the sediment is called the "blowing method", based on the hydrophobic characteristics of ostracod valves,

Fig. 7. Plastic basket and container used for separating living ostracods from the sediment (from Brandorff et al., 2013).

which is also the effect used by the aeration method described in Szlauer-Łukaszewska & Radziejewska (2013). The sediment with the living and subrecent ostracods was put into a container and a small amount of water was added. The container is then repeatedly tilted so that alternating parts of the sediment and ostracods are exposed to air or gently recovered by water. Consequently, ostracods were affixed to the surface of the water via their hydrophobic valve surface. The surficial water is then poured or gently blown over the edge of the container into a net. This procedure has to be repeated until no ostracods can be seen at the surface of the water after tilting the container. Most of the living and subrecent ostracods can be separated from the sediment in a short time (E. Schornikov, pers. comm.).

In February 2014 samples were taken from the town moat of Bremen (Scharf & Viehberg, 2014; site 6) for testing the effectiveness of the described methods in separating living ostracods from the sediment (table II). The water was covered by ice and had an electrical conductivity of 1453 μS cm^{-1}. Compared with previous sample-dates, all species were recaptured, except *Physocypria kraepelini* G. W. Müller, 1903 (table II) (op. cit.). In total, 4922 adult living specimens were detected and additionally adults of *Candona candida* (O. F. Müller, 1776), *Cyprideis torosa* (Jones, 1850) and *Herpetocypris chevreuxi* (Sars, 1896) were present. *Cypria ophtalmica* (Jurine, 1820) was eudominant, *Candona neglecta* Sars, 1887 was dominant and *Cypria subsalsa* Redeke, 1936 and *Ilyocypris bradyi* Sars, 1890 were subdominant. The rest of the identified species were recent or subrecent.

The emigration method is compared with the blowing method (table II). In the experiments, about 68-99% of the living dominant and subdominant adult animals emigrated successfully, whereas 93-94% was separated from the sediment by the

TABLE II
Results of separating ostracods from the sediment experiments

Species		Emigration experiment 1			Emigration experiment 2			Emigration experiment 3			Blowing experiment					
		Emigrated		r.i.s.	Emigrated		r.i.s.	Emigrated		r.i.s.	Eliminated by blowing				r.i.s.	
		Living			Living			Living			Living		Subrecent		Living	Subrecent
		ind.	%	ind.	ind.	%	ind.	ind.	%	ind.	ind.	%	ind.	%	ind.	ind.
		1	2	3	4	5	6	7	8	9	10	11	12	13	14	15
Candona candida (O. F. Müller, 1776)	ad	13	87	2	1	100							4	100	1	
	juv												1	100		
Candona neglecta Sars, 1887	ad	102	75	34	44	42	60	71	66	36	44	100	22	55		18
	juv	14	52	13	2	17	10	3	43	4	15	100	137	84		27
Cyclocypris ovum (Jurine, 1820)	ad	99	99	1	395	99	6	364	98	7	90	94	18	100	6	
	juv															
Cypria ophtalmica (Jurine, 1820)	ad	812	86	131	609	68	288	368	71	148	430	93	24	89	31	3
	juv															
Cypria subsalsa Redeke, 1936	ad	12	86	2	56	30	128	92	66	48	40	98			1	
	juv	6	100		19	76	6			4	2	100				
Cyprideis torosa (Jones, 1850)	ad						1									
	juv															
Cypridopsis vidua (O. F. Müller, 1776)	ad	16	89	2	13	68	6	3	60	2	23	92	20	59	2	14
	juv				1	100					1	100				2
Cytheromorpha fuscata (Brady,1869)	ad	8	100		26	81	6	28	97	1	6	75	1	10	2	9
	juv	2	100		4	100		1	100							
Darwinula stevensoni (Brady & Robertson, 1870)	ad			5	2	8	24	1	13	7	9	90	6	55	1	5
	juv	2	67	1			5	2	50	2	6	86	5	100	1	
Herpetocypris reptans (Baird, 1835)	ad	2	40	3						1	2	100				
	juv	1	14	6	1	14	6	2	50	2	2	100	12	67		6
Herpetocypris chevreuxi (Sars, 1896)	ad										1	100				
	juv															

TABLE II
(Continued)

Species		Emigration experiment 1			Emigration experiment 2			Emigration experiment 3			Blowing experiment					
		Emigrated		r.i.s.	Emigrated		r.i.s.	Emigrated		r.i.s.	Eliminated by blowing				r.i.s.	
		Living			Living			Living			Living		Subrecent		Living	Subrecent
		ind.	%	ind.	ind.	%	ind.	ind.	%	ind.	ind.	%	ind.	%	ind.	ind.
		1	2	3	4	5	6	7	8	9	10	11	12	13	14	15
Ilyocypris bradyi Sars, 1890	ad	10	100		59	97	2	97	96	4	5	56	24	51	4	23
	juv				1	100		3	100						1	
Ilyocypris decipiens Masi, 1905	ad												27	27		73
	juv												54	33		108
Isocypris beauchampi (Paris, 1920)	ad	2	50	2	9	53	8	10	36	18	2	100				
	juv				1	14	6	2	29	5	1	100				
Limnocythere inopinata (Baird, 1843)	ad												10	15		55
	juv															2
Plesiocypridopsis newtoni (Brady & Robertson, 1870)	ad												4	100		
	juv												4	67		2
Prionocypris zenkeri (Chryzer & Toth, 1858)	ad	3	100		18	100		24	100		2	100	3	60		2
	juv	2	100		3	75	1	2	100		2	100	6	38		10
Pseudocandona compressa (Koch, 1838)	ad															
	juv			1							1	100				
Potamocypris variagata (Brady & Norman, 1889)	ad	5	100		11	79	3	5	71	2	4	100	3	60		2
	juv															

In the emigration experiments the percentage is calculated as the part of the living emigrated individuals in the relation to the sum of the found living animals, e.g., in emigration experiment 1 for *Candona candida*: 13 individuals are 87% of 15 (13 + 2 = 15, columns 1-3). In the blowing experiment the percentage is calculated as part of the living individuals, eliminated by blowing, in relation the sum of living ostracods, e.g., for *Ilyocypris bradyi*: 5 individuals correspond to 56% of the living 9 animals found (columns 10 and 14), respectively, for the subrecent ostracods. r.i.s., remains in the sediment; ad, adult; juv, juvenile; ind., individual.

blowing method. In the emigration experiment the sediment with the ostracods is put into the basket and here most of the ostracods are under the surface of the sediment, i.e., they must crawl first to the surface of the sediment to emigrate. *Cyclocypris ovum* (Jurine, 1820) and *Ilyocypris bradyi* seem to be able to do it very well. For *Cypria ophtalmica*, a swimming species, it is more difficult to solve this problem. It is surprising that not all specimens of both *Candona*-species are not able to reach the surface of the sediment or do not emigrate. It is noteworthy that also the blowing experiment does not separate all specimens out of the sediment. Certainly if one has repeated the blowing procedure the percentage would increase. However, with each extraction step a little sediment is transported into the net in which the animals drop and, hence, increases the amount of time to separate the animals from the sediment in the lab. Therefore, one must compromise between time consumption and efficiency of separation. In our case all of the living species are represented with a high percentage by the blowing method. Additionally, subrecent species such as *Ilyocypris decipiens* Masi, 1905 and *Plesiocypridopsis newtoni* (Brady & Robertson, 1870) (summer forms) and *Limnocythere inopinata* (Baird, 1843) could be isolated by their empty valves. According to these results we recommend the blowing method for the separation of ostracods from the sediment. Our emigration method can be compared with the deterioration technique of Szlauer-Łukaszewska & Radziejewska (2013). We use a basket (fig. 7) and they use an hour glass which is filled with sediment. Our blowing experiment is based on the hydrophobic characteristic of the ostracod valves. This is also the basis for their aeration experiment (op. cit.). The results are similar, but Szlauer-Łukaszewska & Radziejewska (2013) studied a great number of experiments so that they could statistically evaluate the results. Aeration experiments were performed with sediment preserved in ethanol, whereas we use unpreserved sediment. The unpreserved sediment has the advantage that most of the animals live after the separation procedure and can be utilized for further experiments or for a genetic analysis.

ACKNOWLEDGEMENTS

B.S. thanks Julia Lorenschat (Technical University Braunschweig, Braunschweig, Germany) and Ursula Beddig, Bremen, for their assistance during the tests of the new method for collecting ostracods in stony sediment of a creek. We thank Raphael Gromig, Cologne, for correcting the English text of the previous version of the manuscript and Lisa Park (University of Akron, Akron, OH, U.S.A.) for the final language editing, and the reviewers for their helpful comments which have improved the manuscript.

REFERENCES

BOU, C. & R. ROUCH, 1967. Un nouveau champ de recherches sur la faune aquatique souterraine. C. R. Hebdomad. Seances Acad. Sci., **265**: 369-370.

BRANDORFF, G. O., W. HOLLWEDEL, B. SCHARF & A. F. VIEHBERG, 2013. Der Bremer Stadtgraben — die derzeitige Besiedlung mit Cladocera, calanoiden Copepoda und Ostracoda (Crustacea), mit einer Betrachtung der historischen Entwicklung des Gewässers. Abh. Naturwiss. Ver. Bremen, **47**: 93-119.

BRUNKE, M. & T. GONSER, 1999. Hyporheic invertebrates — the clinal nature of interstitial communities structured by hydrological exchange and environmental gradients. J. N. Am. Benthol. Soc., **18**: 344-362.

MACAN, T. T., 1958. Methods of sampling the bottom fauna in stony sediments. Int. Ver. Limnol. Mitt., **8**: 1-21.

MARMONIER, P., 1988. Biocenoses interstitielles et circulation des eaux dans le sous-écoulement d'un chenal amenage du Haut-Rhône français. (Thesis, University Claude Bernard, Lyon).

SCHARF, B. & M. BRUNKE, 2013. The colonization of the river Elbe with Ostracoda (Crustacea) after the union of Germany in 1989. Int. Rev. Hydrobiol., **98**: 1-8.

SCHARF, B. & F. A. VIEHBERG, 2014. Living Ostracoda (Crustacea) from the town moat of Bremen, Germany. Crustaceana, **87**: 1124-1135.

SCHWOERBEL, J., 1994. Methoden der Hydrobiologie Sußwasserbiologie (4 ed.): 1-368. (Gustav Fischer, Stuttgart, Jena).

SZLAUER-ŁUKASZEWSKA, A. & T. RADZIEJEWSKA, 2013. Two techniques of ostracod (Ostracoda, Crustacea) extraction from organic detritus-rich sediments. Limnologica, **43**: 272-276.

VIEHBERG, F. A., 2002. A new and simple method for quantitative sampling of meiobenthos-communities. Limnologica, **32**: 350-351.

WILCZEK, S., H. FISCHER, M. BRUNKE & M. T. PUSCH, 2004. Microbial activity within a subaqueous dune in a large lowland river (River Elbe, Germany). Aquat. Microb. Ecol., **36**: 83-97.

First received 1 February 2013.
Final version accepted 23 May 2014.